Lipid Peroxides
in Biology and Medicine

Academic Press Rapid Manuscript Reproduction

Based on the conference
"Lipid Peroxides in Biology and Medicine"
held November 7–8, 1980 in Nagoya, Japan.

Lipid Peroxides
in Biology and Medicine

Edited by

Kunio Yagi

Institute of Biochemistry
Faculty of Medicine
University of Nagoya
Nagoya, Japan

1982

ACADEMIC PRESS, INC.

(Harcourt Brace Jovanovich, Publishers)

Orlando   San Diego   San Francisco   New York   London
Toronto   Montreal   Sydney   Tokyo   São Paulo

ACADEMIC PRESS, INC.
Orlando, Florida 32887

United Kingdom Edition published by
ACADEMIC PRESS, INC. (LONDON) LTD.
24/28 Oval Road, London NW1 7DX

Library of Congress Cataloging in Publication Data
Main entry under title:

Lipid peroxides in biology and medicine

   Proceedings of the International Conference on Lipid
Peroxides in Biology and Medicine, which was held in
Nagoya, Japan, Nov. 7-8, 1980.
   Includes index.
   1. Lipids--Metabolism--Congresses. 2. Peroxides--
Congresses. 3. Peroxides--Toxicology--Congresses.
I. Yagi, Kunio, Date   . II. International
Conference on Lipid Peroxides in Biology and Medicine
(1980 : Nagoya-shi, Japan) [DNLM: 1. Lipid peroxides--
Congresses. QU 85 L7634 1980]
QP751.L553 1982   612'.397   82-16430
ISBN 0-12-768050-0

PRINTED IN THE UNITED STATES OF AMERICA

   84 85   9 8 7 6 5 4 3 2

Contents

Contributors

Numbers in parentheses indicate the pages on which the authors' contributions begin.

Hiroshi Abe (285), *First Department of Medicine, Osaka University Hospital, Osaka, Japan*

Bruce N. Ames (339), *Department of Biochemistry, University of California, Berkeley, California 94720*

Steven D. Aust (23), *Department of Biochemistry, Michigan State University, East Lansing, Michigan 48823*

Anne P. Autor (131), *Department of Pharmacology, University of Iowa, Iowa City, Iowa 52240*

Richard Cathcart (339), *Department of Biochemistry, University of California, Berkeley, California 94720*

Gerald Cohen (199), *Mount Sinai School of Medicine, City University of New York, New York, New York 10029*

Lars Ernster (55), *Department of Biochemistry, Arrhenius Laboratory, University of Stockholm, Stockholm, Sweden*

Leopold Flohé (149), *Center of Research, Grünenthal GmbH, Aachen, Federal Republic of Germany*

David Gelmont (161), *Laboratory of Nuclear Medicine and Radiation Biology, University of California, Los Angeles, California 90024*

Donald D. Gibson (179), *Biomembrane Research Laboratory, Oklahoma Medical Research Foundation, Oklahoma City, Oklahoma 73101*

Yuichiro Goto (295), *Department of Medicine, Tokai University School of Medicine, Bohseidai, Isehara, Japan*

Osamu Hayaishi (41), *Department of Medical Chemistry, Kyoto University Faculty of Medicine, Kyoto, Japan*

Shunsaku Hirai (305), *Institute of Neurology and Rehabilitation, Gunma University School of Medicine, Maebashi, Japan*

Paul Hochstein (81), *Institute for Toxicology, University of Southern California, Los Angeles, California 90024*

Monica C. Hollstein (339), *Department of Biochemistry, University of California, Berkeley, California 94720*

Toshihiko Isobe (243), *Laboratory of Nutrition, Faculty of Agriculture, Tohoku University, Sendai, Japan*

Shuichi Kimura (243), *Laboratory of Nutrition, Faculty of Agriculture, Tohoku University, Sendai, Japan*

John W. Lightsey (1), *Department of Chemistry, Louisiana State University, Baton Rouge, Louisiana 70803*

Yong Y. Lin (89), *Division of Biochemistry, University of Texas Medical Branch, Galveston, Texas 77550*

Paul B. McCay (179), *Biomembrane Research Laboratory, Oklahoma Medical Research Foundation, Oklahoma City, Oklahoma 73101*

Joe M. McCord (123), *Department of Biochemistry, University of South Alabama, Mobile, Alabama 36688*

Ronald N. McElhaney (161), *Department of Biochemistry, University of Alberta, Edmonton, Canada*

Michael F. McGehee (89), *Division of Biochemistry, University of Texas Medical Branch, Galveston, Texas 77550*

James F. Mead (161), *Laboratory of Nuclear Medicine and Radiation Biology, and Department of Biological Chemistry, University of California, Los Angeles, California 90024*

Mitsunori Morimatsu (305), *Institute of Neurology and Rehabilitation, Gunma University School of Medicine, Maebashi, Japan*

Minoru Nakano (107), *College of Medical Care and Technology, Gunma University, Maebashi, Japan*

Kerstin Nordenbrand (55), *Department of Biochemistry, Arrhenius Laboratory, University of Stockholm, Stockholm, Sweden*

Peter J. O'Brien (317), *Department of Biochemistry, Memorial University of Newfoundland, St. John's, Newfoundland, Canada*

Ryohei Ogura (255), *Department of Medical Biochemistry, Kurume University School of Medicine, Kurume, Japan*

Koichi Okamoto (305), *Institute of Neurology and Rehabilitation, Gunma University School of Medicine, Maebashi, Japan*

Minoru Okuma (271), *First Division, Department of Internal Medicine, Faculty of Medicine, Kyoto University, Kyoto, Japan*

Sten Orrenius (55), *Department of Forensic Medicine, Karolinska Institutet, Stockholm, Sweden*

William F. Petrone (123), *Department of Biochemistry, University of South Alabama, Mobile, Alabama 36688*

Donald G. Prier (1), *Department of Chemistry, Louisiana State University, Baton Rouge, Louisiana 70803*

William A. Pryor (1), *Department of Chemistry, Louisiana State University, Baton Rouge, Louisiana 70803*

Catherine Rice-Evans* (81), *Institute for Toxicology, University of California, Los Angeles, California 90024*

*Present address: *Department of Biochemistry, Royal Free Hospital School of Medicine, London, England*

Hiroaki Sai (243), *Laboratory of Nutrition, Faculty of Agriculture, Tohoku University, Sendai, Japan*

Patricia K. Seitz (89), *Division of Biochemistry, University of Texas Medical Branch, Galveston, Texas 77550*

Alex Sevanian (161), *Laboratory of Nuclear Medicine and Radiation Biology, University of California, Los Angeles, California 90024*

Takao Shimizu (41), *Department of Medical Chemistry, Kyoto University Faculty of Medicine, Kyoto, Japan*

Leland L. Smith (89), *Division of Biochemistry, University of Texas Medical Branch, Galveston, Texas 77550*

Erick Sohlberg (161), *Laboratory of Nuclear Medicine and Radiation Biology, University of California, Los Angeles, California 90024*

Robert A. Stein (161), *Laboratory of Nuclear Medicine and Radiation Biology, University of California, Los Angeles, California 90024*

Toshihiko Suematsu (285), *First Department of Medicine and Division of Blood Transfusion, Osaka University Hospital, Osaka, Japan*

Yuji Takahashi (243), *Laboratory of Nutrition, Faculty of Agriculture, Tohoku University, Sendai, Japan*

Hiroshi Takayama (271), *First Division, Department of Internal Medicine, Faculty of Medicine, Kyoto University, Kyoto, Japan*

Al Tappel (213), *Department of Food Science and Technology, University of California, Davis, California 95616*

Jon I. Teng (89), *Division of Biochemistry, University of Texas Medical Branch, Galveston, Texas 77550*

Ming Tien (23), *Department of Biochemistry, Michigan State University, East Lansing, Michigan 48823*

Haruto Uchino (271), *First Division, Department of Internal Medicine, Faculty of Medicine, Kyoto University, Kyoto, Japan*

Guey-Shuang Wu (161), *Laboratory of Nuclear Medicine and Radiation Biology, University of California, Los Angeles, California 90024*

Kunio Yagi (223), *Institute of Biochemistry, Faculty of Medicine, University of Nagoya, Nagoya, Japan*

Preface

Recently, much attention has been focused on the subject of lipid peroxidation, especially in the field of clinical medicine. This is attributed directly to the fact that lipid peroxide increase is related to the pathogenesis of many degenerative diseases such as atherosclerosis, and also, indirectly but more fundamentally, to the biochemical knowledge accumulated in this field. The latter includes the mechanism of formation of lipid peroxides and the development of assay for human samples, thus enabling us to further biomedical research. However, the coordination between biochemical research and clinical investigation has not always been adequate, which is why I proposed a symposium that would provide the opportunity for discussion between scientists engaged in the two disciplines.

The International Conference on "Lipid Peroxides in Biology and Medicine" was organized in collaboration with Drs. L. Ernster (Stockholm), Y. Goto (Isehara), O. Hayaishi (Kyoto), P. B. McCay (Oklahoma), and M. Nakano (Maebashi) and was held in Nagoya on November 7 and 8, 1980. Speakers were asked to submit review articles that included their own data.

Although this book reflects the proceedings of the symposium, the articles, written by experts in the field, have been updated and edited for inclusion in the volume. Recognizing the importance of the control of lipid peroxides in our body for the prevention and treatment of degenerative diseases, it is hoped that this book may contribute to the advancement of this field and thereby promote better health and welfare of man.

I wish to express my sincere thanks to the symposium organizers and all the contributors who made this valuable publication possible.

Kunio Yagi

THE PRODUCTION OF FREE RADICALS IN VIVO FROM THE ACTION OF XENOBIOTICS: THE INITIATION OF AUTOXIDATION OF POLYUNSATURATED FATTY ACIDS BY NITROGEN DIOXIDE AND OZONE

William A. Pryor
John W. Lightsey
Donald G. Prier

Department of Chemistry
Louisiana State University
Baton Rouge, Louisiana

The reactions of ppm levels of nitrogen dioxide with cyclohexene have been studied as a model for the reactions that occur between nitrogen dioxide in polluted air and unsaturated fatty acids in pulmonary lipids. As predicted from the previous studies done at high nitrogen dioxide levels, we find that at concentrations higher than 10,000 ppm, nitrogen dioxide reacts with cyclohexene predominantly by *addition* to the double bond; however, at low ppm levels nitrogen dioxide reacts with this alkene virtually *exclusively by abstraction of allylic hydrogens*. In the presence of air or oxygen, both the addition and abstraction mechanisms initiate alkene autoxidation. We also report some comparable data on the kinetics of the nitrogen dioxide-initiated autoxidation of methyl oleate, linoleate, and linolenate.

Ozone is not a free radical, but some of the pulmonary damage it causes is known to involve radical-mediated reactions. We therefore are using spin-trap methods to study the mechanism of radical production from ozone-alkene reactions. We find that methyl linoleate reacts with ozone at −78°C and that warming to −45°C in the presence of spin traps produces spin adducts of alkoxyl and alkyl radicals. With 2-methyl-2-pentene, a peroxyl radical spin adduct can be observed even at −78°C. We are tentatively hypothesizing that an intermediate is produced, probably a tri-oxygen species such as R–CO–OOOH or ROOOH, that decomposes to produce radicals; these radicals either

are spin-trapped or they react with oxygen or substrate
to generate radicals we trap that are then trapped.

I. INTRODUCTION

There are two important pathways for the production of
radicals in living systems: (1) enzymatically controlled,
one-electron reduction of O_2, and (2) reactions
initiated by xenobiotics. Most of the papers in this
symposium will concentrate on the chemistry and biology of
the reduction of O_2: the mechanism, the enzymes
involved, the reactions of the partially reduced oxygen
species that are formed, and the control of these species.
However, radical production from xenobiotics is equally
important, and in this paper I wish to describe our recent
work in this area.

Xenobiotics can lead to the production of radicals by
several distinctive mechanisms:

(1) Substances can trigger the production of hydrogen
peroxide and superoxide from phagocytic cells; unreactive
materials (polystyrene beads, asbestos) as well as
reactive substances (cigarette smoke) can cause this
effect.

(2) Xenobiotics can be metabolized by radical-
mediated paths; examples are CCl_4 (reduced by
cytochrome P_{450} to Cl^- and $\cdot CCl_3$).

(3) A few toxins are themselves radicals; examples
are NO, NO_2, and organic combustion products (tobacco
smoke and automobile exhaust). These materials can react
with biomolecules to produce radicals without the
intervention of enzymes.

(4) A small group of toxins, while not radicals
themselves, can react to form radicals or radical-
precursor compounds by non-enzymatic paths. This group
includes very reactive species such as ozone and singlet
oxygen.

For several years my group has been interested in the
mechanisms of the damage-producing reactions of nitrogen
dioxide, ozone, and tobacco smoke. In this report, I will
present some data on the first two of these toxins.

II. NITROGEN DIOXIDE

A. Introduction

Nitrogen dioxide is a ubiquitous toxin that occurs in
automobile exhaust and in tobacco smoke (1-5). It is a
free radical that reacts with both alkanes and alkenes at

25°C by free radical mechanisms (6-12). Nitrogen dioxide
alone has been shown to cause pulmonary damage leading to
emphysema (13,14), and the nitrogen dioxide in cigarette smoke
may contribute to emphysema produced by smoking (15-17). Both
in vitro and *in vivo* studies have shown that NO_2 initiates
the autoxidation of unsaturated fatty acids in lipids (18),
and this lipid peroxidation process is known to cause the
destruction of pulmonary lipids (18,19) and lead to membrane
damage and cell death (20-23).

The chemical literature indicates that at ambient temperatures and high concentrations, nitrogen dioxide reacts with
alkenes by addition to the double bond (6-8,12,24) (Eq. 1) and

$$NO_2 \quad + \quad -\overset{|}{\underset{|}{C}}=\overset{|}{\underset{|}{C}}- \quad \underset{k_2}{\overset{k_1}{\rightleftarrows}} \quad O_2N-\overset{|}{\underset{|}{C}}-\overset{|}{\underset{|}{C}}\cdot \qquad (1)$$

$$(I)$$

with compounds such as cyclohexane (11) and toluene (10,25) by
abstraction of hydrogen atoms (Eq. 2). In the liquid phase,

$$NO_2 \quad + \quad R-H \quad \longrightarrow \quad HONO \quad + \quad R\cdot \qquad (2)$$

no studies of the competition between the addition to a
double bond and hydrogen abstraction of an allylic hydrogen
has been reported. In the gas phase at low concentrations of
both nitrogen dioxide and alkene, addition is the only
process reported to occur (26).

It is generally assumed that ppm levels of nitrogen
dioxide in smog initiate lipid autoxidation in the lung by
addition to the double bonds of unsaturated fatty acids, as
shown in equation (1). However, preliminary work in our
laboratory using polyunsaturated fatty acid (PUFA) esters
and ppm levels of nitrogen dioxide indicated that hydrogen
abstraction was the predominant mechanism of initiation. For
this reason, we re-examined the reaction mechanism of the
initiation of autoxidation by nitrogen dioxide (27).

B. Cyclohexene Product Study

To obtain a detailed product analysis, we examined
cyclohexene as a simplified model for an unsaturated fatty
acid. Cyclohexene is a symmetrical alkene with a cis double
bond that gives relatively simple product mixtures. Its

reaction with high concentrations of nitrogen dioxide has been
studied by organic chemists in some detail (6, 7), and the
kinetics of its autoxidation have been well characterized
(28).

Our studies were performed with 6.6 molar solutions of
cyclohexene in hexane; literature evidence indicates that
only radical reactions occur under these conditions (7,8,10).
A carrier gas (nitrogen, air, or oxygen) containing an
equilibrium mixture of both nitrogen dioxide and its dimer,
dinitrogen tetroxide (29), was bubbled through the solutions
at 30°C. This two-phase system was meant to provide a
simplified model for pulmonary exposure conditions.

As indicated by our preliminary results, the principal
initiation mechanism of nitrogen dioxide changes from addition
to a double bond to abstraction of an allylic hydrogen (Eq. 3)

$$NO_2 \; + \; -HC{=}CH{-}CH_2{-} \; \longrightarrow \; HONO \; + \; -\overset{\centerdot}{\overline{CH{-}CH{-}CH}}- \qquad (3)$$
$$(II)$$

as the nitrogen dioxide levels in the carrier gas are
decreased from 30–40% to 50–100 ppm. The two initiation
mechanisms are distinguishable because each forms a unique set
of products. Equation 1 shows that the initial addition of
nitrogen dioxide to a double bond produces radical (I); this
species can react with another nitrogen dioxide to form a
di-nitro or a nitro-nitrite compound, or with oxygen (if air
is present) to ultimately give 2-nitro-hydroperoxides and
other oxygenated compounds. Equation 3 shows that hydrogen
abstraction produces nitrous acid and allylic radical (II).
Radical (II) can combine with nitrogen dioxide to give an
unsaturated nitro or nitrite compound, or with oxygen to give
hydroperoxides and other products of autoxidation. Table I
lists the principle products we have identified from reactions
of nitrogen dioxide and cyclohexene; these compounds account
for 97 ± 14% of the cyclohexene and nitrogen dioxide
consumed. The table shows the mole percentages of products
for selected runs grouped according to the initiation
mechanism responsible for their formation. (The products also
are listed in order of decreasing volatility.) The upper
products are the substituted cyclohexenes that result from
initiation by hydrogen abstraction; the lower compounds result
from initial addition of nitrogen dioxide to the double bond
of cyclohexene and are addition products. (1-Nitrocyclohexene
is a substituted cyclohexene, but it results from elimination
of nitrous acid from 1,2-dinitrocyclohexane (6) and is
classified as an addition product.)

TABLE I. Mole Percentages of Products from Selected Reactions of Cyclohexene and Nitrogen Dioxide at 30°C

Products	Wt. Percent NO_2 + N_2O_4							
	In Nitrogen				In Air			
	37.	0.28	0.071	0.0078[a]	30.	4.7	0.33	0.0076
2-Cyclohexen-1-ol	1	13	25	69	9	12	23	21
2-Cyclohexen-1-hydroperoxide[b]	0	0	0	0	1	13	5	57
2-Cyclohexen-1-nitrate	0	2	5	0	2	19	40	<1
3-Nitrocyclohexene	5	29	39	13	8	2	2	0
2-Cyclohexen-1-one oxime	0	1	2	12	1	0	0	0
Subtotal:	6	45	71	94	21	46	70	78
1-Nitrocyclohexene	33	11	7	2	26	8	<1	0
2-Nitrocyclohexanol[c]	61	38	22	3	47	41	18	22
2-Nitrocyclohexylnitrate	0	3	0	0	2	5	11	0
1,2-Dinitrocyclohexane	<1	3	0	1	4	<1	1	0
Subtotal:	94	55	29	6	79	54	30	22

[a] Ultra Pure Nitrogen (99.999%).
[b] Includes yields of 2-cyclohexen-1-one that are 5-15% of the yields of the 2-cyclohexen-1-hydroperoxide shown.
[c] Includes 2-nitrocyclohexenone and 2-nitrocyclohexyl hydroperoxide.

The weight percentages of nitrogen dioxide (plus dinitrogen tetroxide) in the carrier gases and the identities of the carrier gases are listed across the top of Table I. As expected, addition products are the major products at high nitrogen dioxide concentrations in both carrier gases, but substitution (i.e., hydrogen abstraction) products become the major products as the nitrogen dioxide levels decrease.

More extensive data on product distributions are shown in Figure 1 where the percent addition is plotted against the concentration of nitrogen oxides in the gas phase. In this figure, the mole percentage of addition is calculated as the ratio of the sum of addition products to the sum of all products formed by nitrogen dioxide initiation. In the absence of oxygen, Figure 1A, a linear relationship exists between percent addition and the logarithm of the nitrogen oxide concentration over the concentration range studied. In the presence of oxygen, Figure 1B, the transition from addition to abstraction as the nitrogen oxide concentration is decreased is initially more rapid than in the absence of oxygen. (These NO_2/O_2 data are corrected by subtracting out products that arise solely from autoxidation.) However, at low ppm levels of nitrogen dioxide, more addition occurs in the presence of oxygen than in its absence.

We suggest the following explanation for the increasing amount of hydrogen abstraction observed with decreasing nitrogen dioxide concentration. Addition, the thermodynamically favored process (30), is reversible (Eq. 1) (31,32). In contrast, hydrogen abstraction is not; the nitrous acid formed in equation (3) decomposes to nitric oxide, nitrogen dioxide, and water (Eq. 4) (33,34) preventing

$$2 \text{ HONO} \longrightarrow \text{NO} + \cdot\text{NO}_2 + \text{H}_2\text{O} \tag{4}$$

reversal of equation (3). Therefore, at high concentrations of a radical trapping species (nitrogen dioxide (35), dinitrogen tetroxide, or oxygen), intermediate (I) reacts to give addition products. Equation 5 illustrates this using oxygen

$$\cdot\text{NO}_2 + \overset{|}{-}\text{C}=\overset{|}{\text{C}}- \underset{k_2}{\overset{k_1}{\rightleftharpoons}} \underset{(\text{I})}{\text{O}_2\text{N}-\overset{|}{\underset{|}{\text{C}}}-\overset{|}{\underset{|}{\text{C}}}\cdot} \overset{k_3}{\underset{\text{O}_2}{\longrightarrow}} \underset{(\text{III})}{\text{O}_2\text{N}-\overset{|}{\underset{|}{\text{C}}}-\overset{|}{\underset{|}{\text{C}}}-\text{OO}\cdot} \tag{5}$$

as the radical-trapping species. At low concentrations of trapping species, radical (I) is trapped more slowly and a larger fraction of (I) reverts to reform the alkene; therefore, less addition occurs and hydrogen abstraction becomes favored (36).

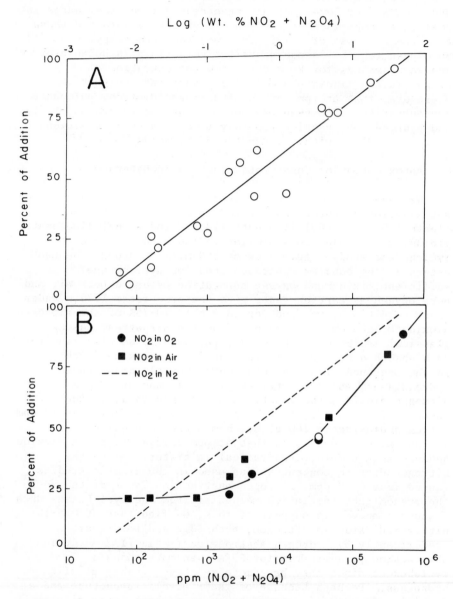

FIGURE 1. *The mole percentage of addition, calculated as the ratio of the sum of addition products to the sum of all products formed by nitrogen dioxide initiation at 30°C, is plotted as a function of the concentration of nitrogen oxides in: (A) Nitrogen and (B) Air or Oxygen.*

Our product study data show that the presence of oxygen in the reaction mixture shifts the reaction mechanism in favor of addition at ppm levels of nitrogen dioxide, but the change is small. A comparison of the kinetic processes involved shows that the reversal of (I) is fast and that the trapping of (I) by oxygen (eq. 5) is not quantitative. (This is because of the large values for k_2 and the low concentration of oxygen in the liquid phase (38).) These data suggest, therefore, that at ppm levels of nitrogen dioxide, pulmonary concentrations of oxygen may not prevent the reversion of (I), and hydrogen abstraction probably predominates in reactions of nitrogen dioxide with unsaturated fatty acids.

C. Autoxidation of Unsaturated Fatty Acid Esters

We have studied the kinetics of nitrogen dioxide-initiated autoxidation of three unsaturated fatty acid esters, methyl oleate (18:1), methyl linoleate (18:2), and methyl linolenate (18:3), under conditions similar to those used in the cyclohexene study. At a flow of 300 ml/min through the neat esters in the bubbler apparatus used for cyclohexene, sufficient mixing and oxygen saturation occur so both air and pure oxygen as carrier gases give the same rates (27). Under these conditions, we find for 18:2 that 70-75% of the nitrogen dioxide molecules are absorbed from the air stream; using classical inhibitor methods (37, pp. 322-29) we also find that each absorbed nitrogen dioxide initiates one kinetic chain. In the presence of inhibitors (α-tocopherol, α-naphthol, hydroquinone, and 2,4,6-trimethylphenol) and at 70 ppm nitrogen dioxide, the kinetic chain length is 3; in the absence of inhibitor, the chain length varies from 7 at 70 ppm nitrogen dioxide to 200 at 0.3 ppm nitrogen dioxide (27).

Our kinetic data show that at ppm levels, nitrogen dioxide behaves as a simple, free-radical initiator. Since the nitrogen dioxide concentrations used in the kinetic studies ranged from 0-70 ppm, hydrogen abstraction is predicted to be the predominant initiation mechanism from the cyclohexene data discussed above. In support of this, we find that at 0.5-1% nitrogen dioxide in nitrogen, both 18:2 and 18:3 react predominantly by hydrogen abstraction (Table II). Hydrogen abstraction predominates for PUFA even more than for cyclohexene since they have more reactive, doubly allylic hydrogens. Perhaps because of their acyclic structure and a resulting increase in steric hinderance about their allylic hydrogens, 18:1 and 1-hexadecene react relatively less by the hydrogen abstraction mechanism than do cyclohexene and 1,5,9-cyclododecatriene (Table II). At lower levels of nitrogen dioxide, however, all three unsaturated fatty acids, like cyclohexene, should shift their reaction mechanisms

TABLE II. Determination of Percent Hydrogen Abstraction Occurring During NO_2-Alkene Reactions [a]

Alkene	% H-Abstraction[b]
Methyl Linolenate (18:3)	71
Methyl Linoleate (18:2)	67[c]
Cyclohexene	56
1,5,9-Cyclododecatriene	52
1-Hexadecene	38
Methyl Oleate (18:1)	36

[a]*Reactions done in sealed vessel at 30°C with 0.5-1% (w/w) NO_2 in nitrogen.*
[b]*The percent hydrogen abstraction is inferred from the yield of water (Eq. 4).*
[c]*Value agrees with the value obtained by product study, Fig. 1A.*

more in favor of hydrogen abstraction. Since hydrogen abstraction produces HONO and does not result in carbon-nitrogen bond formation, the amount of hydrogen abstraction can be determined by elemental analysis of the reisolated PUFA. Using this method we have shown that 18:1 and 18:3 react predominantly by H abstraction at ppm NO_2 levels.

III. OZONE

A. Introduction

Ozone is one of the three most damaging pollutants on a worldwide scale, according to the World Health Organization. This remarkably reactive compound reacts with olefins at extremely rapid rates even at −78°C and reacts with almost every type of organic molecule, even alkanes, at rates that are very fast at ambient temperatures (43). Thus, it is not surprising that breathing ozone-contaminated air such as occurs in photochemical smog produces damage that is predominantly localized in the lungs (22,44,45).

Although ozone is not itself a free radical, it reacts with a variety of types of organic molecules to form radicals (43). A considerable body of inferential evidence indicates that an important, and perhaps the principal, mechanism of the damage-producing reactions of ozone involves radicals (21,46). This evidence includes the production of conjugated dienes and peroxidic materials from the reaction of ozone-air with polyunsaturated fatty acids (PUFA) (47), shortening of the induction period of PUFA autoxidation by ozone in air (20),

the increased production of pentane and ethane from mammals
breathing ozone (48), thiobarbituric acid (TBA) test (48),
thiobarbituric acid (TBA) test results (9,21,49), and effects
of antioxidants on living animals (50) and *in vitro* (51).

The classical mechanism for the reaction of ozone with
olefins in the solution phase was proposed by Criegee (52).
It involves the formation of a 1,2,3-trioxolane that rapidly
undergoes scission (concertedly or stepwise) to give a
carbonyl oxide and a carbonyl compound (43). These two
species either recombine to give the 1,2,4-trioxolane (the
Criegee ozonide) (52), or react in a variety of ways to give
dimeric and polymeric peroxides, polymeric ozonides, ketone
and aldehyde diperoxides, and hydroperoxides (43). Most
authors believe that radicals are not involved in these
liquid-phase processes (43), although the carbonyl oxide has
both dipolar (i.e. zwitterionic) and diradical properties
(53). However, in the gas phase it is generally agreed that
the carbonyl oxide induces a variety of radical reactions that
are important in smog-contaminated air (43,53,54).

For a number of years our group has been interested in
the mechanism through which ozone produces damage in pulmonary
tissue and the extent to which radicals are involved in these
reactions (20,21,45,47,55-58). For our initial study, we have
used PUFA molecules of the type that occur in lung lipids.

B. Observation of Radicals Produced in Low Temperature Ozonation by ESR

ESR Detection of Radicals from Ozone-PUFA Reactions.
Electron spin resonance (ESR) is the most unambiguous method
for the detection of free radicals (59,60). In 1968, it was
reported that bubbling an ozone-oxygen mixture through neat
linoleic acid held in an ESR tube in the cavity of the ESR
spectrometer resulted in an ESR spectrum (61). This important
publication is often cited as conclusive evidence that ozone
does indeed react with PUFA to produce radicals. However, we
doubted the validity of these results for several reasons
(58), and we, therefore, decided to repeat this experiment.
We found that neither the original design nor several
modifications produced ESR signals when ozone-air was bubbled
through either linoleic acid or methyl linoleate at room
temperature (57,58).

Spin Trapping of Radicals from Ozone-Olefin Reactions.
Spin trapping is a technique in which transient radicals are
trapped through their reactions with nitrones or nitroso
compounds to produce nitroxide radicals that are more
conveniently studied by ESR (62,63). Often the trapped

radical can be identified from the ESR parameters of the
nitroxide (64).

Because of our inability to detect radicals directly by
ESR, we have applied the spin-trapping method in an attempt to
identify the radicals produced during initiation of PUFA
oxidation by ozone. We developed the following protocol to
avoid reactions between ozone and the spin trap (58):

(i) An ozone-oxygen or oxygen-free ozone-nitrogen
gaseous mixture is bubbled through a solution of olefin
in pre-ozonized $CFCl_3$ (Freon-11) or hexane in a
bubbler or ESR tube suspended in a dry ice-acetone bath
at $-78°C$.

(ii) After sufficient ozone to produce no more than
10-50% reaction has been bubbled through the solution,
the solution is flushed with nitrogen or argon for about
5 minutes to remove any residual ozone.

(iii) A solution of an appropriate spin trap in
$CFCl_3$ or hexane is then added to give a spin-trap
concentration of 0.1 M. The resulting solution is then
allowed to warm to room temperature.

When this low temperature ozonation and spin-trapping
technique is applied to methyl linoleate using phenyl-*tert*-
butyl nitrone (PBN) as the spin trap, no ESR signal is
obtained while the solution is kept at $-78°C$ either before or
after adding the spin trap. However, upon warming the
solution, a spin-adduct signal appears at about $-45°C$. This
signal grows very fast initially and then much more slowly,
even after the solution has reached room temperature. The
signal consists of a doublet of triplets having $a^N = 13.7$
G an $a^H = 1.8$ G, splittings that are consistent with a
spin trapped alkoxyl radical (64).

We repeated the above experiment using both 5,5-dimethyl-
pyrroline-1-oxide (DMPO) and 2-methyl-2-nitrosopropane (NtB)
as spin traps. With DMPO, upon warming the solution to room
temperature, we observe an ESR spectrum consisting of two
superimposed splitting patterns: a doublet of triplets having
$a^N = 14.3$ G and $a^H = 20.9$ G; and a doublet of triplets
having $a^N = 13.0$ G, $a^H = 6.5$ G, and $a^H = 1.6$ G.
The first signal is consistent with an alkyl radical spin
adduct and the second with an alkoxyl radical spin adduct
(64). With NtB as spin trap, we observed a spin-adduct ESR
spectrum consisting of a doublet of triplets having $a^N =$
15.2 G and $a^H = 1.8$ G, splittings that are consistent with
a carbon-centered radical having an α-hydrogen.

We also have used this spin-trapping protocol to study low
molecular weight olefins and polyenes that are easier to
purify than is methyl linoleate and where ozonation products
can be more easily identified and quantitated (57). In a
series of preliminary experiments, we found that of 9 olefins

tested using our low temperature technique, the tri-substituted
olefin 2-methyl-2-pentene consistently gave the largest yield
of spin-trapped radicals. For example, with PBN as the
spin-trap, reaction of a solution of 2-methyl-2-pentene with
0.1 equivalent of ozone gives a spin adduct concentration of 6
x 10^{-5} M as compared to 8 x 10^{-7} M spin adduct from
methyl linoleate under the same conditions. We therefore have
applied our low temperature and spin-trapping protocol to this
olefin and have also studied the effect of a variety of
modifications of the protocol:

(1) When the procedure described above is followed with
PBN as the spin trap, a weak spin adduct spectrum is observed
while the solution is still at -78°C. The spectrum has
a^N = 13.5 G and a^H = 1.4 G, splittings that are
consistent with a spin-trapped alkylperoxyl radical. When
the solution is allowed to warm to room temperature in 5°
increments ("slow warmup"), the spin adduct signal remains
relatively constant until the temperature reaches about
-45°C, at which time it begins to grow very rapidly. The
peroxyl radical spin-adduct signal reaches a maximum at about
-35°C and then begins to decrease in intensity and be
replaced by a second doublet of triplets having a^N = 14.0
G and a^H = 1.7 G, indicative of spin-trapped alkoxyl
radical. Upon reaching room temperature, only the alkoxyl
spin-adduct signal can be detected.

(2) When the temperature is not controlled and the
reaction mixture containing PBN is allowed to rapidly warm to
room temperature in the cavity of the ESR spectrometer while
the ESR spectra are being obtained ("fast warmup"), we
observe the rapid formation and disappearance of the peroxyl
radical spin adduct, just as before. In this case, however,
this spin adduct signal is replaced by the alkoxyl spin
adduct and an acyl radical spin adduct having a^N = 14.7 G
and a^H = 3.1 G

(3) When the spin trap is nitroso-*tert*-butane, the
fast warmup results in a spin-trapped acyl radical (a^N =
8.1 G) and a spin-trapped primary alkoxyl radical (a^N =
29.2 G and a^H = 1.2 G).

(4) We also obtain spin-adduct spectra even if the
ozonated olefin solution is warmed to room temperature in the
absence of the spin trap. For example, a 0.1 M solution of
2-methyl-2- 2-pentene was ozonated at -78°C, warmed to room
temperature and held there for 30 min without adding PBN, and
then re-cooled to -78°C. When PBN is then added at -78°C and
the solution re-warmed to room temperature in the presence of
oxygen, we observe spin-trapped alkoxyl radicals. When the
solution is warmed to room temperature in the presence of
nitrogen or argon, however, we observe both spin-trapped
alkoxyl and acyl radicals.

(5) We find that radicals are also spin trapped when the ozonated olefin solution is held at −78°C for as long as 17 hours before adding the spin trap and warming to room temperature. Under these conditions when PBN is the spin trap and the solution is warmed to room temperature in the presence of oxygen, we observe the transient peroxyl radical spin adduct and the more stable alkoxyl radical adduct.

Spin Trapping of Radicals from the Ozonation of Aldehydes and Acetals. One hypothesis that could explain our results is that ozonation of 2-methyl-2-pentene or methyl linoleate results in the production of an intermediate that decomposes to produce radicals that are then spin trapped. Decomposition of such an intermediate at a temperature that depends somewhat on the nature of the olefin could account for the appearance of a radical signal at −45°C with methyl linoleate and the rapid growth of the signal at this temperature with 2-methyl-2-pentene. It is known that tetroxides decompose at about −85°C (65) while dioxides are stable at −40°C; however, trioxide species such as R-CO-OOOH, ROOOR, or ROOOH decompose at temperatures between −50°C and −20°C (66-69). Furthermore, aldehydes are produced by ozonation of olefins (43) and are known to react with ozone to produce acyl hydrotrioxides (69-71) that appear to decompose, at least partly (70), to produce radicals. Thus, for example, one might explain radical production from 2-methyl-2-pentene from the reactions shown below:

$$Me_2C=CHEt + O_3 \xrightarrow{-78°} \underset{(IV)}{\underset{Me_2C-\!\!-CHEt}{\overset{O-O-O}{\overset{|\quad|}{}}}} \longrightarrow Me_2C=O-O + EtCHO \quad (6a)$$
$$\underset{(V)}{}$$

$$Me_2C=O-O + EtCHO \longrightarrow \text{Non-radical products} \qquad (6b)$$

$$EtCHO + O_3 \xrightarrow{-78°} \underset{(VI)}{Et\overset{O}{\overset{\|}{-C}}-OOOH} \xrightarrow{Warm} \text{Radicals} \qquad (7)$$

The radicals produced in equations (6) and (7) would then be proposed to initiate the autoxidation of the olefin or diene. This process involves the usual autoxidative chain sequence (40), with alkyl and peroxyl radicals as chain-carrying species.

Initiation:

$$(\text{Primordial radical}) + \text{Olefin} \xrightarrow[\text{H-abstraction}]{\text{Addition or}} \text{R} \cdot \qquad (8)$$

Propagation

$$\text{R} \cdot + \text{O}_2 \longrightarrow \text{ROO} \cdot \qquad (9)$$

$$\text{ROO} \cdot + \text{Olefin} \xrightarrow[\text{H-abstraction}]{\text{Addition or}} \text{R} \cdot \qquad (10)$$

We, therefore, have studied the ozonation of aldehydes by our spin-trap procedure. When propanal is ozonized in $CFCl_3$ at $-78°C$, flushed, and then allowed to warm to room temperature in the presence of PBN, a spin-adduct spectrum with $a^N = 13.2$ and $a^H = 1.4$ G is observed, starting at about $-45°C$, consistent with a spin-trapped peroxyl radical. Repeating this experiment with benzaldehyde as substrate also yielded a spin-trapped radical ($a^N = 13.2$ G and $a^H = 1.3$ G) although, in this case, the temperature at which we observe rapid formation of the spin adduct spectrum is about $-25°C$.

Trioxide species are also known to be formed in the ozonation of acetals (68,69). For example, ozonation of 2-phenyl-1,3- dioxolane produces an alkyl hydrotrioxide that has been identified by low temperature NMR (68). We therefore ozonated 1,3-dioxolane and 2-phenyl-1,3-dioxolane at $-78°C$ in $CFCl_3$. The 2-hydrotrioxide-1,3-dioxolane was produced in each case, as demonstrated by the appearance of an NMR absorbance at 13.1 ppm (66). We then applied the low temperature ozonation and spin-trapping protocol to these compounds using DMPO as the spin trap in $CFCl_3$; spin-adduct spectra are observed on warming that are characteristic of spin-trapped hydroperoxyl ($a^N = 13.0$ G, $a^H = 10.7$ G, and 1.6 G) (72) and alkoxyl radicals ($a^N = 13.0$ G, $a^H = 7.5$ G and = 1.7 G). Applying the same procedure to 2,2-dimethyl-1,3-dioxolane resulted in no spin-adduct spectra.

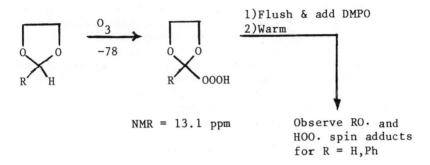

NMR = 13.1 ppm

Observe RO· and
HOO· spin adducts
for R = H,Ph

Detection of Radicals from Ozone-Olefin Reactions Using Radical Scavengers. In our spin-trap studies, it is possible that the spin adducts observed are produced by non-radical reactions that are induced by the reactive nitrone or nitroso spin traps. For this reason, we have performed some preliminary experiments with two radical scavengers that are not nitrone or nitroso compounds.

The stable free radical, 2,2-diphenyl-1-picrylhydrazyl (DPPH) reacts with most radicals and a bleaching of its characteristic violet color is observed (73). We therefore ozonated 3.8 x 10^{-4} moles of methyl linoleate in 5.0 ml of $CFCl_3$ to about 1% reaction using our low temperature protocol. We then flushed with nitrogen and added 1.1 x 10^{-6} moles of DPPH in 2.0 ml of $CFCl_3$ to the ozonated PUFA solution and warmed to room temperature. The DPPH solution became partially decolorized and we determined spectrophotometrically (λ_{max}=530,ε=11,700) that about 3.9 x 10^{-8} moles of DPPH was bleached. This experiment indicates that a solution approximately 6 x 10^{-6} M in radicals was produced; radicals present in that high a concentration could easily be detected by spin trapping methods. Phenolic compounds such as 2,6-di-*tert*-butyl-4-methyl-phenol (BHT) have also been used as traps for free radicals (73). We therefore ozonated a 1.0 M solution of 2-methyl-2-pentene at -78°C to about 10% reaction, flushed, and added a 0.1 M solution of BHT. The resulting solution was then warmed to room temperature in 5° increments in the ESR spectrometer and we observed rapid formation of the phenoxyl radical ESR spectrum at -45°C.

C. Conclusions and Some Mechanistic Speculations

Our results using spin traps leave little doubt that
ozone reacts with mono-olefins and dienes at -78°C in
processes that ultimately yield radicals. We believe these
radicals, in the absence of spin traps, react with the olefin
to initiate autoxidation and are responsible for the
radical-mediated biological effects of ozone.

It must be stressed that the yields of spin-trapped
radicals are small--less than 1% based on the moles of ozone
added. However, since these radicals initiate chain
processes--including the autoxidation of polyunsaturated fatty
acids--their chemical and biological effects can be magnified
(46-48, 56).

The mechanism(s) by which radicals arise is still
unclear, but several possibilities exist. The first
possibility is that ozone abstracts an allylic hydrogen to
produce radicals. Ozone does have diradical character (53),
but it seems unlikely that allylic H abstraction could compete
with addition of ozone to the double bond.

A second possibility is that the radicals result from
some process initiated by the carbonyl oxide. The carbonyl
oxide has diradical character (53) and it has been implicated
in several gas phase (53, 74,75) and solution phase (76)
reactions in which radicals appear to be involved. However,
the evidence for radical production from a carbonyl oxide in
solution is indirect, and there is no evidence that a carbonyl
oxide can initiate autoxidation, particularly in systems in
which aldehyde is present to react with the carbonyl oxide to
produce the Criegee ozonide(76a). There is, however, rather
convincing evidence that there are trioxide bonds in the
"oligomeric" material formed from ozonation of olefins, and it
has been suggested that these trioxide oligomers are formed
from the reaction of the carbonyl oxide with the
1,2,3-trioxolane (77).

A third possibility is that the 1,2,3-trioxolane opens to
a diradical rather than splitting to give the carbonyl oxide and
a carbonyl compound in one synchronous step. If the diradical
from the 1,2,3-trioxolane were to have a finite existence, it
could react either by a β-scission to give the carbonyl oxide
and carbonyl compound or it could undergo an intramolecular
H-transfer reaction (a "backbite" reaction). Although this type
of backbite reaction was thought to be necessary to rationalize
the products from ozonations in the gas phase at one time (43,
54), it now is no longer considered likely (78). The situation
for 1,4-dienes such as the polyunsaturated fatty acids may
be rather special in this regard, however; intramolecular

H-transfer in this case involves abstraction of an allylic
hydrogen, as we have pointed out before (79). It is
interesting that there are relatively few product studies of
the ozonolysis products from 1,4-dienes such as the
polyunsaturated fatty acids. Until such product studies are
done, a possible backbite reaction for 1,4-dienes must be
considered speculative.

A fourth hypothesis that is quite conservative in terms
of the ozonation literature is that a product from the
ozonation of the olefin reacts with more ozone to produce a
species that decomposes to form radicals. As indicated above,
it seems likely for at least some of the olefins we have
studied that the immediate precursor of the radicals is a
trioxygen species. A variety of types of compounds (43) that
are produced in the ozonation of olefins have hydrogens, like
those in acetals and aldehydes (66,68-71), that could react
with ozone to give hydrotrioxides. A collection of these
compounds is shown below.

$$\tag{11}$$

It seems possible that some of these compounds could react
with ozone to give trioxide species that give either
spin-trapped radicals or initiate olefin autoxidation.
Radical production by this indirect mechanism will only be
significant from a compound that is produced in non-trivial
yields from the given olefin and that reacts with ozone at a
rate that is fast enough to compete with the very fast
ozone-olefin reaction. For olefins that give aldehyde
products (such as methyl linoleate and 2-methyl-2-pentene),
most of the radicals may come from the acyl hydrotrioxide, as
indicated in eqs. 6-10 above. The reaction of ozone with
aldehydes may be sufficiently rapid to allow their ozonation
even in the presence of olefins: preliminary kinetics data
(80) indicate that propanal, the aldehyde formed from
2-methyl-2-pentene, reacts with ozone at room temperature
approximately 1000 times slower than does the olefin itself.
If this factor is not very different at -78°C, then ozonation
of a 0.1 M solution of 2-methyl-2-pentene to 10% conversion at
-78°C could yield acyl hydrotrioxide concentrations of

approximately (10%) (1/1000) (0.1M)=10^{-5}M. This appears
consistent with the detection of spin adducts in our
spin-trapping experiments, since we calculate the radicals we
trap are present at concentrations from 10^{-5} to 10^{-7} M.

Thus, we can summarize our current hypothesis for the
initiation of autoxidation of olefins by ozone as follows:
most of the reaction proceeds by way of the usual Criegee path
to give ozonides, carbonyl oxide, oligomers, polymers, and
carbonyl compounds. In a slower pathway, an unstable trioxide
species is produced that decomposes on warming to -50° to -20°
C (depending on the olefin) to give radicals that we spin trap
and that could initiate autoxidation. For olefins such as
methyl linoleate and 2-methyl-2-pentene, the trioxide may be
produced by subsequent ozonation of an aldehyde intermediate
to the acyl hydrotrioxide. However, some of the other
pathways outlined above must also play a role in radical
production since we find that tetramethyl-ethylene, an olefin
that cannot give an aldehyde on ozonation, also gives spin
adducts in our procedure. The nature of the mechanisms that
may operate in the absence of aldehyde products is one subject
of our continuing research.

ACKNOWLEDGMENT

This research was supported in part by grants from the
National Science Foundation and the National Institutes of
Health, Grant HL-16029.

REFERENCES

1. Neurath, G. B., and Dunger, M., in Int. Agency Res. Cancer,
    Sci. Publ. 9, p. 177. I. A. R. C., Lyon, (1974).
2. Bokhoven, D., and Niessen, H. J., Nature (London) 192, 458
    (1961).
3. Newsome, J. R., and Keith, C. H., Tob. Sci. 12, 216 (1978).
4. Melia, R. J. W., Florey, C. duV., Darkey, S. C., Palmers,
    E. D., and Goldstein, B. D., Atmos. Environ. 12, 1397
    (1978).
5. Wade, W. A., Cole, W. A., and Yocom, J. E., J. Air Pollut.
    Control Assoc. 25, 93 (1975).
6. Baldock, H., Levy, N., and Scaife, C. W., J. Chem. Soc.
    1949, 2627.
7. Brand, J. C. D., and Stevens, I. D. R., J. Chem. Soc. 1958,
    629.
8. Shechter, H., Rec. Chem. Prog. 25, 55 (1964)
9. Menzel, D. B., in "Free Radicals in Biology," Vol. II,
    (W. A. Pryor, ed.), p. 181. Academic Press, New York,
    (1976).

10. Titov, A. I., Tetrahedron 19, 557 (1963).
11. A. V. Topchiev, "Nitration of Hydrocarbons" pp. 226–68. Pergamon Press, New York, (1959).
12. Bonetti, G. A., DeSavigny, C. B., Michalski, C., and Rosenthal, R., Amer. Chem. Soc., Div. Petrol. Chem., Preprints 10, 135 (1965).
13. Juhos, L. T., Green, D. P., Furiosi, N. J., and Freeman, G., Am. Rev. Resp. Dis. 121, 541 (1980).
14. Freeman, G., Crane, S. C., Stephens, R. J., and Furiosi, N. J., Am. Rev. Resp. Dis. 98, 429 (1968).
15. Janoff, A., Carp, H., Lee, D. K., and Drew, R. T., Science 206, 1313 (1979).
16. Carp, H., and Janoff, A., Am. Rev. Resp. Dis. 118, 617 (1978).
17. Pryor, W.A., and Dooley, M.M., Biochem. Biophysics. Res. Comm. 106, 981 (1982).
18. Thomas, H. V., Mueller, P. K., and Lyman, R. L., Science 159, 532 (1968).
19. Mead, J. P., in "Free Radicals in Biology" Vol. I, (W. A. Pryor, ed.), p. 51. Academic Press, New York, (1976).
20. Pryor, W. A., Stanley, J. P., Blair, E., and Cullen, G. B., Arch. Environ. Health 31, 201 (1976).
21. Pryor, W. A., in "Molecular Basis of Environmental Toxicity" (R. S. Bhatnagar, ed.) p. 3. Ann Arbor Science Publishers, Ann Arbor, (1980).
22. Pryor, W. A., in "Environmental Health Chemistry" (J. D. McKinney, ed.) p. 445. Ann Arbor Science Publishers, Ann Arbor, (1980).
23. W. A. Pryor, in "Medicinal Chemistry" Vol. V (J. Mathieu, ed.) p. 331. Elsevier, Amsterdam, (1977).
24. Sosnovsky, G., "Free Radical Reactions in Preparative Organic Chemistry" pp. 255–269. The MacMillan Co., New York, (1964).
25. Yoshida, T., Yamamoto, F., and Namba, K, Kogyo Kagaku Zasshi 73, 519 (1970).
26. Akimoto, H., Sprung, J. L., and Pitts, J. N., Jr., J. Amer. Chem. Soc. 94, 4850 (1972).
27. Pryor, W. A., and Lightsey, J. W., Science, 214, 435 (1981); Pryor, W.A., Lightsey, J.W., and Church, D.F., J. Amer. Chem. Soc., in press.
28. Howard, J. A., and Ingold, K. U., Can. J. Chem. 44, 1119 (1966).
29. Redmond, T. F., and Wayland, B. B., J. Chem. Phys. 72, 1626 (1968).
30. Based on calculations using thermochemical data in: Benson, S. W., "Thermochemical Kinetics" pp. 271–310. John Wiley and Sons, New York, (1976).
31. Khan, N. A., J. Chem. Phys. 23, 2447 (1955).

32. Sprung, J. L., Akimoto, H., and Pitts, J. N., Jr., J. Amer.
     Chem. Soc. 96, 6549 (1974).
33. Ashmore, P. G., and Tyler, B. J., J. Chem. Soc. 1961, 1017.
34. Asquith, P. L., and Tyler, B. J., Chem. Comm. 1970, 744.
35. Nitrogen dioxide traps alkyl radicals faster than does
     oxygen: Washiba, N., and Bayes, K. D., Int. J. Chem.
     Kinet. 8, 777 (1976); Hochanadel, C. J., Ghormley, J. A.,
     Boyle, J. W., and Ogren, P. J., J. Phys. Chem. 81, 3
     (1977); Glanzer, K., and Troe, J., previously unpublished
     data prepared for "Reaction Rate and Photochemical Data
     for Atmospheric Chemistry - 1977" (N.B.S. Special Pub. No.
     513) (R. F. Hampson, Jr., and D. Garvin, ed.) p. 43. U.S.
     Gov't. Printing Office, Washington, (1978).
36. This explanation was originally used to rationalize the
     fact that free radical reaction of halogens with alkenes
     gives allylic substitution products at low halogen
     concentrations and addition products at higher
     concentrations (37).
37. Pryor, W. A., "Free Radicals" pp. 180–197. McGraw-Hill
     Book Co., New York, (1966).
38. To rationalize this small effect of oxygen at low nitrogen
     dioxide levels, we suggest that the rate of capture of (I)
     by oxygen (Eq. 5) may not be sufficiently rapid to prevent
     (I) from reverting to starting materials. The rate of
     trapping of (I) by oxygen is dependent on the rate
     constant, $k_3 = 3 \times 10^8$ (39) to 10^9 (40), the
     concentration of (I), and the oxygen concentration.
     Similarly, the rate of reversal of (I) to starting
     materials is dependent on k_2 (5×10^4 to 5×10^5
     in the gas phase) (32) and on the concentration of (I).
     Under the steady-state conditions of autoxidation, the
     estimated oxygen concentration in our reaction solutions
     is 10^{-3} to 10^{-4} molar (41,42). Therefore, the value
     of the ratio: (the rate of reversal of (I) to starting
     materials)/(the rate of trapping of (I)) lies between
     0.05–17, indicating that up to 94% of (I) can revert to
     starting material even in the presence of oxygen.
39. Hasegawa, K., and Patterson, L. K., Photochem. Photobiol.
     28, 817 (1978).
40. Ingold, K. U., Accounts Chem. Res. 2, 1 (1969).
41. Thomsen, E. S., and Gjaldbaek, J. Chrom., Acta Chemica
     Scan. 17, 127 (1963)
42. Osburn, J. O., and Markovic, P. L., Chem. Eng. 76, 105
     (1969).
43. Bailey, P. S. "Ozonations in Organic Chemistry," Vol. 1,
     Academic Press, New York, (1978).

44. Dungworth, D. L., Cross, C. E., Gillespie, J. R., and Plopper, C. G. in "Ozone Chemistry and Technology: A Review of the Literature", (J. S. Murphy and J. R. Orr, eds.), p. 27. The Franklin Institute Press, Philadelphia, (1975).

45. Jaffe, L. S., Arch. Environ. Health. 16, 241 (1968).

46. Menzel, D. B., Ann. Rev. Pharm. 10, 379 (1970).

47. Pryor, W. A. in "Free Radicals In Biology", Vol. 1, (W. A. Pryor ed.), p. 1. Academic Press, New York, (1976).

48. Tappel, A. L. in "Free Radicals In Biology", Vol. IV, (W. A. Pryor, ed.), Academic Press, New York, (1980).

49. Roehm, J. N., Hadley, J. C., and Menzel, D. B. Arch. Environ. Health. 24, 237 (1972).

50. Goldstein, D. B., Buckley, R. D., Cardenas, R., and Balchum, O. J., Science. 169, 605 (1970).

51. Pryor, W. A., and Stanley, J. P., J. Org. Chem. 40, 3615 (1975).

52. Criegee, R., Ann. Chem. 583, 1 (1953).

53. Wadt, W. R., and Goddard, W. A., J. Amer. Chem. Soc. 97, 3004 (1975).

54. O'Neal, G. E., and Blumstein, C., Int. J. Chem. Kinetics. 5, 397 (1973).

55. Pryor, W. A., and Kurz, M. E., Tetrahedron Letters. 698 (1978).

56. Pryor, W. A., Stanley, J. P., Blair, E., and Cullen, G. B., Arch. Environ. Health. 31, 201 (1976).

57. Pryor, W. A., Prier, D. G., and Church, D. F., Environ. Res. 24, 42(1981); idem, J. Amer. Chem. Soc., to be submitted.

58. Pryor, W. A., Prier, D. G., Lightsey, J. L., and Church, D. F., in "Autoxidation in Food and Biological Systems" (M. G. Simic, ed.), Plenum Publishing Corp., New York, in press.

59. Borg, D. C., in "Free Radicals in Biology" Vol. I, (W.A. Pryor, ed.), P. 69. Academic Press, New York, (1976).

60. Swartz, H. M., Bolton, J. R., and Borg, D. C., "Biological Applications of Electron Spin Resonance", John Wiley and Sons, New York, (1972).

61. Goldstein, B. D., Balchum, O. J., Demopoulos, H. B., and Duke, P. S., Arch. Environ. Health. 17, 46 (1968).

62. Janzen, E. G., Acct. Chem. Res. 4, 31 (1971).

63. Janzen, E. G. in "Free Radicals In Biology" Vol. IV, (W. A. Pryor, ed.), p. 116. Academic Press, New York, 1980. 64. Forrester, A. R., in "Magnetic Properties of Free Radicals" Vol. 9 Chapter 1 (K.-H. Hellwege, ed.), p. 192. Springer Verlag, New York, (1979).

65. Bartlett, P. D., and Guaraldi, G., J. Amer. Chem. Soc. 89, 4799 (1967).

66. Stary, F. E., Emge, D. E., and Murray, R. W., J. Amer. Chem. Soc. 96, 5671 (1974).

67. Bartlett, P. D., and Gunther, P., J. Amer. Chem. Soc. 88, 3288 (1966).

68. Kovac, F. and Plesnicar, B., J. Amer. Chem. Soc. 101, 2677 (1979).

69. Deslongchamps, P., Atlani, P., Frehel, D., Malaval, A., and Moreau, C., Can. J. Chem. 52, 3651 (1974).

70. White, H. M., and Bailey, P. S., J. Org. Chem. 30, 3037 (1965).

71. Syrov, A. A., and Tsykovskii, V. K., Z. Org. Khim. 6, 1392 (1970).

72. Harbour, J. R., and Hair, M. L., J. Phys. Chem. 82, 1397 (1978).

73. Forrester, A. R., Hay, J. M., and Thomson, R. H. in "Organic Chemistry of Stable Free Radicals", p. 138. Academic Press, London (1968).

74. Hull, L.A., Hisatsune, I.C., and Heicklen, J., Can. J. Chem. 51, 1504 (1973).

75. Mathia, E., Sanhueza, E., Hisatsune, I.C., and Heicklen, J., Can. J. Chem. 52, 3852 (1974).

76. Jerina, J.W., Jerina, D.M., and Witkop, B., Experimentia 28, 1129 (1972).

76a. Pryor, W.A., and Govindan, C.K., J. Amer. Chem. Soc. 103, 7681 (1981).

77. Greenwood, F.L., and Rubinstein, H., J. Org. Chem. 32, 3369 (1967).

78. Niki, H. in "Chemical Kinetic Data Needs for Modeling the Lower Atomosphere" (J.T. Henson, R.S. Huie, and J.A. Hodgeson, eds.), NES Special Publication 557, U.S. Government Printing Office, Washington, D.C.

79. Pryor, W.A., Photochem. Photobiol. 28, 787 (1978).

80. Pryor, W.A., and Ohto, J. Amer. Chem. Soc., submitted for publication.

COMPARATIVE ASPECTS OF SEVERAL MODEL LIPID PEROXIDATION SYSTEMS

Ming Tien
Steven D. Aust

Department of Biochemistry
Michigan State University
East Lansing, Michigan

The ground state of polyunsaturated fatty acids (PUFA) is of the singlet multiplicity. Their reaction with O_2 is spin forbidden since ground state O_2 is of triplet multiplicity. However, the uncontrolled, deleterious oxidation of PUFA does occur. The reaction of PUFA with O_2 to form lipid hydroperoxides (LOOH) must involve a mechanism that circumvents the spin barrier. One mechanism is the activation of triplet O_2 to singlet oxygen ($^1O_2$) which is highly reactive with diene bonds of PUFA to form LOOH [1,2]. Another mechanism is reaction via free radical mechanisms and indeed, lipid peroxidation is a free radical process.

If the initiation of lipid peroxidation is by methylene hydrogen abstraction from PUFA, the lipid radical is formed. The lipid radical reacts with O_2 to form the lipid peroxy radical [3]. The lipid peroxy radical abstracts a methylene hydrogen from another PUFA to form a LOOH and a second lipid radical. The lipid peroxy radical can also cyclize to form a five member endoperoxide radical. The lipid radical and the lipid endoperoxide radical can react with O_2 and other PUFA to continue the radical chain reaction. The breakdown of LOOH and endoperoxides leads to the formation of numerous products including malondialdehyde (MDA) [4]. The quantitation of MDA is a common assay for lipid peroxidation [5]. LOOH can also be quantitated by its reaction with KI [6].

Due to the toxic implications of lipid peroxidation to aerobic life, research into the mechanism of lipid peroxidation has been quite intense during the last fifteen years. One area of lipid peroxidation being intensely investigated is initiation via enzymatic systems. A common aspect of lipid peroxidation initiated by an enzymatic system is the passage of electrons from substrate to a system containing O_2 and chelated iron. We have investigated the mechanism of lipid peroxidation initiated by NADPH-cytochrome P450 reductase, which reduces chelated iron, and by xanthine oxidase, which reduces O_2. This paper reviews lipid peroxidation initiated by these two systems in addition to lipid peroxidation catalyzed by reductive cleavage of peroxides by Fe^{+2}.

I. NADPH-Dependent Lipid Peroxidation

Lipid peroxidation initiated by an enzymatic process was first demonstrated with rat liver microsomes. NADPH-dependent microsomal lipid peroxidation was shown to require ADP or pyrophosphate chelated iron for maximal activity (7-15). In the absence of iron or NADPH, no lipid peroxidation occurs in rat liver microsomes (Table I). The chelation of Fe^{+3} by ADP greatly stimulated activity while addition of both ADP-Fe^{+3} and EDTA-Fe^{+3} resulted in maximal activity. While other investigators have utilized EDTA as an inhibitor of lipid peroxidation (10,16), Pederson et al. (17) showed that EDTA-Fe^{+3} can stimulate NADPH-dependent

TABLE I. NADPH-Dependent Microsomal Lipid Peroxidation

	MDA
	nmol/min/ml
Control	0.03
NADPH	0.15
ADP-Fe^{+3}	0.41
EDTA-Fe^{+3}	0.49
NADPH, EDTA-Fe^{+3}	0.54
NADPH, ADP-Fe^{+3}	3.87
NADPH, ADP-Fe^{+3}, EDTA-Fe^{+3}	7.23
NADPH, ADP-Fe^{+3}, EDTA	0.19

Reaction mixture contained 0.5 mg of microsomal protein/ml in 50 mM Tris-Cl, pH 7.5 at 37°C. The following additions were made as indicated: 0.1 mM NADPH, ADP-Fe^{+3} (1.7 mM ADP, 0.1 mM $FeCl_3$), EDTA-Fe^{+3} (0.11 mM EDTA, 0.1 mM $FeCl_3$), and 0.11 mM EDTA. Reactions were initiated by the addition of NADPH and sampled periodically to assay for MDA (6).

lipid peroxidation. As shown in Table I, EDTA-Fe^{+3} cannot catalyze NADPH-dependent lipid peroxidation. It can enhance the rate of lipid peroxidation only in the presence of ADP-Fe^{+3}. However, the addition of EDTA, instead of EDTA-Fe^{+3} resulted in inhibition of activity, presumably by competing with ADP for the chelation of Fe^{+3}. This result is consistent with the proposal that initiation of NADPH-dependent lipid peroxidation is dependent on ADP-Fe^{+3}. Due to its Fe^{+3} chelating ability, EDTA would decrease the concentration of Fe^{+3} chelated by ADP and consequently decrease the rate of lipid peroxidation.

The role of NADPH-cytochrome P-450 reductase in lipid peroxidation was initially established, in part, by the ability of an antibody to the reductase to inhibit NADPH-dependent microsomal lipid peroxidation (18). Subsequent investigations demonstrated that NADPH-dependent lipid peroxidation could be reconstituted utilizing a purified protease solubilized NADPH-cytochrome P450 reductase (19). As shown in Table II, no lipid peroxidation of liposomes occurs in the absence of NADPH cytochrome P450 reductase (52 units/mg protein). The liposomes were made from rat liver microsomal lipid extracted anaerobically to minimize autoxidation (18). Both NADPH and ADP-Fe^{+3} are required for activity. The addition of EDTA-Fe^{+3} to this system greatly stimulated activity. This enhancement was initially observed by

TABLE II. NADPH-Dependent Liposomal Lipid Peroxidation

	MDA	LOOH
	nmol/min/ml	
Control	0.00	0.0
NADPH, ADP-Fe^{+3}	0.00	0.0
NADPH, Reductase	0.00	0.0
NADPH, Reductase, ADP-Fe^{+3}	0.30	1.8
NADPH, Reductase, ADP-Fe^{+3}, EDTA-Fe^{+3}	3.40	19.5
NADPH, Reductase, ADP-Fe^{+3}, cytochrome P450	1.51	6.6

Reaction mixtures contained 1 µmol of lipid phosphate/ml in 50 mM Tris-Cl, pH 7.5 at 37°C. The following additions were made as indicated: 0.1 mM NADPH, ADP-Fe^{+3} (1.7 mM ADP, 0.1 mM $FeCl_3$), EDTA-Fe^{+3} (0.11 mM EDTA, 0.1 mM $FeCl_3$), 0.1 unit NADPH-cytochrome P450 reductase/ml, and 0.3 nmol ferric cytochrome P450/ml. Reactions were initiated by the addition of NADPH and sampled periodically for MDA and lipid hydroperoxides (6).

Pederson and Aust (17,19), who suggested that EDTA-Fe^{+3} has catalytic activity similar to an endogenous microsomal component that participates in lipid peroxidation. Svingen and coworkers (20) reconstituted NADPH-dependent lipid per-oxidation with purified cytochrome P450. As shown in Table II ferric cytochrome P450 (13.2 nmol/mg protein), like EDTA-Fe^{+3}, enhances both MDA and LOOH formation. In addition, drug substrates that are metabolized by cytochrome P450 were found to inhibit NADPH-dependent microsomal lipid peroxidation and EDTA-Fe^{+3} could reverse this inhibition. This observation in addition to the reconstitution results led these investigators to propose that cytochrome P450 is involved in NADPH-dependent lipid peroxidation and that EDTA-Fe^{+3} had similar catalytic properties. It is impor-tant to point out that the enhancement of activity by cyto-chrome P450 is by the ferric form of the cytochrome since the protease solubilized reductase used in these experiments cannot reduce cytochrome P450 (21).

Like microsomes, the effect of EDTA on the rate of lipid peroxidation is exceedingly sensitive to whether Fe^{+3} is associated with the added EDTA. The results shown in Fig. 1 were from assays containing NADPH, ADP-Fe^{+3}, NADPH- cyto-chrome P450 reductase and EDTA-Fe^{+3} (1.1 EDTA to 1 Fe^{+3} molar ratio). Low amount of MDA is formed in the absence of EDTA-Fe^{+3}. Increasing EDTA-Fe^{+3} concen-tration up to 50 µM resulted in a corresponding increase in activity. Above 50 µM, EDTA-Fe^{+3} approaches saturation. Instead of having saturation kinetics, EDTA without iron has a peak of activity at 50 µM. At high concentrations, EDTA inhibits lipid peroxidation presumably due to its ability to chelate Fe^{+3} away from ADP. This again demonstrates the requirement for ADP chelated Fe^{+3} for initiation of NADPH-dependent lipid peroxidation. This result could explain the discrepancies in the literature concerning the effects of EDTA on lipid peroxidation.

II. Superoxide-Dependent Lipid Peroxidation

Enzymatic lipid peroxidation was also observed using xan-thine oxidase to generate superoxide (O_2^{-}). Interest in O_2^{-} mediated toxicity began with the realization that O_2^{-} was produced in many common biochemical reactions and that other enzymes catalytically scavenged this radical. O_2^{-} has been shown to be produced by the autoxidation of flavins (22) and ferrodoxins (23) in addition to certain enzymatic reactions (24,25). Superoxide dismutase (SOD), scavenging O_2^{-} at rates limited by diffusion (26,27), has been shown

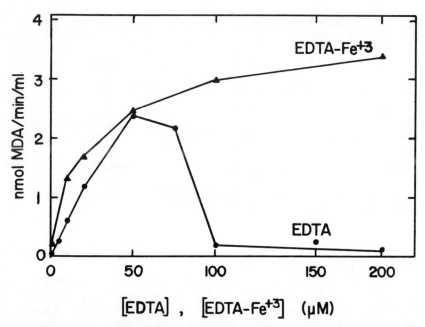

FIGURE 1.   Effect of EDTA and EDTA-Fe^{+3} on Lipid
Peroxidation.   Reaction mixtures contained liposomes (0.5
μmol lipid phosphate/ml), ADP-Fe^{+3} (2 mM ADP, 0.12 mM
Fe^{+3}), 0.2 mM NADPH, 0.28 mg NADPH-cytochrome P450
reductase/ml and the amount of EDTA (●——————●) or EDTA-Fe^{+3}
(▲——————▲) indicated in 0.25 M Tris-Cl, pH 6.8 at 37°C.
Reactions were initiated by the addition of NADPH and sampled
periodically for MDA (6).

to be ubiquitous throughout respiring organisms (28). O_2^-
mediated toxicity has been implicated in toxicity of paraquat
(29,30), alloxan (31), 6-hydroxydopamine (32), dialuric acid
(33) and adriamycin (34).
    Table III demonstrates the components required to cata-
lyze O_2^--dependent lipid peroxidation of liposomes made
from extracted rat liver microsomal lipid. The generation of
O_2^- and H_2O_2 by xanthine oxidase in the absence of iron
did not result in lipid peroxidation. These results confirm
the conclusion that neither O_2^- nor H_2O_2 nor a reaction
product of the two can initiate lipid peroxidation.   The
addition of ADP-Fe^{+3} resulted in activity as evidenced by
low levels of MDA (0.18 nmol/min/ml) and lipid hydroperoxide
(1.6 nmol/min/ml) formation. High rates of lipid peroxidation are attained only when either EDTA-Fe^{+3} or cytochrome P450 is included in the reaction mixture.

TABLE III.   Superoxide-Dependent Lipid Peroxidation

	MDA	LOOH
	nmol/min/ml	
Control	0.00	0.0
ADP-Fe^{+3}, EDTA-Fe^{+3}	0.02	0.2
Xanthine Oxidase	0.20	0.8
Xanthine Oxidase, ADP-Fe^{+3}	0.71	5.2
Xanthine Oxidase, EDTA-Fe^{+3}	0.02	0.9
Xanthine Oxidase, ADP-Fe^{+3}, EDTA-Fe^{+3}	4.17	13.6
Xanthine Oxidase, ADP-Fe^{+3},		
cytochrome P-450	5.10	24.1

Reaction mixtures contained liposomes (1 µmol lipid phosphate/ml) and 0.33 mM xanthine in 50 mM Tris-Cl, pH 7.5 at 37°C. The following additions were made as indicated; ADP-Fe^{+3} (1.7 mM ADP, 0.1 mM FeCl$_3$), EDTA-Fe^{+3} (0.11 mM EDTA, 0.1 mM FeCl$_3$), 0.1 unit xanthine oxidase/ml and 0.3 nmol cytochrome P-450/ml. Reactions were initiated by the addition of xanthine oxidase and assayed periodically for MDA and lipid hydroperoxides (6).

III. Lipid Hydroperoxide-Dependent Lipid Peroxidation

Lipid peroxidation can be thought of as a two-step process. The first involves a mechanism for the formation of free radicals and LOOH starting with peroxide-free PUFA. The second is the degradation of these lipid hydroperoxides to free radical products which can participate in the propagation of additional lipid peroxidation. Once formed, lipid hydroperoxides are relatively stable to uncatalyzed unimolecular decomposition (35). The calculated half-life for the uncatalyzed unimolecular homolysis of H_2O_2 at 30°C is 10^{11} years (3). However, in the presence of transition metals, hydroperoxides are readily decomposed by a redox mechanism generating free radicals (36,37,38). For example, iron readily catalyzes the decomposition of H_2O_2 in a Fenton-type reaction (39):

$$Fe^{+2} + H_2O_2 \longrightarrow Fe^{+3} + \cdot OH + {}^-OH \tag{1}$$

Iron has also been shown to catalyze the decomposition of lipid hydroperoxides to yield free radical products by a redox mechanism (35). In the presence of unsaturated lipid these free radicals can participate in the propagation of lipid peroxidation (37,40). It is well known that in the presence of lipid hydroperoxides, lipid peroxidation can be catalyzed by numerous heme compounds (41). This mechanism, which we term lipid hydroperoxide-dependent lipid peroxidation, can also be demonstrated by the addition of lipid or other organic peroxides to liver microsomes. In this case, cytochrome P450 reacts with the lipid hydroperoxides to promote peroxidation of microsomal lipid (42). The ability of cytochrome P450 to participate in hydroperoxide dependent-lipid peroxidation has also been demonstrated by reconstitution of lipid peroxidation utilizing purified cytochrome P450 (20). As shown in Table IV the action of soybean lipoxygenase on detergent-treated liposomes resulted in the formation of lipid hydroperoxides (0.5 nmol/min/ml). The inclusion of cytochrome P450 (0.3 nmol/ml) in this reaction mixture resulted in the formation of MDA and a 3-fold increase in rates of lipid hydroperoxide formation. In any system lacking cytochrome P450 or heme to participate in lipid hydroperoxide-dependent lipid peroxidation, the rate of lipid peroxidation would be much less. Lipid hydroperoxide-dependent lipid peroxidation can be added to any system by the inclusion of EDTA-Fe^{+3} and a system to reduce it.

EDTA-Fe^{+2} has been demonstrated to react with H_2O_2 to produce the hydroxyl radical ($\cdot$OH) (43):

$$\text{EDTA-}Fe^{+2} + HOOH \longrightarrow \text{EDTA-}Fe^{+3} + \cdot OH + {}^-OH \qquad (2)$$

TABLE IV. Effect of Cytochrome P-450 on Lipid Peroxidation

	MDA	LOOH
	nmol/min/ml	
Control	0.02	0.1
Cytochrome P-450	0.05	0.3
Lipoxygenase	0.08	0.5
Lipoxygenase, cytochrome P-450	0.88	1.4

Liposomal reaction mixtures contained 1 μmol lipid phosphate/ml, and 0.04% sodium deoxycholate in 50 mM Tris-Cl, pH 7.5 at 37°C. Ferric cytochrome P-450 (0.3 nmol/ml) and 100 μg of lipoxygenase/ml were added where indicated. Reactions were initiated by the addition of lipoxygenase and sampled periodically for MDA and lipid hydroperoxides (6).

There is good evidence to believe that lipid hydroperoxides can undergo a similar reaction with EDTA-Fe^{+2} to produce the lipidoxy radical (44):

$$EDTA\text{-}Fe^{+2} + LOOH \longrightarrow EDTA\text{-}Fe^{+3} + LO\cdot + {}^-OH \qquad (3)$$

To investigate the effect of different iron chelates on lipid hydroperoxide-dependent lipid peroxidation, lipid hydroperoxides were first generated by preincubation of detergent treated liposomes with soybean lipoxidase for 30 minutes (44). At the end of this period total lipid was extracted (45). These lipid hydroperoxides were mixed with unperoxidized phospholipid to make liposomes with a known content of lipid hydroperoxides for the studies shown in Table V. The ADP and EDTA ferric chelates have very little if any catalytic activity. Of the ferrous chelates, only the addition of EDTA-Fe^{+2} causes a significant enhancement in the rate of lipid peroxidation. Although ADP-Fe^{+2} does have some catalytic ability, it is apparent that its ability to promote lipid hydroperoxide-dependent lipid peroxidation is minimal

TABLE V.   Role of EDTA-Fe^{+3} in Lipid Hydroperoxide-Dependent Lipid Peroxidation

	MDA	LOOH
	nmol/min/ml	
Control	0.01	0.6
ADP-Fe^{+3}	0.02	0.6
EDTA-Fe^{+3}	0.01	1.1
ADP-Fe^{+2}	0.60	1.3
EDTA-Fe^{+2}	1.16	9.3
NADPH, Reductase	0.21	0.4
NADPH, Reductase, EDTA-Fe^{+3}	1.11	5.5
Xanthine Oxidase	0.20	0.8
Xanthine Oxidase, EDTA-Fe^{+3}	0.77	4.4

Liposomal reaction mixtures contained 1 μmol lipid phosphate/ml, and 0.1 μmol lipid hydroperoxide per μmol lipid phosphate in 50 mM Tris-Cl, pH 7.5 at 37°C. The following additions were made as indicated: ADP-Fe^{+3} (1.7 mM ADP, 0.1 mM FeCl$_3$), ADP-Fe^{+2} (1.7 mM ADP, 0.1 mM FeCl$_2$), EDTA-Fe^{+3} (0.11 mM EDTA, 0.1 mM FeCl$_3$), EDTA-Fe^{+2} (0.11 mM EDTA, 0.1 mM FeCl$_2$), 0.1 mM NADPH, 0.1 unit NADPH-cytochrome P-450 reductase/ml, and 0.1 unit xanthine oxidase/ml.

compared to that of EDTA-Fe^{+2}.   It is important to point out that EDTA-Fe^{+2} cannot initiate lipid hydroperoxide free lipid peroxidation (44).   It is also important to point out that these studies were carried out at pH 7.5.   At acidic pH the decompositions of lipid hydroperoxides with ferric iron can be significant.

Other investigators have proposed that lipid hydroperoxide-dependent lipid peroxidation can also be promoted by O_2^- in the absence of iron (46).

$$LOOH + O_2^- \longrightarrow LO\cdot + {}^-OH + O_2 \qquad (4)$$

This reaction, which is the lipid hydroperoxide equivalent to the Haber-Weiss reaction (47), has been proposed to

$$O_2^- + H_2O_2 \longrightarrow \cdot OH + {}^-OH + O_2 \qquad (5)$$

initiate O_2^- dependent lipid peroxidation.   However, the results shown in Table V indicate that no lipid peroxidation occurs in the absence of iron.   This demonstrates that O_2^- generated by xanthine oxidase or NADPH-cytochrome P450 reductase is not capable of promoting lipid hydroperoxide-dependent lipid peroxidation.   Lipid peroxidation does occur when EDTA-Fe^{+3} is added to the reaction mixture.   The rates of lipid peroxidation catalyzed by the enzymatic reduction of EDTA-Fe^{+3} are comparable to the rates observed by the direct addition of EDTA-Fe^{+2}.   In the xanthine oxidase system, EDTA-Fe^{+3} is reduced by O_2^- (48).   In systems reconstituted with NADPH-cytochrome P450 reductase, the reductase directly reduces EDTA-Fe^{+3} (20).

More recent results from this laboratory show that the rate of lipid hydroperoxide-dependent lipid peroxidation promoted by unchelated FeCl$_2$ is much higher than that of EDTA-Fe^{+2}.   This result suggests that chelation of Fe^{+2} by EDTA does not impart any special characteristics to Fe^{+2} in promoting lipid hydroperoxide-dependent lipid peroxidation.   There is, however, good reason to believe that chelation of Fe^{+3} by EDTA does facilitate the reduction of Fe^{+3} by O_2^- and by NADPH-cytochrome P450 reductase.   This would account for the enhancement in lipid peroxidation in both O_2^- and NADPH-dependent lipid peroxidation upon chelation of Fe^{+3} by EDTA.

IV. Initiation of Lipid Peroxidation.   Is $\cdot$OH Involved?

The past publications of this laboratory have all doubted the participation of $\cdot$OH in lipid peroxidation.   This doubt is contrary to the predominant view. Fong and coworkers (49) studied NADPH-dependent and O_2^--dependent lipid peroxidation and concluded that both systems generated $\cdot OH$ by an iron catalyzed Haber-Weiss reaction. SOD stimulated the system,

$$O_2^- + ADP\text{-}Fe^{+3} \longrightarrow O_2 + ADP\text{-}Fe^{+2} \tag{6}$$

$$2O_2^- + 2H^+ \longrightarrow H_2O_2 + O_2 \tag{7}$$

$$ADP\text{-}Fe^{+2} + H_2O_2 \longrightarrow ADP\text{-}Fe^{+3} + \cdot OH + {}^-OH \tag{8}$$

presumably by increasing the production of H_2O_2, while catalase and $\cdot OH$ traps inhibited activity. Kellogg and Fridovich (50,51) studied O_2^--dependent lipid peroxidation in a system which had no added iron. SOD and catalase both inhibited lipid peroxidation. Inclusion of $\cdot OH$ traps had very little effect but singlet oxygen traps were very effective inhibitors. They therefore concluded that $\cdot OH$ and singlet oxygen, both produced by the Haber-Weiss reaction (5) were responsible for initiating lipid peroxidation.

Lai and Piette (52) proposed that initiation of NADPH-dependent and O_2^- dependent lipid peroxidation occurs via the $\cdot OH$ on the basis of EPR spin trapping techniques. Utilizing the spin trap 5,5-dimethyl-1-pyrroline-1-oxide

TABLE VI. Mechanism of NADPH-Dependent Lipid Peroxidation

	MDA	LOOH
	nmol/min/ml	
Complete System	3.40	19.5
Complete System, SOD	0.50	3.5
Complete System, DPF	3.10	16.3
Complete System, BHT	0.00	0.0
Complete System, Benzoate	3.45	20.1

The complete system contained liposomes (1 μmol lipid phosphate/ml), 0.1 unit NADPH-cytochrome P-450 reductase/ml, ADP-Fe^{+3} (1.7 mM ADP, 0.1 mM $FeCl_3$), 0.1 mM NADPH, and EDTA-Fe^{+3} (0.11 mM EDTA, 0.1 mM $FeCl_3$) in 50 mM Tris-Cl, pH 7.5 at 37°C. The following additions were made where indicated: 1 unit SOD/ml, 0.2 mM DPF, 1 mM BHT and 40 mM benzoate. Reactions were initiated by the addition of NADPH and sampled periodically for MDA and lipid hydroperoxide (6).

(DMPO), these investigators detected a DMPO-OH radical adduct signal from a microsomal incubation mixture containing NADPH and EDTA-Fe^{+2}. The same signal was detected from a xanthine oxidase incubation mixture containing xanthine and EDTA-Fe^{+2}. These investigators proposed that ·OH is formed via the iron catalyzed Haber-Weiss reaction.

Our results with both NADPH-dependent and O_2^{-}-dependent lipid peroxidation are not in accord with the results of these laboratories. First of all, liver microsomes are heavily contaminated with catalase (53), yet microsomes very effectively promote NADPH-dependent lipid peroxidation (Table I). This would be unlikely if ·OH was generated by either the Haber-Weiss reaction or an iron catalyzed Haber-Weiss reaction. The mechanism of initiation is free radical since butylated hydroxy toluene (BHT) caused complete inhibition of activity (Table VI). The $^1O_2$ trap 2,5-diphenylfuran (DPF) did not cause significant inhibition of activity indicating that $^1O_2$ involvement in initiation is minimal. The involvement of ·OH also appears to be minimal due to lack of inhibition by 40 mM benzoate. SOD, however, caused 83% inhibition in both MDA and LOOH formation.

O_2^{-}-dependent lipid peroxidation was investigated by similar techniques (Table VII). SOD almost totally inhibited

TABLE VII. Mechanism of Superoxide-Dependent Lipid Peroxidation

	MDA	LOOH
	nmol/min/ml	
Complete System,	2.36	22.8
Complete System, SOD	0.03	5.0
Complete System, DPF	1.67	19.9
Complete System, BHT	0.01	0.1
Complete System, Benzoate	2.64	23.9

The complete system contained liposomes (1 μmol lipid phosphate/ml), ADP-Fe^{+3} (1.7 mM ADP, 0.1 mM FeCl$_3$), EDTA-Fe^{+3} (0.11 mM EDTA, 0.1 mM FeCl$_3$), 0.33 mM xanthine and 0.1 unit/ml xanthine oxidase in 50 mM Tris-Cl, pH 7.5 at 37°C. The following additions were made as indicated: 1 unit SOD/ml, 0.2 mM DPF, 1 mM BHT and 40 mM benzoate. Reactions were initiated by the addition of xanthine oxidase and sampled periodically for MDA and lipid hydroperoxides (6).

O_2^-- dependent lipid peroxidation presumably by blocking the reduction of iron. The slight inhibition of activity by DPF indicates that $^1O_2$ participation is minimal. Complete inhibition by BHT demonstrates that the mechanism involves free radicals. Catalase had no effect on activity suggesting again that H_2O_2 is not involved. If ·OH is not generated from H_2O_2, it does not appear to be generated by any other mechanism since benzoate had no effect on activity.

Our EPR spin trapping results show no evidence for ·OH formation in either the NADPH dependent or O_2^- dependent lipid peroxidation (54). As Finkelstein and coworkers (55) have pointed, great caution should be exercised in interpretation of EPR spectra. These workers have pointed out that the DMPO-OH radical EPR signal can be obtained by mechanisms not involving ·OH. The DMPO-O_2^- radical adduct can decompose to yield the DMPO-OH radical adduct. It is also important to point out that activities published by Lai and Piette (52) on NADPH-dependent microsomal lipid peroxidation are not expressed per mg protein. The maximal rates given were 1.5 nmol MDA/5 min in a incubation containing 2.1 mg protein/ml. If the sample size was 1 ml, then the rate would be 0.15 nmol MDA/min/mg protein. However these rates are not much greater than rates obtained by autoxidation of microsomes (Fig. 1). The rates we typically observe for NADPH-dependent microsomal lipid peroxidation are 20 to 30 times greater.

V. Lipid Peroxidation Promoted by a Fenton-Type Reaction

In order to further investigate the role of the ·OH in lipid peroxidation, a system of lipid peroxidation dependent upon the hydroxyl radical was developed. This system consists of Fe^{+2} and H_2O_2 interacting in a Fenton-type reaction (39). The generation of ·OH was confirmed by EPR spin trapping. No lipid peroxidation occurred in the absence of Fe^{+2} or H_2O_2 (Table VIII) or in the presence of catalase or mannitol. The chelation of iron by ADP decreased activity somewhat but more importantly, the activity was no longer sensitive to ·OH traps such as mannitol. One possible explanation is that free ·OH is never formed and that the ferryl ion, which may be equated to the ·OH is formed:

$$ADP\text{-}Fe^{+2} + H_2O_2 \longrightarrow ADP\text{-}[FeO]^{+2} + H_2O_2 \qquad (9)$$

Perhaps chelation of iron with ADP favors the ferryl ion as opposed to dissociation to free ·OH. If the ADP chelated

TABLE VIII. Hydroxyl Radical—Dependent Lipid Peroxidation

	MDA
	nmol/min/ml
Control	0.01
Fe^{+2}	0.15
H_2O_2	0.10
Fe^{+2}, H_2O_2	5.20
Fe^{+2}, H_2O_2, Catalase	0.25
Fe^{+2}, H_2O_2, Mannitol	0.12
ADP-Fe^{+2}	0.20
ADP-Fe^{+2}, H_2O_2	5.10
ADP-Fe^{+2}, H_2O_2, Mannitol	4.85

Liposomes, 1 μmol lipid phosphate/ml, were incubated in 30 mM NaCl, pH 7.5 at 37°C. The following additions were made as indicated: 0.2 mM $FeCl_2$, 0.1 mM H_2O_2, ADP-Fe^{+2} (1 mM ADP, 0.2 mM $FeCl_2$), 10 mM mannitol and 1 unit catalase/ ml. Reactions were initiated by the addition of iron and sampled periodically to obtain rates of MDA formation (6).

ferryl ion is not reactive with mannitol but capable of initiating lipid peroxidation, this would result in a mechanism appearing not to be dependent on the ·OH.

A comparison of the ·OH dependent lipid peroxidation system with our results on O_2^- dependent and NADPH-dependent lipid peroxidation leads us to believe that ·OH is not responsible for initiation in these latter model systems. We have previously proposed that initiation in both these model systems occurred by the same mechanism and that this mechanism has more biological significance than the ·OH. In our proposed model, ADP-Fe^{+3} is reduced by either NADPH- cytochrome P450 reductase in the case of microsomes, or by O_2^- to form ADP-Fe^{+3} - O_2^- $\longleftrightarrow$ ADP-Fe^{+2} - O_2. This complex then reacts with PUFA to initiate lipid peroxidation (Fig. 2). However, it is possible that this complex is further reduced to the ferryl ion which then initiates lipid peroxidation.

$$ADP\text{-}Fe^{+2} - O_2 \xrightarrow{\ e^-\ } ADP\text{-}Fe^{+2} - O_2^- \qquad (10)$$

$$ADP\text{-}Fe^{+2} - O_2^- \cdot \xrightarrow{\ e^-\ } [ADP\text{-}Fe^{+2} - O_2^=] \qquad (11)$$

$$[ADP\text{-}Fe^{+2} - O_2^=] \longrightarrow ADP\text{-}[FeO]^{+2} + H_2O \qquad (12)$$

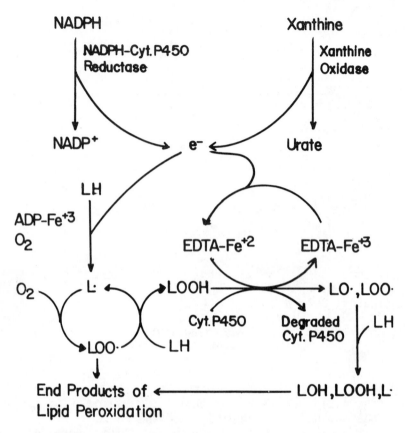

FIGURE 2.   Proposed Scheme of Lipid Peroxidation.

This mechanism could explain the lack of inhibition by catalase.   If ADP-Fe^{+3} and $O_2^=$ have a high binding constant, this would explain why free peroxide is not involved
in the mechanism.   This mechanism would also explain the lack
of inhibition by mannitol since the ADP-Fe^{+3} $O_2^=$
complex could be equated to the ADP-Fe^{+2} and H_2O_2
system which initiates lipid peroxidation but is not
inhibited by mannitol.

     In our scheme of lipid peroxidation, free radical forma-
tion in PUFA occurs via an activated iron oxygen complex.
This activated complex is generated by providing reducing
equivalents to a system containing ADP-Fe^{+3} and O_2.   In
NADPH-dependent lipid peroxidation, reducing equivalents are
provided by the NADPH-cytochrome P450 reductase catalyzed
reduction of iron by NADPH.   In $O_2^{\cdot-}$-dependent lipid peroxidation, reducing equivalents are provided by xanthine oxidase

catalyzed reduction of O_2 by xanthine. Previous publications from this laboratory have proposed that the activated complex is $ADP\text{-}Fe^{+3} - O_2^- \longleftrightarrow ADP\text{-}Fe^{+2} - O_2$ (44). However, more recent results from this laboratory suggest that this complex is precursor to a more reactive species which is responsible for initiation. There is good evidence to believe that an oxygen bridged di-iron complex ($ADP\text{-}Fe^{+2}O_2Fe^{+3}ADP$) intermediate is involved in the mechanism. The nature of the initial lipid free radical formed by this complex is yet to be determined. One of the free radicals formed would be the lipid free radical ($L\cdot$). This radical reacts with O_2 to form the lipid peroxyl radical ($LOO\cdot$) which can react with another PUFA to form the lipid free radical and lipid hydroperoxide. Lipid hydroperoxides can be decomposed to free radical products by $EDTA\text{-}Fe^{+2}$ or cytochrome P450. These radical can participate in the propagation reaction of lipid peroxidation and also breakdown to numerous products.

REFERENCES

1. Foote, C.S., Accounts Chem. Res. 1, 104 (1968).
2. Fenical, W., Kearns, D.R., and Radlick, P., J. Am. Chem. Soc. 91, 3396 (1969).
3. Pryor, W.A., Fed. Proc. 32, 1862 (1973).
4. Gardner, H.W., Kleiman, R., and Weisleder, D., Lipids 9, 696 (1974).
5. Bernheim, F., Bernheim, L.C., and Wilbur, K.M., J. Biol. Chem. 174, 257 (1948).
6. Buege, J.A., and Aust, S.D., in "Methods in Enzymology, Biomembranes C-Biological Oxidations", (Fleischer, S., and Packer, L., eds.), p. 302. Academic Press, New York (1978).
7. Wills, E.D., Biochem. J. 113, 315 (1969).
8. Wills, E.D., Biochem. J. 113, 325 (1969).
9. Beloff-Chain, A., Catanzaro, R., and Serlupi-Cescenzi, G., Nature 198, 351 (1963).
10. Hochstein, P., and Ernster, L., Biochem. Biophys. Res. Commun. 12, 388 (1963).
11. Hochstein, P., Nordenbrand, K., and Ernster, L., Biochem. Biophys. Res. Commun. 14, 323 (1964).
12. Beloff-Chain, A., Serlupi-Crescenzi, G., Cantanzaro, R., Venettacci, D., and Balliano, M., Biochim. Biophys. Acta 97, 416 (1965).
13. May, H.E., and McCay, P.B., J. Biol. Chem. 243, 2288 (1968).

14. Poyer, J.L., and McCay, P.B., J. Biol. Chem. 246, 263 (1971).
15. Ernster, L., and Nordenbrand, K., Methods Enzymol. 10, 547 (1967).
16. Noguchi, T., and Nakano, M., Biochim. Biophys. Acta 368, 446 (1974).
17. Pederson, T.C., and Aust, S.D., Biochim. Biophys. Acta 385, 232 (1975).
18. Pederson, T.C. Buege, J.A., and Aust, S.D., J. Biol. Chem. 248, 7134 (1973).
19. Pederson, T.C., and Aust, S.D., Biochem. Biophys. Res. Commun. 48, 789 (1972).
20. Svingen, B.A., Buege, J.A., O'Neal, F.O., and Aust, S.D., J. Biol. Chem. 254, 5892 (1979).
21. Coon, M.J., Strobel, H.W., and Boyer, R.F., Drug Metab. Disp. 1, 92 (1973).
22. Misra, H.P., and Fridovich, I., J. Biol. Chem. 247, 188 (1972).
23. Misra, H.P., and Fridovich, I., J. Biol. Chem. 246, 6886 (1971).
24. Fridovich, I., and Handler, P., J. Biol. Chem. 233, 1581 (1958).
25. Brady, F.O., Forman, H.J., and Feigelson, P., J. Biol. Chem. 246, 7119 (1971).
26. McCord, J.M., and Fridovich, I., Fed. Proc. 28, 346 (1968).
27. McCord, J.M., and Fridovich, I., J. Biol. Chem. 244, 6049 (1969).
28. McCord, J.M., Keele, B.B., and Fridovich, I., Proc. Nat. Acad. Sci., U.S.A. 68, 1024 (1971).
29. Bus, J.S., Aust, S.D., and Gibson, J.E., Biochem. Biophys. Res. Commun. 58, 479 (1974).
30. Bus, J.S., Aust, S.D., and Gibson, J.E., Res. Commun. Chem. Path. Pharmacol. 11, 31 (1975).
31. Cohen, G., and Heikkila, R.E., J. Biol. Chem. 249, 2447 (1974).
32. Heikkila, R.E., and Cohen, G., Science 181, 456 (1973).
33. Heikkila, R.E., and Cohen, G., Ann. N.Y. Acad. Sci. 258, 221 (1975).
34. Goodman, J., and Hochstein, P., Biochem. Biophys. Res. Commun. 77, 797 (1977).
35. O'Brien, P.J., Can. J. Biochem. 47, 485 (1969).
36. O'Brien, P.J., and Little C., Can. J. Biochem. 47, 493 (1969).
37. Sosnovsky, G., and Rawlinson, D.J., in "Organic Peroxides", Vol. II, (Swern, D. ed.), p. 153. Wiley-Interscience, New York (1971).

38. Uri, N., in "Autoxidation and Antioxidants", Vol. I, (Lundberg, W.O., ed.), p. 55. Interscience Publishers, New York (1961).
39. Fenton, H.J.H., J. Chem. Soc. 65, 899 (1894).
40. Demopoulos, H.B., Fed. Proc. 32, 1903 (1973).
41. Maier, V.P. and Tappel, A.L., J. Am. Oil Chemists' Soc. 36, 8 (1959).
42. O'Brien, P.J., and Rahimtula, A., J. Agric. Food Chem. 23, 154 (1975).
43. Walling, C., Partch, R.E., and Weil, T., Proc. Nat. Acad. Sci. U.S.A. 72, 140 (1975).
44. Svingen, B.A., O'Neal, F.O., and Aust, S.D., Photochem. Photobiol. 28, 803 (1978).
45. Pederson, T.C., and Aust, S.D., Biochem. Biophys. Res. Commun. 52, 1071 (1973).
46. Thomas, M.J., Mehl, K.S., and Pryor, W.A., Biochem. Biophys. Res. Commun. 83, 927 (1978).
47. Haber, F., and Weiss, J., Proc. R. Soc. Edinb. Sect. A. Math. Phys. Sci. 147, 332 (1934).
48. Ilan, Y.A., and Czapski, G., Biochim. Biophys. Acta 498, 386 (1977).
49. Fong, K., McCay, P.B., Poyer, J.L., Keele, B.B., and Misra, H., J. Biol. Chem. 248, 7792 (1973).
50. Kellogg, E.W. III, and Fridovich, I., J. Biol. Chem. 250, 8812 (1975).
51. Kellogg, E.W. III, and Fridovich, I., J. Biol. Chem. 252, 6721 (1977).
52. Lai, C.S., and Piette, L.H., Arch. Biochem. Biophys. 190, 27 (1978).
53. Welton, A.F., and Aust, S.D., Biochim. Biophys. Acta 373, 197 (1974).
54. Tien, M., Svingen, B.A., and Aust, S.D., Fed. Proc. (in press).
55. Finkelstein, E., Rosen, G.M., and Rauckman, E.J., Arch. Biochem. Biophys. 200, 1 (1980).

METABOLIC AND FUNCTIONAL SIGNIFICANCE
OF PROSTAGLANDINS IN LIPID PEROXIDE RESEARCH

Osamu Hayaishi
Takao Shimizu

Department of Medical Chemistry
Kyoto University Faculty of Medicine
Kyoto

Thiobarbituric acid (TBA) reaction has been widely used to determine the amount of so-called lipid hydroperoxides in the blood and various tissues (1-3). Although the exact mechanism of the reaction is yet to be elucidated, it has been generally assumed that lipid hydroperoxides undergo decomposition through several possible intermediates to produce malondialdehyde (MDA) (4,5) (Fig. 1). MDA thus produced reacts with TBA to yield a pink coloured TBA pigment with characteristic absorption and fluorescence spectra (3). It has also been known that during prostaglandin biosynthesis in the seminal vesicle, this TBA-reacting materials were produced significantly (6). TBA reactive material apparently decomposes under the conditions of the TBA test to produce MDA which then

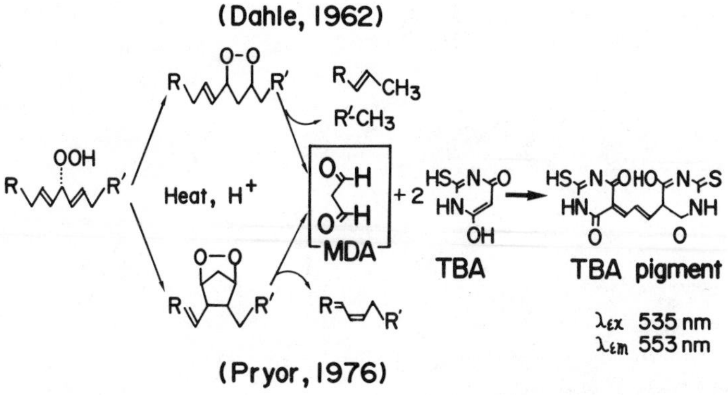

Figure 1. Lipid hydroperoxides and TBA reaction.

LIPID PEROXIDES IN BIOLOGY AND MEDICINE

41

reacts with TBA but the identity of these compounds has not
yet been well established. In this paper, we describe our
recent experiments indicating that prostaglandin (PG) endoperoxides, namely PGG and H, are TBA positive and also that when
they are decomposed either enzymatically or nonenzymatically,
MDA is produced in significant quantities both in vitro and in
vivo (7-9).

In Fig. 2 so-called arachidonate oxygenation pathways are
illustrated. Unsaturated fatty acids such as arachidonic acid
are known to be converted enzymatically to various hydroperoxy
acids such as 5, 12, and 15 hydroperoxy arachidonic acids (9-
12). It was recently shown by Samuelsson and co-workers that
5-hydroperoxy arachidonic acid is further converted to leuko-
triene C, or D, so-called slow reacting substance(SRS-A) (13).
These hydroperoxy unsaturated fatty acids exhibit various biological functions such as chemotaxis (14) and vasodilation,
cause some deteriorative reactions in various degenerative
disorders (15,16), and have generally been considered to be as
the major TBA reactive materials in various tissues and blood.
During the past ten years or so, rapid progress has been made
in the studies of PG and related compounds (17). Arachidonic
acid was shown to be oxygenated by cyclooxygenase to produce
PG endoperoxides such as G_2 and H_2. These unstable endoperoxides decompose spontaneously to MDA and hydroxy-heptadeca-
trienoic acid (HHT) (18,19). PG endoperoxides also undergo
enzymatic conversion to various so-called primary PG such as
E_2, D_2, and $F_{2\alpha}$. Subsequently two new important members of
this group, namely thromboxane A_2 and prostacyclin (PGI_2) were
discovered and these highly biologically important but

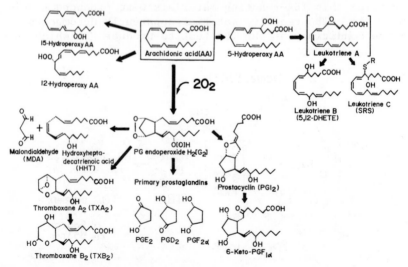

Figure 2. Arachidonate oxygenation pathways.

extremely unstable compounds were converted to thromboxane B_2 and 6-keto-$PGF_{1\alpha}$, respectively. We have purified the enzymes which catalyze isomerization of PG endoperoxides to these highly biologically active compounds and found that the production of MDA and HHT was accerelated by both thromboxane synthetase (7) and prostacyclin synthetase (8). Furthermore, we demonstrated that the nonenzymatic decomposition is greatly accelerated by the presence of both heme and thiol compounds (9). First the enzymatic formation of MDA during the biosynthesis of thromboxane A and PGI and then, nonenzymatic formation catalyzed by glutathione and heme compounds are described. Finally some of the in vivo studies indicating that these compounds are also responsible for the TBA reactive materials in vivo are discussed.

As shown in Fig. 3, when PGH_2 was incubated with thromboxane synthetase highly purified from blood platelets, thromboxane A_2 was produced which was converted non-enzymatically to a more stable compound thromboxane B_2 (7). However, during the course of this reaction HHT and MDA were also produced almost to the same extent. The data supporting this interpretation are presented in Fig. 4.

$[^{14}C]$ Carboxyl-labeled PGH_2 was incubated with thromboxane synthetase. Then the product of the reaction was extracted with organic solvents and was subjected to thin layer chromatography (TLC). When PGH_2 was the substrate, almost stoichiometric amounts of HHT and MDA were produced concomitant with the formation of thromboxane B_2. MDA was not radioactive, did not appear on TLC and was determined separately. However, when PGH_1 was used as substrate, much less thromboxane B_1 was produced and HHD (hydroxy-heptadecadienoic acid) became the major product of the reaction together with MDA. When the enzyme was inhibited by derivatives of imidazole or pyridine, both of these activities decreased to the same extent

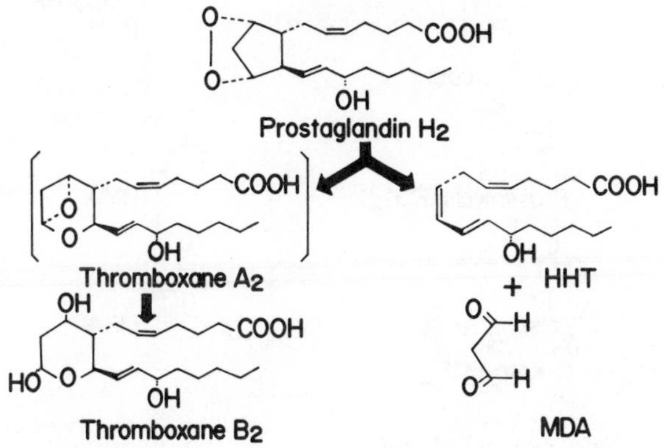

Figure 3. Conversion of PGH_2 to thromboxanes, HHT and MDA.

indicating that both reactions are catalyzed by the same enzyme.   Similar but more striking difference was observed when
prostacyclin synthetase was used instead of thromboxane

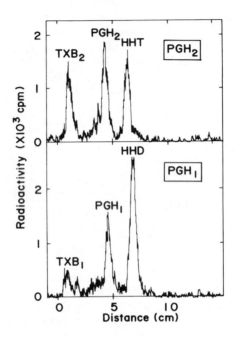

Figure 4.   Malondialdehyde formation by thromboxane synthetase.

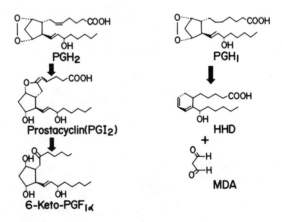

Figure 5.   Conversion of PGH$_2$ and H$_1$ to prostacyclin (PGI$_2$),
HHD and MDA.

synthetase.

When prostacyclin synthetase purified from rabbit aorta was used as the enzyme source, PGH_2 was predominantly converted to prostacyclin that decomposed to 6-keto-$PGF_{1\alpha}$ (Fig. 5). However, when PGH_1 was used as substrate, PGI_1 formation was hardly observed, and MDA and HHD became the major products of the reaction (8). Such data are shown in Fig. 6. When PGH_2 was used as substrate for PGI synthetase, the major product was the decomposition product of prostacyclin, namely 6-keto-$PGF_{1\alpha}$, but HHT formation was hardly observed. However, when PGH_1 was used as substrate, the formation of prostacyclin was almost negligible and HHD, the dienoic acid, became the major product of the reaction indicating that the structure of the substrate rather than the nature of the enzyme determines the quality and quantity of the products produced under these conditions. The formation of 6-keto-$PGF_{1\alpha}$ and HHD is inhibited by 15-hydroperoxy arachidonic acid, an inhibitor of PGI synthetase, indicating that both of these reactions are catalyzed by the same enzyme. Above findings were also observed by other investigators (20-22).

HHT was then isolated, derivatized and the structure determined by gas chromatography-mass spectrometric procedure as shown in Fig. 7. MDA concomitantly formed was identified by

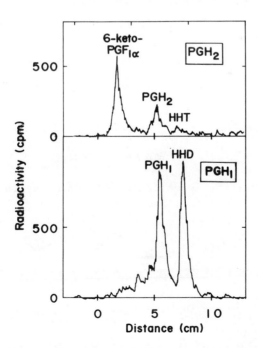

Figure 6.   Malondialdehyde formation by PGI synthetase.

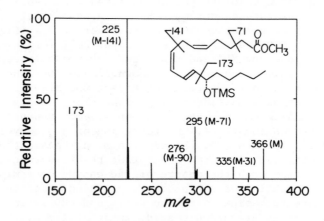

Figure 7.   Identification of the reaction product as HHT (12-
hydroxy-5,8,10-heptadecatrienoic acid) by GC-MS.

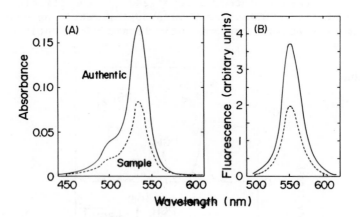

Figure 8.   Identification of malondialdehyde by TBA reaction.

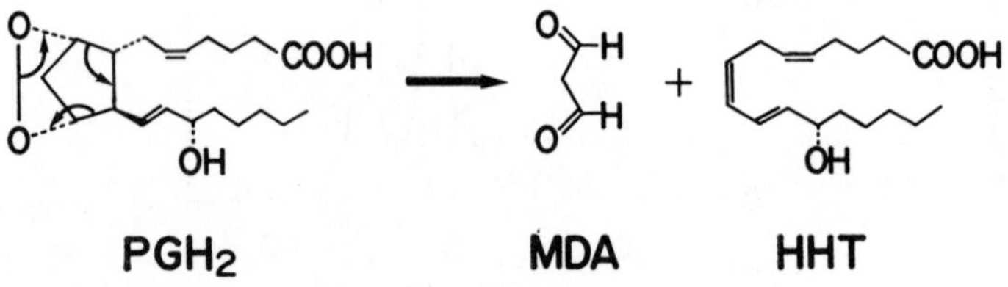

Figure 9.   Syntheses of malondialdehyde (MDA) and 12L-hydroxy-
5,8,10-heptadecatrienoic acid (HHT).

TBA reaction as shown in Fig. 8.

In panel (A), the solid line represents the absorption spectrum of the authentic sample and the dotted line represents that of the product of the reaction. In panel (B), are shown the fluorescence spectra of the authentic and isolated samples as described by Yagi (3) further supporting its identity as MDA. It is clear from these studies that during the conversion of the PG endoperoxides by either thromboxane or prostacyclin synthetase, MDA and HHT or HHD are produced concomitantly with the formation of either thromboxanes or prostacyclins. However, it has been known for some time that PG endoperoxides undergo nonenzymatic decomposition to produce MDA and HHT as shown in Fig. 9. As can be seen, PGH_2 undergoes spontaneous decomposition to produce MDA and HHT. This reaction is accelerated by acid and heat and therefore, under the conditions of TBA reaction, PGH_2 itself give positive TBA reaction. However, it has been reported from a number of laboratories that this reaction is accelerated by the boiled homogenates of different tissues (23,24). We have recently isolated materials responsible for this conversion from various tissues, purified them by column chromatography and identified them to be heme compounds and thiol compounds (9).

As can be seen in Fig. 10, when radiactive PGH_2 was incubated in the presence of heme and GSH, a radioactive compound, which was identified to be HHT on TLC, was produced with the concomitant formation of MDA. When either GSH or heme was omitted from the reaction mixture, this conversion was not observed to a significant extent. In Fig. 11 is shown the dependency of this conversion on various concentrations of GSH. The solid triangles indicate the HHT formation in the

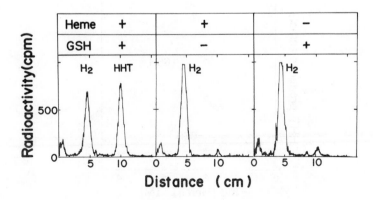

Figure 10. Conversion of PGH_2 to HHT in the presence of heme and GSH.

presence of hematin and the open triangles in its absence. As
can be seen, 10^{-3} to 10^{-5} molar concentrations of GSH was required under these conditions. Oxidized GSH was completely
inactive. NADH, NADPH, and ascorbic acid were all inactive
whereas other thiol compounds were as active as GSH (9). Fig.
12 shows the concentration dependency on heme in the presence
of GSH and its absence. 10^{-4} to 10^{-6} Molar concentrations of
heme are required for this conversion. Hemoproteins were also
active but inorganic iron was almost completely inactive.

Fig. 13 shows the TBA reactivity of MDA, various PG endoperoxides and hydroperoxy acids. As compared to authentic MDA,

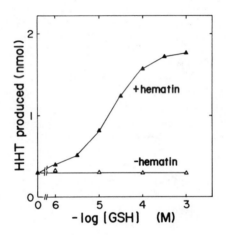

Figure 11.   Effect of GSH on $PGH_2 \longrightarrow HHT$.

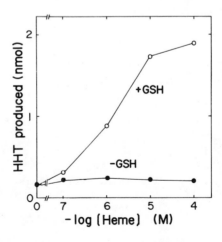

Figure 12.   Effect of heme on $PGH \longrightarrow HHT$.

PGG_2 is about 50% as reactive, followed by PGH_2 (30%). 15-Hydroperoxy arachidonic acid is much less reactive (6%) and primary PG such as E_2, D_2, $F_{2\alpha}$ are almost completely inactive (9). The low reactivity of lipid hydroperoxides was also pointed out by others (25). These results do indicate that MDA is produced during the process of enzymatic and nonenzymatic decomposition of PGG_2 and H_2 but also PGG_2 and H_2 are TBA reactive materials by themselves.

In order to determine what percentage of TBA reactive materials in the serum was due to lipid hydroperoxide and PG endoperoxide and their derivatives, we have carried out the following experiment. As shown in Fig. 14, unsaturated fatty acids can be converted by lipoxygenases to these hydroperoxides. On the other hand, arachidonic acid is converted by the prostaglandin synthetase system to various prostaglandin derivatives via PG endoperoxides such as G_2 and H_2 as common intermediates. The cyclooxygenase reaction is known to be inhibited specifically by nonsteroidal anti-inflammatory agents such as aspirin (26,27). However, lipoxygenases are generally not inhibited by aspirin and the conversion of the hydroperoxides to hydroxy compounds are inhibited by aspirin (28). It is, therefore, reasonable to assume that when aspirin is given to animals or humans, then the production of PG endoperoxides is inhibited whereas that of the lipid hydroperoxides is not inhibited. In fact, it may be expected to increase because the further conversion of hydroperoxides to hydroxy compounds is inhibited by aspirin. We, therefore, took rabbits, gave aspirin intravenously (50 mg/kg) and determined the amount of TBA reactive materials in the serum.

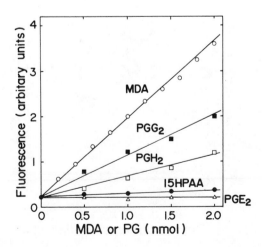

Figure 13. Reaction of PG endoperoxides with TBA reagent.

If lipid peroxides are the major TBA reactive materials in the
blood, then the TBA reactive materials should increase after
aspirin administration.   On the other hand, if PG endoperoxides and their derivatives are the major TBA reactive materials in the blood, then the administration of aspirin should
lead to decrease of TBA values in the serum.   The results of
such an experiment are illustrated in Fig. 15.
      We have taken 6 rabbits and determined serum TBA values

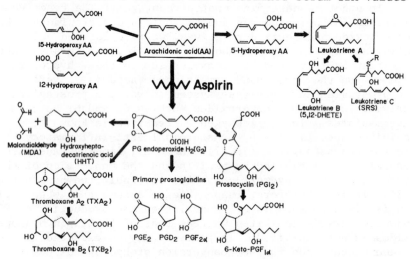

Figure 14.   Arachidonate oxygenation pathways.

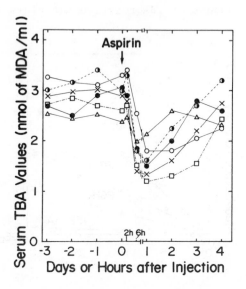

Figure 15.   Effect of aspirin on serum TBA values.

according to Yagi's procedure. As you can see, 6 hours after aspirin administration the TBA values went down by 25 to 60%, which then gradually increased after several days to return to the normal values. We also determined the amount of 13,14-dihydro-15-keto-PGF$_{2\alpha}$, the major serum metabolites of PGF$_{2\alpha}$, which also decreased to about 25% of the normal values after aspirin treatment concomitantly with the decrease in TBA values.

The results of inhibitor experiments must of course be interpreted with great caution but these results indicate that 25 to 60% of the so-called TBA reactive materials in the blood may be due to PG endoperoxides and their derivatives rather than unsaturated hydroperoxy fatty acids as has been generally believed.

In summary, it has been generally believed that serum thiobarbituric acid values represent lipid hydroperoxide levels namely conventional hydroperoxy unsaturated fatty acids, in the blood and tissues. However, PG endoperoxides such as G$_2$ and H$_2$ are also TBA positive and during the enzymatic conversion of these endoperoxides to prostacyclins and thromboxanes, a significant amount of MDA was produced concomitantly with HHT or HHD. Furthermore, these endoperoxides undergo rapid and non-enzymatic conversion to malondialdehyde in the presence of heme and reduced glutathione. After intravenous administration of aspirin, a cyclooxygenase inhibitor, the thiobarbituric acid reactive materials in the rabbit serum decreased by 25 to 60% with a concomitant decrease of serum PG levels. These results taken together strongly indicate that thiobarbituric acid values considered to be an indicator of lipid hydroperoxides are, to a significant extent, due to prostanoid hydroperoxides namely PG endoperoxides and their derivatives.

ACKNOWLEDGMENTS

We are grateful to Dr. N. Ohishi and Prof. K. Yagi of Nagoya University for critical discussion. Part of this work was carried out in collaboration with Dr. K. Kondo in our laboratory.

This work was supported in part by Grant-in-Aid for Scientific Research from the Ministry of Education, Science and Culture of Japan, and by grants from the Japanese Foundation on Metabolism and Diseases, Research Foundation for Cancer and Cardiovascular Diseases, and Japan Heart Foundation 1979.

REFERENCES

1. Bernheim, F., Bernheim, M. L. C., and Wilbur, K. M.
    (1948) J. Biol. Chem. 174, 257.
2. Niehaus, W. G. Jr., and Samuelsson, B. (1968) Eur. J.
    Biochem. 6, 126.
3. Yagi, K. (1976) Biochem. Med. 15, 212.
4. Dahle, L. K., Hill, E. G., Holman, R. T. (1962) Arch.
    Biochem. Biophys. 98, 253.
5. Pryor, W. A., Stanley, J. P., and Blair, E. (1976) Lipids
    11, 370.
6. Flower, R. J., Cheung, H. S., and Cushman, D. W. (1973)
    Prostaglandins 4, 325.
7. Yoshimoto, T., Yamamoto, S., Okuma, M., and Hayaishi, O.
    (1977) J. Biol. Chem. 252, 5871.
8. Watanabe, K., Yamamoto, S., and Hayaishi, O. (1979)
    Biochem. Biophys. Res. Commun. 87, 192.
9. Shimizu, T., Kondo, K., and Hayaishi, O. (1981) Arch.
    Biochem. Biophys. 206, 271.
10. Hamberg, M., and Samuelsson, B. (1967) J. Biol. Chem.
    242, 5336.
11. Nugteren, D. H. (1975) Biochim. Biophys. Acta 380, 299.
12. Borgeat, P., Hamberg, M., and Samuelsson, B. (1976) J.
    Biol. Chem. 251, 7816.
13. Örning, L., Hammarström, S., and Samuelsson, B. (1980)
    Proc. Natl. Acad. Sci. USA 77, 2014.
14. Goetzl, E. J., and Gorman, R. R. (1978) J. Immunol. 120,
    526.
15. Desai, I. D., and Tappel, A. L. (1963) J. Lipid Res. 4,
    204.
16. Tappel, A. L. (1973) Fed. Proc. 32, 1870.
17. Samuelsson, B., Goldyne, M., Granström, E., Hamberg, M.,
    Hammarström, S., and Malmsten, C. (1978) Ann. Rev.
    Biochem. 47, 997.
18. Hamberg, M., and Samuelsson, B. (1974) Proc. Natl. Acad.
    Sci. USA 71, 3400.
19. Nugteren, D. H., and Hazelhof, E. (1973) Biochim. Biophys.
    Acta 326, 448.
20. Hammarström, S., and Farlardeau, P. (1977) Proc. Natl.
    Acad. Sci. USA 74, 3691.
21. Wlodawer, P., and Hammarström, S. (1979) FEBS. Lett. 97,
    32.
22. Wlodawer, P., and Hammarström, S. (1980) Prostaglandins
    19, 969.
23. Shimizu, T., Yamamoto, S., and Hayaishi, O. (1979) J.
    Biol. Chem. 254, 5222.
24. Pace-Asciak, C. K., and Nashat, M. (1976) J. Neurochem.
    27, 551.

25. Ohkawa, H., Ohishi, N., and Yagi, K. (1978) J. Lipid Res. 19, 1053.
26. Vane, J. R. (1971) Nature New Biol. 231, 232.
27. Miyamoto, T., Ogino, N., Yamamoto, S., and Hayaishi, O. (1976) J. Biol. Chem. 251, 2629.
28. Siegel, M. I., McConnel, R. T., and Cuatrecasas, P. (1979) Proc. Natl. Acad. Sci. USA 76, 3774.

MICROSOMAL LIPID PEROXIDATION:
MECHANISM AND SOME BIOMEDICAL IMPLICATIONS

Lars Ernster
Kerstin Nordenbrand

Department of Biochemistry
Arrhenius Laboratory
University of Stockholm
Stockholm, Sweden

Sten Orrenius

Department of Forensic Medicine
Karolinska Institutet
Stockholm, Sweden

I. INTRODUCTION

The occurrence of enzymically induced peroxidation of lipids in rat liver microsomes was first reported in 1963 by
Hochstein and Ernster (1). The reaction required, besides molecular oxygen, the presence of NADPH, and was greatly enhanced
by nucleoside di- or triphosphates, as well as inroganic pyrophosphate, in combination with ferrous or ferric iron (2). The
properties of the reaction were compared with those of the non-
enzymic peroxidation of lipids induced by ascorbate (Fig. 1)
(1-3). Evidence was presented (4-7) that the enzymically induced lipid peroxidation involved the microsomal flavoprotein

---

Abbreviations: DPPD, diphenyl-*p*-phenylenediamine; DPTA, diethy-
lenetriamine pentaacetic acid; MA, malonaldehyde; pOHMB, *p*-hyd-
roxymercuribenzoate; PP_i, inorganic pyrophosphate; SKF-525A,
β-diethylaminoethyl-diphenylpropylacetate; TCA, trichloroacetic acid.

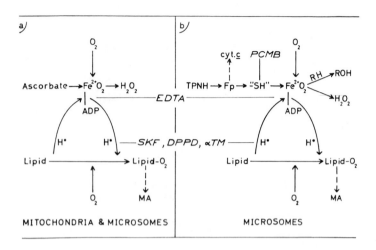

Fig. 1. Schematic representation of lipid peroxidation linked to ascorbate or TPNH (NADPH) oxidation in microsomes and mitochondria.   From Hochstein and Ernster (1).

known as NADPH-cytochrome c reductase, subsequently identified as NADPH-cytochrome P-450 reductase. Drugs undergoing hydroxylation by way of the cytochrome P-450-linked monooxygenase system were shown to inhibit NADPH-linked lipid peroxidation (4). Data were also reported (3) which indicated that lipid peroxidation drastically alters the structure and enzymic activities of the microsomes. Various aspects of microsomal lipid peroxidation have been the subject of extensive studies in many laboratories over the past years (8-32).

This paper summarizes our present state of knowledge of the microsomal lipid-peroxidation system, based partly on the above information and partly on unpublished results that have been obtained in our laboratory. Data concerning the qualitative and quantitative requirements of the reaction are presented, together with information regarding its stoichiometry, kinetics, inhibitors, and its relationship to drug metabolism. The results are discussed in relation to current proposals concerning the mechanism of enzymically induced microsomal lipid peroxidation. The effects of lipid peroxidation on microsomal structure and function have also been studied in some detail. Some biomedical implications of these effects are considered.

II. EXPERIMENTAL

Sprague-Dawley rats of both sexes, weighing 150-250 g, were used. The animals were starved overnight before sacrifice. Liver microsomes were prepared from homogenates made in 0.25 M sucrose as described by Ernster *et al.* (33). The microsomal pellets were rinsed 2-3 times with 0.15 M KCl in order to remove sucrose (which interferes with the colorimetric assay for malonaldehyde). The preparation was finally suspended in 0.15 M KCl to contain 20 mg microsomal protein per ml. The biuret method was used to determine microsomal protein.

Lipid peroxidation was followed by measuring O_2 consumption polarographically with a Clark electrode, NADPH oxidation fluorometrically with an Eppendorf fluorimeter, and malonaldehyde colorimetrically with the thiobarbituric acid reaction (34). The incubation medium consisted of 25 mM Tris-Cl buffer, pH 7.5, and 0.15 M KCl. Concentrations of microsomes, NADPH and other additions are specified in the figure and table legends.

Light-scattering changes of the microsomes were followed by measuring the optical density at 520 nm. Assays of NADH- and NADPH-cytochrome *c* reductase (33), glucose-6-phosphatase (33) and IDPase (35) activities of the microsomes were made as previously described.

Fe^{2+} was determined by the method of Diehl and Smith (36), using bathophenanthroline sulfonate. Fe^{3+} was estimated by the same method, after reduction with hydroxylamine. When inorganic pyrophosphate or a nucleoside di- or triphosphate was present, it was necessary to hydrolyze these compounds (by treatment with 1 M HCl in boiling water-bath for 10 minutes) before adding the bathophenanthroline reagent.

All chemicals were commercial products. Special care was taken to use batches of nucleoside di- and triphosphates as free as possible from contaminating iron (*cf.* ref. 2).

III. RESULTS

A. *Parameters, kinetics, stoichiometry*

As reported earlier (1-5), NADPH-linked lipid peroxidation in rat liver microsomes can suitably be followed by measuring (a) O_2 consumption; (b) NADPH disappearance; (c) malonaldehyde formation. Results of an experiment involving the measurement of all three parameters are shown in Fig. 2. Addition of 0.05 mM NADPH to a microsomal suspension resulted in low rates of

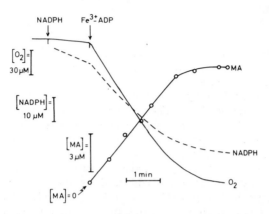

Fig. 2. NADPH- and ADP-Fe³⁺-induced lipid peroxidation in rat liver microsomes. The reaction mixture contained 25 mM Tris-Cl, pH 7.5, 150 mM KCl, and 3.8 mg microsomal protein in a final volume of 3 ml. Additions: 50 μM NADPH and 1 mM ADP + 10 μM Fe³⁺. Temperature, 25°C.

O_2 consumption and NADPH disappearance and no formation of malonaldehyde. Upon the addition of 10 μM $FeCl_3$ and 1 mM ADP (ADP-Fe^{3+}) there was a marked rise in the rates of O_2 consumption and NADPH disappearance, and malonaldehyde was formed at a linear rate. When all NADPH was consumed, the O_2 uptake and malonaldehyde formation ceased, and could be re-initiated by the addition of NADPH or of an NADPH-regenerating system (not shown).

The ADP-Fe^{3+}-induced increase in O_2 consumption and NADPH oxidation as well as the accompanying malonaldehyde formation showed Michaelis-Menten kinetics with respect to NADPH concentration (K_m = 0.55 μM) and were competitively inhibited by $NADP^+$ (K_i = 7.2 μM), consistent with the involvement of NADPH-cytochrome P-450 reductase (37). In contrast, the NADPH oxidation occurring in the absence of ADP-Fe^{3+} exhibited nonhyperbolic kinetics, required relatively high concentrations of NADPH (half-maximal rate at 0.1 mM), and was not inhibited by $NADP^+$. This reaction, which probably is identical with the "NADPH oxidase" originally described by Gilette *et al.* (38) evidently is unrelated to NADPH-cytochrome P-450 reductase and lipid peroxidation.

The molar ratio, O_2/NADPH, that can be deduced from Fig. 2 based on the *increments* in the rates of O_2 consumption and NADPH oxidation following the addition of ADP-Fe^{3+}, is approximately 6. This means that for each molecule of NADPH oxidized by O_2 via NADPH-cytochrome P-450 reductase and ADP-Fe^{3+} (lead-

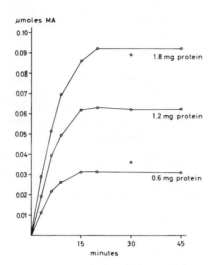

*Fig. 3. Lipid peroxidation with varying amounts of micro-
somes. The reaction mixture contained 25 mM Tris-Cl, pH 7.5,
150 mM KCl, 0.5 mM NADP⁺, 10 mM isocitrate, 0.1 mM MnCl₂, 40
mM nicotinamide, 1 unit of isocitrate dehydrogenase and 10 µM
PPᵢ + 10 µM Fe³⁺ in a final volume of 2 ml. Temperature, 30°C.*

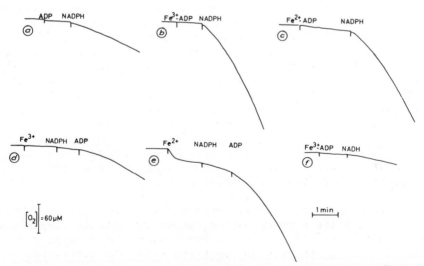

*Fig. 4. Comparison of the effects of NADPH and NADH and of
Fe³⁺ and Fe²⁺ on lipid peroxidation.*
*The reaction mixture contained 25 mM Tris-Cl, pH 7.5, 150
mM KCl, 0.3 mM NADPH or NADH, 1 mM ADP, 12 µM Fe²⁺ or Fe³⁺,
and 3.6 mg microsomal protein in a final volume of 3 ml. Temperature 30°C.*

ing to the formation of H_2O_2; see below), approximately 5 molecules of O_2 are incorporated into lipids to form lipid peroxides.

The formation of H_2O_2 as a reaction product was demonstrated by performing the lipid peroxidation in the presence of catalase and methanol, and measuring the amount of formaldehyde formed. Although the microsomal preparation contained some catalase, it was necessary to add rather large amounts of catalase to ensure efficient removal of H_2O_2. In agreement with earlier reports (25,28), this did not inhibit lipid peroxidation.

The molar ratio, O_2/malonaldehyde, as deduced from Fig. 2, is appr. 23. When corrected for the amount of O_2 accounted for by NADPH oxidation (by way of both "NADPH oxidase" and ADP-Fe^{3+}-induced NADPH-cytochrome P-450 reductase, assuming that both reactions lead to H_2O_2 formation, i.e., to the consumption of one mole of O_2 per mole of NADPH) the O_2/malonaldehyde ratio becomes appr. 16. This is close to the value reported in the literature (39) for arachidonic acid, which has been found by May and McCay (13) to be the quantitatively preponderant component decreasing in level during microsomal lipid peroxidation.

Both the rate and the extent of lipid peroxidation were proportional to the amount of microsomal protein (Fig. 3). The maximal extent of lipid peroxidation resulted in the formation of about 5 nmoles of malonaldehyde, corresponding to the consumption of about 1 µmole of O_2, per mg protein.

Omission of ADP or Fe^{3+}, or replacement of NADPH by NADH, resulted in little or no lipid peroxidation (Fig. 4). Fe^{2+} was able to substitute for Fe^{3+} in conjunction with ADP. When added alone, Fe^{2+} induced a transient O_2 uptake (and malonaldehyde formation), which was independent of NADPH (see further Fig. 11).

B. *Comparison with ascorbate-induced lipid peroxidation. Effects of various iron chelates*

As reported earlier (2), replacement of NADPH by ascorbate resulted in a lipid peroxidation, which was likewise dependent on added ADP-Fe^{2+} (Fig. 5). The ascorbate-induced lipid peroxidation was nonenzymic and occurred also with heat-denatured microsomes as well as with mitochondria.

Various nucleoside di- and triphosphates activated NADPH-linked lipid peroxidation in a fashion similar to ADP; nucleoside monophosphates were inactive (Fig. 6). Maximal activities were reached at nucleotide concentrations of 0.2-1 mM. Inorganic pyrophosphate (PP$_i$) also activated the NADPH-linked lipid peroxidation, but maximal activity occurred at about

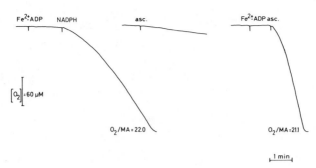

Fig. 5. Comparison of the effects of NADPH and ascorbate on lipid peroxidation. The reaction mixture contained 25 mM Tris-Cl, pH 7.5, 150 mM KCl, 0.3 mM NADPH or 1 mM ascorbate, 1 mM ADP + 12 µM Fe²⁺, and 3.2 mg microsomal protein in a final volume of 3 ml. Temperature, 30°C.

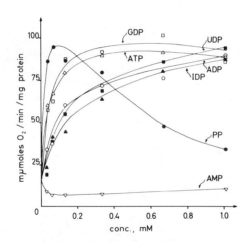

Fig. 6. Effects of various nucleotides and of inorganic pyrophosphate on NADPH-linked lipid peroxidation. From Hochstein et al. (2).

0.05–0.1 mM, above which concentration PP_i was inhibitory. In all cases, Fe^{2+} and Fe^{3+} were equally efficient.

The optimal concentration of Fe^{2+} (or Fe^{3+}) was 0.04 mM in the case of ADP (Fig. 7) and with other nucleotides. In the case of PP_i, the optimal iron concentration varied with the concentration of PP_i as shown in Fig. 8. In the range of 0.05–0.1 mM PP_i, which gave maximal activity, the optimal iron concentration was 0.012 mM. Iron oxalate also activated NADPH-

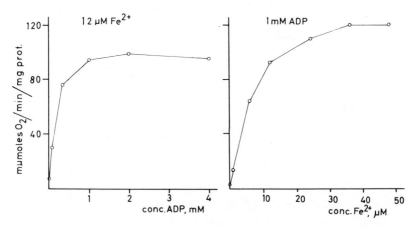

Fig. 7. Effects of varying concentrations of ADP and Fe^{2+} on NADPH-linked lipid peroxidation.   From Hochstein et al. (2).

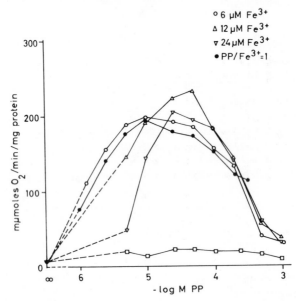

Fig. 8. Effect of varying concentrations of pyrophosphate and Fe^{3+} on NADPH-linked lipid peroxidation.
The reaction mixture contained 25 mM Tris-Cl, pH 7.5, 0.15 M KCl, 0.15 mM NADPH, 2.8 mg microsomal protein, and pyrophosphate + Fe^{3+} as indicated. Final volume, 3 ml. Temperature 23°C.

linked lipid peroxidation (Fig. 9); 0.02 mM iron and 1 mM oxalate gave maximal activity. In all cases, Fe^{2+} and Fe^{3+} were equally efficient.

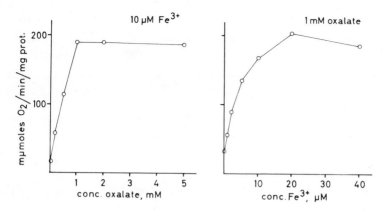

Fig. 9. *Effect of varying concentrations of oxalate and* Fe^{3+} *on NADPH-linked lipid peroxidation. Conditions were as in Fig. 8, except that oxalate + Fe^{3+} were added as indicated.*

Table I compares the capacities of various iron chelates to support NADPH- and ascorbate-linked lipid peroxidation in liver microsomes. The nucleoside di- and triphosphate (NDP and NTP) and the oxalate complexes of iron were efficient activators of lipid peroxidation with both types of electron donor. Iron-PP_i was only active in the case of NADPH; in the case of ascorbate PP_i was inactive at any concentration tested (0.007--1 mM) and it also inhibited ascorbate-linked lipid peroxidation as induced by other iron chelates. Conversely, iron-malonate was highly active in the case of ascorbate but only slightly active in the case of NADPH. The iron chelate of 8-hydroxyquinoline was active with NADPH; the activity with ascorbate was not tested. The cyanide, EDTA and 1,10-phenanthroline chelates of iron were inactive with both types of electron donor. These compounds were also able to inhibit lipid peroxidation as induced by other iron complexes. The inhibitory effect of these complexes was dependent on their concentration in relation to the concentration of activating iron complex. As a general rule, all iron chelates capable of supporting NADPH- or ascorbate-induced lipid peroxidation seem to be in the redox potential range of +0.1 to -0.15 V. Ferredoxins, cytochrome c, cytochrome b_5, hemoglobin, methemoglobin, catalase, horseradish peroxidase and ferritin neither activated nor inhibited NADPH- or ascorbate-linked lipid peroxidation in microsomes. Cu^{2+} did not replace iron in connection with any of the chelating agents studied.

Also indicated in Table I are the effects of Fe^{2+} and Fe^{3+} alone, without any added chelating agent, on the NADPH- and

Table 1. Iron Chelates Inducing Microsomal Lipid Peroxida-tion[a]

Ligand	E'_0	NADPH	Ascorbate
	Volts	nmoles O_2/min.	
1,10-Phenanthroline	1.06	1	15
H_2O	0.77	1[b]	15[b]
CN	0.36	1	19
EDTA	0.25	1	15
Malonate	0.1	22	232
Oxalate	-0.01	208	289
NDP, NTP		186	400
PP_i	-0.10	214	4
8-OH-Quinoline	-0.15	135	
Ferredoxin (bact.)	-0.42	1	11
Ferredoxin (spinach)	-0.42	8	

[a]*The reaction mixtures contained 25 mM Tris-Cl, pH 7.5, 150 mM KCl, 0.3 mM NADPH or 1 mM ascorbate, and 4.3 mg microsomal protein in a final volume of 3 ml. Fe^{2+} or Fe^{3+} were added in final concentrations of 12 µM together with the ligands indicated. The concentrations of the ligands were as follows: 1,10-phenanthroline, 1 mM; EDTA, 1 mM; ferro-or ferricyanide, 12 µM; malonate, 1 mM; oxalate, 1 mM; NDP or NTP, 1 mM; PP_i, 0.1 mM; 8-OH-quinoline, 40 µM, ferredoxin, 70 µg (bact.) or 53 µg (spinach). Temperature, 30°C. E'_0 values quoted from refs. 40 and 41.*

[b]*These values refer to Fe^{3+}, added prior to NADPH or ascorbate. If Fe^{3+} is added after ascorbate, the rate is 255. Fe^{2+} alone initiates lipid peroxidation which however ceases shortly. Subsequent addition of NADPH or ascorbate reinitiates the reaction at rates of 19 and 268, respectively.*

ascorbate-linked lipid peroxidation. Fe^{3+} was inactive in the case of NADPH and slightly active in the case of ascorbate. In contrast, Fe^{2+} showed a low activity with NADPH and a high activity with ascorbate, a pattern similar to that found with iron-malonate. This phenomenon is further illustrated by the polarographic traces in Fig. 10. As already shown (*cf.* Fig. 4), addition of Fe^{2+} alone to the microsomes gave rise to a transient O_2 uptake. The fact that subsequent addition of NADPH or ascorbate elicited reponses different from those found in the case of Fe^{3+}, indicated that the cessation of O_2 uptake in the case of Fe^{2+} could not be due simply to an oxidation to Fe^{3+}.

Fig. 11 a shows the effect of varying amounts of added

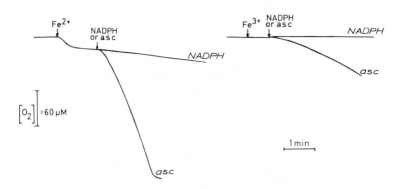

Fig. 10. Comparison of the effects of Fe^{2+} and Fe^{3+} in the absence of chelating agents on NADPH- and ascorbate-linked lipid peroxidation. Conditions as in Fig. 5.

Fe^{2+} on O_2 consumption and malonaldehyde formation. There was an increasing lag in the initiation of O_2 uptake with increasing Fe^{2+} concentration, and maximal rate occurred when O_2 consumption reached about one-half of its final extent. This suggests that an Fe^{2+}/Fe^{3+} ratio of about 1 was necessary to obtain maximal rate of O_2 uptake with any given concentration of added Fe^{2+}. This maximal rate was proportioned to the total amount of Fe^{2+} added. The total amount of O_2 consumed was also dependent on the amount of Fe^{2+} added; the molar ratio of O_2/Fe^{2+} varied between 1.5 and 1.9. Malonaldehyde was formed in the course of the O_2 uptake, with a molar ratio of O_2/malonaldehyde varying between 13 and 20.

Repeated additions of equal amounts of Fe^{2+} to the microsomes resulted in a repeated transient O_2-uptake, with a slight progression in the amount of O_2 consumed for each addition (Fig. 11 b). When twice the amount of microsomes was used this did not increase the amount of O_2 taken up or malonaldehyde formed per Fe^{2+} added. The response to added Fe^{2+} was also observed with microsomes that had been stored for several days in the cold as well as with heat-denatured or TCA-precipitated preparations. When the TCA-precipitated microsomes were extracted once with absolute ethanol at room temperature, an O_2-uptake upon the addition of Fe^{2+} was found in the ethanol extract but not in the residue. When microsomes were extracted with TCA after repeated additions of Fe^{2+}, the bulk of the added iron was recovered in the precipitate. If lipid peroxidation was inhibited by DPPD, all the added iron was recovered in the TCA extract.

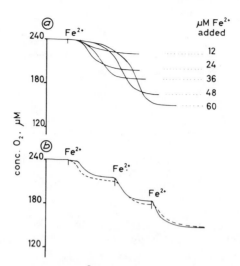

Fig. 11. Effect of Fe^{2+} on lipid peroxidation in the absence of NADPH or ascorbate.

The reaction mixture consisted of 25 mM Tris-Cl, pH 7.5, and 0.15 M KCl. Final volume, 3 ml. Temperature, 24°C.
(a) Effect of varying Fe^{2+} concentrations.
(b) Effect of repeated additions of 12 μM Fe^{2+}. 3.2 (solid line) or 6.4 (dotted line) mg microsomal protein.

These findings indicate that lipid peroxidation induced by added Fe^{2+} alone is nonenzymic and can be reproduced with isolated microsomal lipids. It appears, furthermore, that this lipid peroxidation leads to a binding of iron to the microsomes, in a form that is similar to iron-malonate (iron-malonaldehyde?).

C. Inhibitors. *Relationship to drug metabolism*

The antioxidant DPPD is a very potent inhibitor of NADPH-linked peroxidation in microsomes (Fig. 12). Strong inhibitions are also obtained with vitamins E and A. With all three inhibitors, malonaldehyde formation is completely abolished whereas O_2 consumption and NADPH disappearance are only partially inhibited. In the case of DPPD and vitamin A, the inhibition of lipid peroxidation was found to be transient, and the duration of the inhibition was dependent on the concentration of the compound. This phenomenon is probably due to the conversion of these compounds into a noninhibitory form during the incubation.

NADPH-linked lipid peroxidation is also blocked by *p*-hyd-

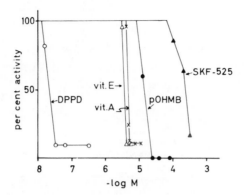

*Fig. 12. Effect of inhibitors on NADPH-linked lipid peroxi-
dation. Conditions as in Fig. 2.*

roxymercuribenzoate (*cf.* Fig. 12); in this case the inhibition
is complete also with regard to O_2 consumption and NADPH disappearance. A depression of NADPH-linked lipid peroxidation
was also observed with SKF-525A, a known inhibitor of drug metabolism. DPPD, vitamin E and vitamin A inhibited ascorbate-
linked lipid peroxidation as well, whereas pOHMB and SKF-525A
acted only on the NADPH-linked process.

Drugs undergoing cytochrome P-450-linked monooxygenation
have been shown to inhibit NADPH-induced (but not ascorbate-
induced) microsomal lipid peroxidation (4). The inhibition
required active drug metabolism and was relieved by carbon monoxide or by aging of the microsomes leading to an inactivation of cytochrome P-450. Data in Fig. 13 further illustrate
this effect. Microsomes suspended in 0.15 M KCl were preincubated with codeine or aminopyrine, recentrifuged and washed.
The microsomes containing bound drug were then incubated with
NADPH and ADP-Fe^{3+}, and O_2 consumption was measured. As may be
seen in Fig. 13, the drugs inhibited O_2 uptake initially, but
the degree of inhibition decreased with time, most probably
due to a removal of the drugs by metabolism. The inhibition
was relieved by carbon monoxide, or by preincubation of the
drug-containing microsomes for 15 minutes with NADPH prior to
the addition of ADP-Fe^{3+}.

An inhibition of lipid peroxidation was also observed with
liver microsomes isolated from animals after the administration of phenobarbital *in vivo*. As has been shown elsewhere
(42), there occurs a rapid binding of the injected drug to the
liver microsomes, which lasts for several hours and is not released even by extensive washing of the isolated microsomes

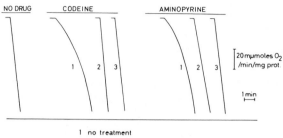

1 no treatment
2 40% CO, 4% O$_2$, 56% N$_2$
3 preincub. with NADPH

Fig. 13. Inhibition of NADPH-linked lipid peroxidation by preincubation of microsomes with drugs, and its relief by carbon monoxide.

Microsomes were pretreated for 15 min. at 37°C in the absence or presence of 5 mM codeine or aminopyrine, centrifuged, washed once with and resuspended in 0.15 M KCl. O$_2$ consumption was measured in a reaction mixture as in Fig. 1. When indicated the drug-treated microsomes were preincubated with NADPH for 15 min. at 37°C, centrifuged, resuspended in 0.15 M KCl and then assayed for lipid peroxidation as described in Fig. 2.

with Tris-KCl. Fig. 14 shows the extent of inhibition of microsomal lipid peroxidation *in vitro* at various intervals after the intraperitoneal injection of a single dose of phenobarbital. As expected, the inhibition was relieved by carbon monoxide. Like in the case of the microsomes treated with drugs

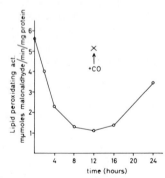

Fig. 14. Effect of phenobarbital administration to rats on the NADPH-linked lipid peroxidation of liver microsomes.

Rats were injected intraperitoneally with a single dose of 100 mg/kg phenobarbital. Liver microsomes were prepared after various intervals and tested for NADPH-linked lipid peroxidation as described in Fig. 2.

in vitro (*cf.* Fig. 13), the inhibition of lipid peroxidation
was only transient and was relieved by preincubation of the
microsomes for a suitable time interval with NADPH and O_2
prior to the addition of ADP-Fe^{3+} (not shown).

D. *Effects on microsomal structure and functions*

   As already reported briefly (3), lipid peroxidation causes
lysis of the microsomes, with a concomitant inactivation of
certain microsomal enzymes. Results of a more complete experiment are shown in Fig. 15. Microsomes were incubated in the
presence of NADPH and O_2. Samples were removed at suitable intervals and supplemented with DPPD to interrupt lipid peroxidation. A control series was run with DPPD present from the
beginning. Light-scattering at 520 nm was used as a measure of
lysis. It may be seen that the light-scattering decreased with
time in the sample exhibiting lipid peroxidation, while it
slightly increased in the control. The difference in optical
density between the two samples paralleled the formation of
malonaldehyde.
   Among the microsomal enzymes investigated, nucleoside di-
phosphatase showed a substantial activation in the course of
lipid peroxidation, while glucose-6-phosphatase revealed a
transient activation followed by a virtually complete inactivation. The NADH-cytochrome *c* reductase activity gave a varying response, showing no change in some experiments, and a
slight decrease or increase in others. The NADPH-cytochrome *c*
reductase activity remained constant during lipid peroxidation.
These responses of the various enzyme activities to progressive lysis caused by lipid peroxidation are similar to those
earlier found to occur upon treatment of microsomes with increasing concentrations of deoxycholate (33,35,43).
   Lipid peroxidation also resulted in a conversion of cytochrome P-450 into the P-420 form (not shown; *cf.* ref. 29).
Parallel to this, there was a progressive decrease in the capacity of the microsomes to carry out NADPH-linked monooxygenation.
   Further aspects of the effect of lipid peroxidation on
microsomal structure and function have been studied by Högberg
et al. (29).

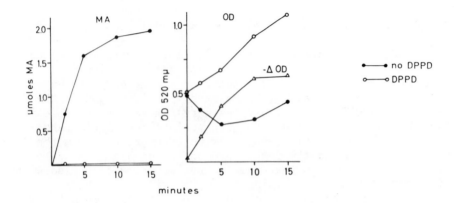

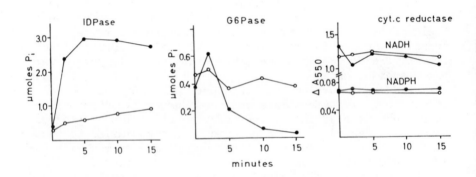

Fig. 15. *Effect of lipid peroxidation on microsomal structure and enzymes.*
    *The reaction mixture contained 25 mM Tris-Cl, pH 7.5, 150
mM KCl, 0.3 mM NADPH, 10 μM PP$_i$-Fe^{2+}, and 30 mg microsomal
protein in a final volume of 30 ml. When indicated, 1 μg/ml
DPPD was added. Temperature, 30°C. Aliquots were removed and
lipid peroxidation in the samples not containing DPPD was
stopped by the addition of 1 μg/ml DPPD.*
    *Malonaldehyde, optical density and enzyme activities were
determined as indicated in EXPERIMENTAL.*

IV. CONCLUSIONS AND COMMENTS

A. *Mechanism*

The present data confirm and extend previous conclusions (1-7) regarding the mechanism of NADPH-linked microsomal lipid peroxidation. From the available information we vizualize the process to involve the following reactions:

$$NADPH + H^+ + O_2 \rightarrow NADP^+ + H_2O_2 \tag{1}$$

$$LH + O_2 \rightarrow LOOH \tag{2}$$

where LH is a lipid undergoing peroxidation.

Reaction 1 may involve the following steps:

$$NADPH + 2XPP\text{-}Fe^{3+} \rightarrow NADP^+ + 2XPP\text{-}Fe^{2+} + H^+ \tag{1a}$$

$$2XPP\text{-}Fe^{2+} + 2O_2 \rightarrow 2XPP\text{-}Fe^{2+}\text{-}O_2 \tag{1b}$$

$$2XPP\text{-}Fe^{2+}\text{-}O_2 \rightarrow 2XPP\text{-}Fe^{3+}\text{-}O_2^{\overline{}} \tag{1c}$$

$$2XPP\text{-}Fe^{3+}\text{-}O_2^{\overline{}} \cdot + 2H^+ \rightarrow 2XPP\text{-}Fe^{3+} + H_2O_2 + O_2 \tag{1d}$$

Sum: $\quad NADPH + H^+ + O_2 \rightarrow NADP^+ + H_2O_2 \tag{1}$

where XPP-Fe is a suitable Fe chelate (XPP standing for an organic or inorganic pyrophosphate). Reaction 1a is catalyzed by the flavoenzyme NADPH-cytochrome P-450 reductase, whereas reactions 1b, 1c and 1d are nonenzymic. The dismutation according to reaction 1d may consist of several steps, *e.g.* a dissociation of $XPP\text{-}Fe^{3+}\text{-}O_2^{\overline{}} \cdot$ into $XPP\text{-}Fe^{3+}$ and $O_2^{\overline{}} \cdot$ and a subsequent disproportionation of $O_2^{\overline{}}$. The latter reaction is relatively slow at physiological pH in the absence of a catalyst (44,45), and, thus, reaction 1d probably is the rate-limiting step of reaction 1. This enables $XPP\text{-}Fe^{3+}\text{-}O_2^{\overline{}} \cdot$ to initiate lipid peroxidation and also explains why superoxide dismutase inhibits this process.

The initiation of lipid peroxidation (reaction 2) may be envisaged as involving the following steps:

$$LH + XPP\text{-}Fe^{3+}\text{-}O_2^{\overline{}} \cdot \rightarrow L\cdot + XPP\text{-}Fe^{2+}\text{-}O_2H \tag{2a}$$

$$L\cdot + O_2 \rightarrow LOO\cdot \tag{2b}$$

$$LOO\cdot + XPP\text{-}Fe^{2+}\text{-}O_2H \rightarrow LOOH + XPP\text{-}Fe^{3+}\text{-}O_2^{\overline{}}\cdot \tag{2c}$$

Sum: $\quad LH + O_2 \rightarrow LOOH \tag{2}$

The fact that NADPH- and ascorbate-induced lipid peroxidations have similar requirements for iron chelate suggests that it is reaction 2 rather than reaction 1 that determines this requirement.

Several investigators have proposed (21,22,26) that microsomal lipid peroxidation occurs via an ADP-Fe^{2+}-facilitated
Haber-Weiss reaction (46), with the hydroxyl radical (OH·) as
the initiating species. The latter would be generated from
H_2O_2 according to the reaction:

$$ADP-Fe^{2+} + H_2O_2 \rightarrow ADP-Fe^{3+} + OH^- + OH· \qquad (3)$$

However, the involvement of this mechanism has been questioned on several grounds (31). One important objection is that
microsomal lipid peroxidation is not inhibited by catalase, in
concentrations that efficiently remove H_2O_2. Our results confirm this conclusion. On the other hand, OH· may be involved
in other instances of lipid peroxidation, *e.g.* that induced
by xanthine oxidase (47).

In addition to reaction 2, lipid peroxidation is known to
proceed also by propagation, *i.e.* through a reaction sequence
which is initiated by a lipid containing a fatty acid peroxide
free radical (LOO·) extracting a hydrogen atom from an adjacent lipid (L'H) and subsequent oxygenation of the latter
according to the reactions:

$$LOO· + L'H \qquad \rightarrow LOOH + L'· \qquad (4a)$$

$$L'· + O_2 \qquad \rightarrow L'OO· \qquad (4b)$$

Sum: $LOO· + L'H + O_2 \rightarrow LOOH + L'OO·$ $\qquad (4)$

A mechanism for the initiation of NADPH-induced lipid peroxidation similar to that outlined above has in recent years
been proposed by Aust and associates (24,30,31). Their mechanism differs from ours, however, in one important respect.
According to these authors (*cf.* ref. 31) the ADP-perferryl ion
($XPP-Fe^{2+}-O_2 \leftrightarrow XPP-Fe^{3+}-O_2^-·$ using the present symbolism)
transfers oxygen directly to LH, without the concomitant formation of H_2O_2. Such a mechanism does not account for the des-
tiny of NADPH. Under the conditions employed by Aust and associates (*cf.* refs. 30,31), using a reconstituted liposomal system, initiation accounted for a minor part of the observed lipid peroxidation, over 90% occurring via propagation. In our
experiments, NADPH was continuously needed for lipid peroxidation, and H_2O_2 was formed as a product of NADPH oxidation.
The maximal number of lipid peroxide molecules was 5 for each
molecule of NADPH oxidized, *i.e.* 2.5 for each ADP-perferryl
ion formed. This means that maximally 1.5 lipid peroxide was
formed by propagation for each lipid peroxide formed by ADP-
perferryl-induced initiation. This low extent of propagation
as compared to the liposomal system may be explained by the
fact that the phospholipid bilayer of the microsomal membrane
is interrupted by many proteins which may limit lipid-lipid

interaction.

In a recent paper Svingen et al. (31) have arrived at the interesting conclusion that the propagation of lipid peroxidation under their conditions may require the participation of cytochrome P-450. They found that the EDTA and DPTA chelates of iron can replace cytochrome P-450 in this function and can also relieve the inhibition of lipid peroxidation by various drugs, earlier described in our laboratory (4). Under our conditions, the need for cytochrome P-450 in lipid peroxidation seems unlikely, in view of the findings (4) that lipid peroxidation is insensitive to carbon monoxide and that it proceeds uninhibited after the inactivation of cytochrome P-450 occurring in the course of lipid peroxidation of during aging of the microsomes. Whether this lack of involvement of cytochrome P-450 in microsomal lipid peroxidation is due to the relatively minor role played by propagation under the conditions of our studies, as discussed above, is not known. It should be noted, however, that if a peroxidase action of cytochrome P-450, of the type described by Hrycay and O'Brien (48,49) and involving lipid peroxides as substrates, would play a major role in microsomal lipid peroxidation, in accordance with the propagation mechanism proposed by Svingen et al. (31), then one would expect to find significant amounts of hydroxy fatty acids among the reaction products. This does not seem to be the case (27).

Our earlier conclusion (4) that the inhibition of NADPH-linked lipid peroxidation by drugs undergoing cytochrome P-450-linked monooxygenation is due to a competition for the supply of reducing equivalents via NADPH-cytochrome P-450 reductase has recently been challenged by Miles et al. (32). Our conclusion was based on the findings that these drugs did not inhibit ascorbate-induced lipid peroxidation, and that the inhibition of the NADPH-linked lipid peroxidation was dependent on an active monooxygenation and was relieved by carbon monoxide or by aging of the microsomes leading to an inactivation of cytochrome P-450. Additional evidence supporting this conclusion it reported in the present paper. According to Miles et al. (32), the inhibition by various drugs is also observed with ascorbate-induced lipid peroxidation and is not relieved by metyrapone, an inhibitor of drug metabolism. They conclude that the inhibition is due to an antioxidant effect of the drugs. However, the conditions used by Miles et al. (32), were different from ours in that no iron chelate was added to activate lipid peroxidation and, consequently, the activities measured were very low. Furthermore, the inhibition to occur required rather high drug concentrations and long periods of incubation. It is quite possible that, under those conditions, the drugs tested do act as antioxidants. The simplest (although not necessarily the only) explanation for our findings

still seems to be that during active monooxygenation, cytochrome P-450 diverts electrons from the reductase, thus preventing the reduction of the added iron chelate and thereby
the initiation of lipid peroxidation.

B. *Biomedical aspects*

Lipid peroxidation is one of several types of microsomal
oxidations with important biomedical implications that are
catalyzed by the flavoenzyme NADPH-cytochrome P-450 reductase.
Other reactions of interest in this context are the oxidation
of certain xenobiotics to epoxides, and the one-electron reduction of quinones to autoxidable semiquinones (Fig. 16).
Lipid peroxidation results in membrane damage and lysis, whereas epoxides and semiquinones may bind to nucleophilic groups
in proteins and nucleic acids and cause cell death or muta-
tions leading to transformation of normal cells into cancer
cells. In concluding this paper it may be appropriate to briefly summarize information concerning the relationship between
these reactions and to consider conceivable physiological devices that control their activities.

Microsomal lipid peroxidation is efficiently prevented by
naturally occurring antioxidants, *e.g.*, vitamins E and A, as
well as by superoxide dismutase and glutathione peroxidase by
removing $O_2^-\cdot$ and H_2O_2. Once formed, microsomal lipid peroxides may be removed, in principle, by way of glutathione
peroxidase or cytochrome P-450, but to what extent these mechanisms are operative is not known.

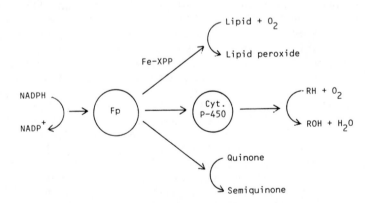

*Fig. 16. Reactions catalyzed by NADPH-cytochrome P-450
reductase.*

In the case of epoxides, the conversion of these to phenols, dihydrodiols and glutathione conjugates constitute the
chief physiological control devices (50). Both the formation
and the removal of epoxides are subject to regulatory mechanisms involving enzyme induction (51).

Mechanisms involved in the formation and removal of quinones represent a highly complex interplay between microsomal
quinone reductases and oxygenases, the importance of which
only recently began to emerge. The scheme in Fig. 17 may
serve as a framework for the following discussion.

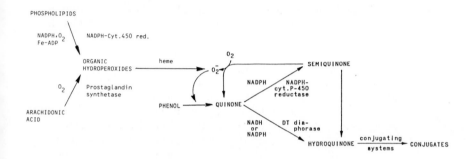

Fig. 17. Relationship of microsomal peroxidations and
quinone reductases.

Quinones, both naturally occurring and of xenobiotic origin, can be reduced by a number of enzymes. Some of these, *e.g.*
NADPH-cytochrome P-450 reductase, are one-electron transfer
enzymes (52) and give rise to semiquinones which are readily
autoxidized with the formation of O_2^-. For example, as recently shown by Bachur *et al.* (53), the highly cytotoxic anticancer drug adriamycin and related agents of quinone nature are
potent $O_2^-\cdot$ generators following reduction by way of NADPH-cytochrome P-450 reductase. Once formed, $O_2^-\cdot$ may in turn promote quinone formation from certain phenols (*e.g.* from 6-hyd-
roxybenzpyrene [54,55]) and thereby the generation of more $O_2^-\cdot$
in an autocatalytic fashion. Benzpyrene quinone formation is
also promoted by organic hydroperoxides such as cumene hydroperoxide (56), or by the hydroperoxy endoperoxide, PGG_2, which
is an intermediate in prostaglandin biosynthesis from arachidonic acid (57). Incidentally, PGG_2 has recently also been
shown to trigger the oxygenation of benzpyrene-7,8-dihydrodiol
to the highly mutagenic 9,10-epoxide (58).

Other quinone reductases are two-electron transfer enzymes

and produce hydroquinones which are relatively stable and
ready to undergo conjugation reactions. A quinone reductase
of this type is the flavoprotein DT diaphorase (NAD(P)H:qui-
none oxidoreductase), an enzyme that has been extensively studied in our laboratory over the last 20 years (59-62). When
both one- and two-electron transferring quinone reductases are
present, the latter type would be expected to limit autoxidation and concomitant O_2^-· formation, by converting all available quinone into the stable hydroquinone. Indeed, such a relationship can easily be demonstrated (62) by incubating microsomes, which contain both NADPH-cytochrome P-450 reductase
and DT diaphorase, with NADPH and a limiting amount of a quinone - *e.g.* menadione - and comparing the extents of NADPH
oxidation in the absence and presence of dicoumarol, which
selectively inhibits DT diaphorase (59-62). When DT diaphorase
is active, the extent of NADPH oxidation is essentially stoichiometric with the amount of quinone added, indicating that
practically no autoxidation has taken place. This relationship is especially pronounced with microsomes from rats
treated with 3-methylcholanthrene, which is a potent inducer
of DT diaphorase (63,64). In contrast, when DT diaphorase is
inhibited, NADPH oxidation continues beyond stoichiometry, due
to autoxidation of the semiquinone formed via NADPH-cytochrome
P-450 reductase. This relationship explains why DT diaphorase
is the preferred quinone reductase in connection with conjugation reactions (65), despite the high quinone reductase activity of NADPH-cytochrome P-450 reductase (66,67). Of great
interest in this connection is a recent report by Benson *et al.*
(68) demonstrating a striking increase in DT diaphorase activity of various tissues of mice fed with butylhydroxyanisole
(BHA), a widely used antioxidant food additive and protective
agent against carcinogenesis and toxicity. These findings
strengthen the view (62,69) that DT diaphorase functions as
an important cellular control device for O_2^-· formation and
peroxidative tissue damage. While antioxidant vitamins, superoxide dismutases and glutathione peroxidases prevent lipid
peroxidation by promoting the removal of O_2^-·, DT diaphorase
may achieve the same purpose by preventing O_2^-· formation.

ACKNOWLEDGMENTS

Work quoted from the authors' laboratories was supported
by the Swedish Cancer Society, Swedish Medical and Natural-Sci-
ence Research Councils, Swedish Tobacco Company, and the National Cancer Institute, National Institutes of Health, U.S.A.
(Contract No. NO1 CP 33363 and Grant No. 1RO 1CA 26261-01).

REFERENCES

1. Hochstein, P., and Ernster, L., *Biochem. Biophys. Res. Commun. 12*, 388 (1963).
2. Hochstein, P., Nordenbrand, K., and Ernster, L., *Biochem. Biophys. Res. Commun. 14*, 323 (1964).
3. Hochstein, P., and Ernster, L., *in* Ciba Foundation Symp. on Cellular Injury (A.V.S. de Reuck and J. Knight, eds.) p. 123, Churchill Ltd, London (1964).
4. Orrenius, S., Dallner, G., and Ernster, L., *Biochem. Biophys. Res. Commun. 14*, 329 (1964).
5. Ernster, L., and Nordenbrand, K., *Meth. Enzymol. 10*, 574 (1967).
6. Ernster, L., Nordenbrand, K., Orrenius, S., and Das, M.L., *Hoppe-Seyler's Z. Physiol. Chem. 349*, 1604 (1968).
7. Nilsson, R., Orrenius, S., and Ernster, L., *Biochem. Biophys. Res. Commun. 17*, 303 (1964).
8. Beloff-Chain, A., Serlupi-Crescenzi, G., Catanzaro, R., Venettacci, D., and Balliano, M., *Biochim. Biophys. Acta, 97*, 416 (1965).
9. Robinson, J.D., *Arch. Biochem. Biophys 112*, 170 (1965).
10. Wills, E.D., *Biochem. J. 99*, 667 (1966).
11. Marks, F., and Hecker, E., *Hoppe-Seyler's Z. Physiol. Chem. 348*, 727 (1967).
12. Utley, H.G., Berhheim, F., and Hochstein, P., *Arch. Biochem. Biophys. 118*, 29 (1967).
13. May, H.E., and McCay, P.B., *J. Biol. Chem. 243*, 2288 (1968).
14. Slater, T.F., *Biochem. J. 106*, 155 (1968).
15. Christophersen, B.O., *Biochem. J. 106*, 515 (1968).
16. Marks, F., and Hecker, E., *Hoppe-Seyler's Z. Physiol. Chem. 349*, 523 (1968).
17. Wills, E.D., *Biochem. J. 113*, 315 (1969).
18. Wills, E.D., *Biochem. J. 113*, 325 (1969).
19. Wills, E.D., *Biochem. J. 113*, 333 (1969).
20. Poyer, J.L., and McCay, P.B., *J. Biol. Chem. 246*, 263 (1971).
21. McCay, P.B., Pfeifer, P.M., and Stripe, W.M., *Ann. N.Y. Acad. Sci. 203*, 62 (1972).
22. Fong, K.-L., McCay, P.B., Poyer, J.L., Keele, B.B., and Misra, H., *J. Biol. Chem. 248*, 7792 (1973).
23. Pederson, T.C., and Aust, S.D., *Biochem. Biophys. Res. Commun. 48*, 789 (1972).
24. Pederson, T.C., Buege, J.A., and Aust, S.D., *J. Biol. Chem. 248*, 7134 (1973).
25. Pederson, T.C., and Aust, S.D., *Biochem. Biophys. Res. Commun. 52*, 1071 (1973).

26. King, M.M., Lai, E.K., and McCay, P.B., *J. Biol. Chem.* *250*, 6496 (1975).

27. McCay, P.B., Gibson, D.D., Fong, K.-L., and Hornbrook, K.R., *Biochim. Biophys. Acta*, *431*, 459 (1976).

28. Tyler, D.D., *FEBS Lett. 51*, 180 (1975).

29. Högberg, J., Bergstrand, A., and Jakobsson, S.V., *Eur. J. Biochem. 37*, 51 (1973).

30. Svingen, B.A., O'Neal, F.O., and Aust, S.D., *Photochem. Photobiol. 28*, 803 (1978).

31. Svingen, B.A., Buege, J.A., O'Neal, F.O., and Aust, S.D., *J. Biol. Chem. 254*, 5892 (1979).

32. Miles, P.R., Wright, J.R., Bowman, L., and Colby, H.D., *Biochemical Pharmacology*, *29*, 565 (1980).

33. Ernster, L., Siekevitz, P., and Palade, G., *J. Cell. Biol. 15*, 541 (1962).

34. Bernheim, F., Bernheim, M.L., and Wilbur, K.M., *J. Biol. Chem. 174*, 257 (1948).

35. Ernster, L., and Jones, L.C., *J. Cell. Biol. 15*, 563 (1962).

36. Diehl, H., and Smith, G.F., *"The Iron Reagents"*, publ: by The G. Frederick Smith Chemical Co, Columbus, Ohio (1960).

37. Ernster, L., and Orrenius, S., *Fed. Proceed 24*, 1190 (1965).

38. Gilette, J.R., Brodie, B.B., and La Du, B.N., *J. Pharm. Exp. Therap. 119*, 532 (1957).

39. Dahle, L.K., Hill, E.G., and Holman, R.T., *Arch. Biochem. Biophys. 98*, 253 (1962).

40. Michaelis, L., and Smythe, C.V., *J. Biol. Chem. 94*, 329 (1931).

41. Buckingham, D.A., and Sargeson, A.M., *in* Chelating Agents and Metal Chelates (F.P. Dwyer and D.P. Mellor, eds.) p. 237, Academic Press, New York and London (1967).

42. Orrenius, S., Kupfer, D., and Ernster, L., *FEBS Lett. 6*, 249 (1970).

43. Dallner, G., *Acta Pathol. Microbiol. Scand. Suppl.*, 166 (1963).

44. Czapski, G., *Ann. Rev. Phys. Chem. 22*, 171 (1971).

45. Halliwell, B., *FEBS Lett. 56*, 34 (1975).

46. Haber, F., and Weiss, J., *Proc. R. Soc. Edinb. Sect. A. Math. Phys. Sci.*, *147*, 332 (1934).

47. Kellog, E.W. III, and Fridovich, I., *J. Biol. Chem. 252*, 6721 (1977).

48. Hrycay, E.G., and O'Brien, P.J., *Arch. Biochem. Biophys. 147*, 14 (1971).

49. Hrycay, E.G., and O'Brien, P.J., *Arch. Biochem. Biophys. 147*, 28 (1971).

50. DePierre, J.W., and Ernster, L., *Biochim. Biophys. Acta*, *473*, 149 (1978).

51. Seidegård, J., Morgenstern, R., DePierre, J.W., and Ernster, L., *Biochim. Biophys. Acta, 586,* 10 (1980).
52. Iyanagi, T., and Yamazaki, J., *Biochim. Biophys. Acta, 216,* 282 (1970).
53. Bachur, N.R., Gordon, S.L., Gee, M.V., and Kon, H., *Proc. Natl. Acad. Sci. USA, 76,* 954 (1979).
54. Lorentzen, R.J., Caspary, W.J., Lesko, S.A., and Ts'o, P. O.P., *Biochemistry, 14,* 3970 (1975).
55. Lorentzen, R.J., and Ts'o, P.O.P., *Biochemistry, 16,* 1467 (1977).
56. Capdevila, J., Estabrook, R.W., and Prough, R.A., *Arch. Biochem. Biophys. 200,* 186 (1980).
57. Marnett, L.J., and Reed, G.A., *Biochemistry, 18,* 2923 (1979).
58. Marnett, L.J., Abstr. 10th Linderstrøm-Lang Conference, Skokloster, Sweden (1980).
59. Ernster, L., Ljunggren, M., and Danielson, L., *Biochem. Biophys. Res. Commun. 2,* 88 (1960).
60. Ernster, L., Danielson, L., and Ljunggren, M., *Biochim. Biophys. Acta, 58,* 171 (1962).
61. Ernster, L., *Meth. Enzymol 10,* 309 (1967).
62. Lind, C., Hochstein, P., and Ernster, L., *in* Third Internatl. Symp. on Oxidases and Related Oxidation Reduction Enzymes (T.E. King, H.S. Mason, and M. Morrison, eds.) Pergamon Press, New York, in press.
63. Huggins, C., Ford, E., and Fukunishi, R., *J. Exp. Med. 119,* 943 (1964).
64. Lind, C., and Ernster, L., *Biochim. Biophys. Res. Commun. 56,* 392 (1974).
65. Lind, C., Vadi, H., and Ernster, L., *Arch. Biochem. Biophys. 190,* 97 (1978).
66. Capdevila, J., Estabrook, R.W., and Prough, R., *Biochem. Biophys. Res. Commun. 83,* 1291 (1978).
67. Chen, A.L., Fahl, W.E., Wrighton, S.A., and Jefcoate, C. R., *Cancer Res. 39,* 4123 (1979).
68. Benson, A.M., Hunkeler, M.J., and Talalay, P., *Proc. Natl. Acad. Sci. USA, 77,* 5216 (1980).
69. Ernster, L., Abstr. 10th Linderstrøm-Lang Conference, Skokloster, Sweden (1980).

LIPID PEROXIDATION AND MEMBRANE ALTERATIONS IN ERYTHROCYTE SURVIVAL[1]

Paul Hochstein
Catherine Rice-Evans[2]

Institute for Toxicology
University of Southern California
Los Angeles, California

In the body, human erythrocytes with a diameter of approximately 8 μm circulate through a network of vessels whose diameters range from 25 mm to less than 5 μm. The principal thrust of investigations in blood rheology in recent years has been to understand the factors which contribute to the flow behavior of cells in the smallest of those vessels. It is apparent that such factors also contribute to the sequestration of erythrocytes during normal aging as well as in accelerated aging. This accelerated aging is observed in several genetic abnormalities and after the administration of certain drugs. The capacity of the erythrocyte to deform in the microcirculation under these circumstances is a determinant of its survival.

Erythrocyte deformability is a function of cell shape as well as of a variety of events which effect surface/volume ratios (S/V), the internal viscosity of the cells, and intrinsic membrane microviscosity (1). More than one of these factors may operate simultaneously to influence the mechanical properties of cells, to alter their flow behavior, and to

[1]This work was supported in part by grants from the National Institutes of Health (AG-00471) and from the American Heart Association, Greater Los Angeles Affiliate (537-IG4).
[2]Visiting Research Scholar, Permanent Address: Department of Biochemistry, Royal Free Hospital School of Medicine, London, England.

lead to their removal from the circulation. In this paper
we summarize some of the experimental evidence that the
peroxidation of endogenous membrane lipids may be a contributing factor in decreasing cellular deformability. These
experiments suggest that lipid peroxidation may alter membrane
constituents and their interactions in such a way that membrane microviscosity is increased and cellular rigidity
enhanced.

I. LIPID PEROXIDATION IN NORMAL ERYTHROCYTE AGING

The accumulation of fluorescent chromolipids in the
tissues of older animals is well known (2). The occurrence
of these pigments is presumed to be a consequence of the
breakdown of peroxidized fatty acids to yield malonyldialdehyde (MDA). This agent has the capacity to cross-link,
through Schiff's base formation, the amino groups of both
phospholipids and proteins to form fluorescent derivatives (3).
Erythrocyte membranes also contain such derivatives and it is
of special interest that they appear to be present in highest
quantities in membranes prepared from the oldest cells in the
circulation (4). Furthermore, chromolipids with identical
fluorescent characteristics (maximal excitation 390–400 nm,
maximal emission 460 nm) can be formed, in amounts proportional to the concentration of MDA (between 1 and 20 µM),
by mixing authentic MDA with human erythrocytes.

In addition to containing more fluorescent pigment, the
membranes of the older normal cells also contain increased
amounts of high molecular weight protein (HMWP) (4). These
polymers are presumably derived from spectrin (bands 1 and 2)
since their occurrence is associated with decreased amounts
of this membrane protein. No changes in other membrane proteins are detected after SDS gel electrophoresis and staining
with Coomassie Brilliant Blue of PAS. The occurrence of HMWP
is not prevented by agents such as DTE (dithioerythritol)
which suggests that disulfides are not primarily involved in
their formation. Moreover, the addition of MDA (1.0 to
10 µM) to erythrocytes causes a similar decrease in spectrin
and accumulation of HMWP.

These phenomena are consistent with a view that the peroxidation of unsaturated fatty acids results in their breakdown to form an agent (MDA) with the capacity to polymerize
phospholipids (e.g., PE) and at least one essential membrane
protein, spectrin.

Such a view suggests that during normal aging in erythrocytes the autoxidation of hemoglobin (Fe^{+2}) may be a critical
event. The formation of superoxide anions (O_2^-) by autoxidation of hemoglobin would also yield H_2O_2 after the subsequent dismutation. These species have the capacity to
interact to form still other potential initiators of lipid
peroxidation, e.g., ·OH. In this connection, it is possible
that hemoglobin may also act as a Fenton reagent to enhance
the formation of ·OH from H_2O_2 and that the association of
hemoglobin with the membranes of older cells or in disease
states might enhance lipid peroxidation. Alternatively,
iron-nucleotide complexes in erythrocytes (5) might serve a
similar function in generating radical species with the
capacity to initiate lipid peroxidation (6). The central
role of H_2O_2 in these events would be in keeping with the
diminished capacity of older cells to generate NADPH through
glucose-6-phosphate dehydrogenase activity. Such a restriction would result in a failure to maintain glutathione (GSH)
levels for GSH peroxidase activity (7).

II. PHENYLHYDRAZINE-INDUCED LIPID PEROXIDATION IN ERYTHROCYTES

The defense mechanisms (GSH, GSH Peroxidase, Superoxide
Dismutase) which normally serve to protect all but the oldest
cells from radical-induced lipid peroxidation may be over-
whelmed or bypassed by many chemical agents. These substances
interact with oxygen to generate initiators of peroxidation
(O_2^-, H_2O_2, ·OH) in amounts in excess of the capacity of
detoxification systems or, equally likely, in membrane sites
where the detoxification mechanisms may not function adequately. Among these substances is phenylhydrazine. This agent
is known to react with oxygen to form both O_2^- and H_2O_2 (8-
10). Its administration to animals results in severe
hemolytic anemia accompanied by dramatic elevations in the
number of circulating reticulocytes.

These young cells may also be produced in animals in
response to bleeding. As would be expected, the reticulocytes formed after bleeding animals have a longer half life
than do the cells in a population of adult erythrocytes
(16 vs. 11 days). Paradoxically, the reticulocytes formed in
response to phenylhydrazine have a half life, as measured by
chromium-labeling, of less than 2 days (11). Recent evidence
suggests that in vivo lipid peroxidation might be a factor
responsible for the premature removal of these "young" cells
from the circulation by the spleen (12).

Thus, the reticulocytes in the circulation in response
to phenylhydrazine contain four-fold more fluorescent chromolipids than do those produced in response to bleeding. It
appears that the membranes of the former cells have undergone
extensive peroxidation. Electrophoresis of the membrane
proteins of these erythrocytes also reveals a striking
decrease in spectrin content (-30%) and a corresponding
increase in HMWP. There appear to be few other distinguish-
ing biochemical or morphological features between the two
types of reticulocytes. It seems reasonable to posit that
the formation of lipid polymers (fluorescent chromolipids),
and the accumulation of high molecular weight protein
polymers at the expense of spectrin might contribute to the
altered membrane mechanical properties which are ultimately
expressed in enhanced cellular rigidity and sequestration.

III. PHENYLHYDRAZINE-INDUCED ALTERATIONS IN MEMBRANE
     FLUIDITY

The treatment of human erythrocytes with phenylhydrazine,
in fact, results in a marked decrease in bulk lipid fluidity
(13). This is illustrated in Table I. After in vitro
exposure to this agent, membranes were labeled with 1,6-
diphenyl hexatriene (DPH) and membrane microviscosity
parameters determined. It may be seen that the peroxidation
of membrane lipids leads to an increased lipid microviscosity

TABLE I. Lipid peroxidation, relative microviscosity
parameters, and ANS fluorescence of human erythrocyte
membranes isolated from cells treated with phenylhydrazine.

$PhNH-NH_2$ (mM)	MDA (A_{532}/mg prot.)	Microviscosity[a] $\left[\dfrac{r^o}{r} - 1\right]^{-1}$	ANS (rel. fluor. & em. max.)
0	0.069	1.64	— (465 nm)
0.50	0.516	3.24	-32% (472 nm)

[a]The values are the "microviscosity parameter" where r^o is
the limiting anisotropy for DPH and r is the fluorescence
anisotropy transposed from observed polarization values.

as measured by fluorescence polarization techniques with DPH. Such an increase might reflect an increase in lipid-lipid interaction and lipid packing density as well as a diminution in lipid-protein interaction (14). It is noteworthy that other experiments with the negatively charged fluorescent probe 1-anilino-8-naphthalene sulfonate (ANS) reveal decreased and red-shifted fluorescence intensity after phenylhydrazine treatment of cells (Table I). These results suggest a lesser penetration of the probe into hydrophobic regions of the membrane and possibly a change in the orientation of phospholipids at the surface of the membrane.

IV. MALONYLDIALDEHYDE AND CELLULAR DEFORMABILITY

We have previously demonstrated that treatment of cells with an agent which induces lipid peroxidation, tertiary butyl hydroperoxide, also causes a marked increase in cellu- lar deformability (15). That these changes in deformability result from the capacity of MDA itself to polymerize membrane constituents is suggested by the experiments illustrated in Table II. In these studies (16), the addition of MDA to cells resulted in an increase in cellular rigidity which is particularly evident at low sheer forces. Deformability was measured directly with a counter-rotating, cone-plate rheo- scope mounted on an inverted microscope. The cells were subjected to sheer stresses that ranged from 0 to 500 dynes/ cm^2. Deformation is defined as the ratio L-W/L+W where L is the length of the cell in the direction of the flow and

TABLE II.   Sheer-stress dependent deformability of human erythrocytes exposed to malonyldialdehyde.

	τ (dynes/cm^2)	D/D_o
Control RBC	–	1.0
RBC + MDA (20 μM)	2.5	0.52
	7.5	0.68
	75.0	0.97

W is the width of the cell perpendicular to the flow. D increases with increasing cell deformation approaching a limiting value at high sheer stresses. D/D_0 is the ratio of deformation of the treated cells (D) to the untreated cells (D_0). The action of MDA in reducing cellular deformability was observed in the absence of measurable effects on cell volume. Comparable concentrations of methanol, formaldehyde or glutaraldehyde did not alter deformability.

V. SYNOPSIS OF THE EFFECTS OF LIPID PEROXIDATION IN ERYTHROCYTES

We view the senescence of erythrocytes as affected by the reactions outlined below:

AUTOXIDATION ($HbFe^{+2}$, Cu^{+2}, PHENOLIC & HYDRAZO COMPOUNDS)
   ↓
   ↓
 O_2^- + H_2O_2 (DETOXIFICATION: SOD, GSH, GSH PEROXIDASE)
        ↓
        ↓ ($HbFe^{+2}$, $NPP-Fe^{+2}$)
        ↓
       •OH (DETOXIFICATION: RADICAL SCAVENGERS)
        ↓
        ↓
MEMBRANE PHOSPHOLIPIDS (PE, PS)
        ↓   ↑
        ↓   ↑
     LIPID-OOH → → → → RADICALS & VOLATILE HYDROCARBONS
        ↓
        ↓
       MDA → → → →   FLUORESCENT CHROMOLIPIDS
        ↓   ↓
        ↓   ↓
SPECTRIN → SPECTRIN POLYMERS   ↓
              ↓   ↓
              ↓   ↓
        _____
           ↑LIPID-LIPID INTERACTION
           ↓LIPID-PROTEIN INTERACTION
           ↑MEMBRANE MICROVISCOSITY
           ↓CELLULAR DEFORMABILITY
           ↑CELLULAR SEQUESTRATION

To a greater degree than most other cells, erythrocytes are exposed to radicals with the capacity to initiate the peroxidation of membrane phospholipids. For example, these radicals may be generated subsequent to the autoxidation of intracellular hemoglobin or a variety of chemicals which find their way into the blood, e.g., metals such as Cu^{+2}, and phenolic and hydrazo compounds. The failure of systems to adequately detoxify O_2^- and H_2O_2 may result in their inter-action to form hydroxy ($\cdot OH$) radicals. Alternatively, the decomposition of H_2O_2 in the presence of endogenous hemo-globin or nucleotide pyrophosphate-Fe (NPP-Fe) complexes may yield $\cdot OH$. In either event the formation of lipid hydroperoxides and their ultimate decomposition result in the formation of several volatile products, hydrocarbon radicals, and MDA. This dialdehyde has the capacity to cross-link amino phospholipids and membrane spectrin. We suggest that these polymerization reactions contribute to the observed alterations in membrane bulk lipid fluidity, to the decreased deformability of affected cells, and to the eventual entrapment of cells in the microcirculation of the spleen.

Spectrin plays a central role in maintaining erythrocyte membrane stability through its interaction with other mem-brane proteins and lipids. Although we have emphasized the role of MDA in forming spectrin polymers, similar aggregates might be formed during lipid peroxidation either through disulfide formation or by direct radical attack. In any of these events the state of spectrin phosphorylation and function might be altered. However, the association of fluorescent chromolipids with spectrin polymers is consistent with the concept that MDA is a key participant in the series of reactions which contribute to both normal and accelerated aging in erythrocytes.

REFERENCES

1. Mohandas, N., Phillips, W. M., and Besis, M. Semin. Hematol. 16, 95 (1979).
2. Wolman, M. Israel J. Med. Sci. 11 (Supplement), 1 (1975).
3. Tappel, A. L. Fed. Proc. 32, 1870 (1973).
4. Jain, S. K., and Hochstein, P. Biochem. Biophys. Res. Comm. 92, 247 (1980).
5. Bartlett, G. Biochem. Biophys. Res. Comm. 70, 1063 (1976).
6. Hochstein, P. Israel J. Chem. 21, 52 (1981).
7. Cohen, G., and Hochstein, P. Science 134, 1574 (1961).

8. Rostorfer, H. H., and Cormier, M. J. Archiv. Biochem.
    Biophys. 71, 235 (1957).
9. Cohen, G., and Hochstein, P. Biochem. 3, 895 (1964).
10. Jain, S. K., and Hochstein, P. Biochim. Biophys. Acta
    586, 128 (1979).
11. Jain, S. K., and Subrahmanyam, D. Indian J. Exp. Biol.
    16, 255 (1978).
12. Jain, S. K., and Hochstein, P. Archiv. Biochem. Biophys.
    201, 683 (1980).
13. Rice-Evans, C., and Hochstein, P. Biochem. Biophys. Res.
    Comm. 100, 1537 (1981).
14. Shinitzky, M., and Barenholz, Y. Biochim. Biophys. Acta
    515, 367 (1978).
15. Corry, W. D., Meiselman, H. J., and Hochstein, P.
    Biochim. Biophys. Acta 597, 224 (1980).
16. Pfafferott, C., Meiselman, H. J., and Hochstein, P.
    Blood 59, 12 (1982).

LIPID PEROXIDATIONS OF CHOLESTEROL[1]

Leland L. Smith
Jon I. Teng
Yong Y. Lin
Patricia K. Seitz[2]
Michael F. McGehee

Division of Biochemistry
The University of Texas Medical Branch
Galveston, Texas

1. INTRODUCTION

There is increasing recognition of the broad range of biological activities of oxidized cholesterol derivatives in intact animals and in cultured cells, such activities including
cytotoxic effects, stimulatory or inhibitory effects on specific enzymes, and cell membrane effects. Accordingly, interest
in the generation of oxidized cholesterol derivatives *in vivo*
and the presence of such compounds in foodstuffs as possible
influences on human health is developing.

There are presently recognized at least four distinct
modes of oxidation of cholesterol:strictly chemical oxidation,
specific enzyme oxidation, autoxidation, and generalized lipid
peroxidation. Lipid peroxidation is considered here to be
distinct from autoxidation although the same oxidation products are formed. Lipid peroxidation involves an enzyme component providing reductants (NADPH, $Fe(II)$, O_2^-, etc) and
molecular oxygen, both atoms of which are incorporated into
hydroperoxides as initial stable products. Although lipid
peroxidation is usually defined for unsaturated fatty acyl

[1] Financial support of these studies from the Robert A. Welch
Foundation, Houston, Texas and the U.S. Public Health Service (grants HL-10160 and ES-00944) is gratefully acknowledged.
[2] Robert A. Welch Post-doctoral Fellow, 1978-1980.

derivatives, the phenomenon is also observed for cholesterol
and related unsaturated sterols.

Cholesterol may be oxidized *in vivo* chemically, enzymically by specific enzymes, and perhaps by lipid peroxidation, a
matter addressed here. Pure cholesterol and cholesterol in
tissue *post mortem* are subject to autoxidation, autoxidation
being viewed as apparently uncatalyzed oxidation by the oxygen of the air. The details of cholesterol autoxidation processes are now fairly well known (1,2).

Initial stable cholesterol autoxidation products are the
epimeric cholesterol 7-hydroperoxides, the quasiequatorial
7β-hydroperoxide predominating (3,4), formed by attack of
ground-state molecular oxygen ($^3O_2$) on a C-7 cholesteryl radical, the existence of which is supported by electron spin
resonance data. Formal reduction of product 7-hydroperoxides
yields the corresponding 7-alcohols cholest-5-ene-3β,7α-diol
and cholest-5-ene-3β,7β-diol; formal dehydration yields the
7-ketone 3β-hydroxycholest-5-en-7-one (5-7). Dehydration of
the 7-ketone gives cholesta-3,5-dien-7-one. The 7-hydroper-
oxides also oxidize cholesterol to give the isomeric cholesterol 5,6-epoxides, the 5β,6β-epoxide predominating (8).
Hydration of either 5,6-epoxide gives 5α-cholestane-3β,5,6β-
triol. Other autoxidation reactions occur, including attack
in the side chain to form hydroperoxides, chief among which
is cholesterol 25-hydroperoxide from which cholest-5-ene-3β,
25-diol is derived (5,9).

Lipid peroxidation of cholesterol follows the same course
of 7-hydroperoxide formation (10-13) and subsequent deriva-
tion of the cholest-5-ene-3β,7-diols and 3β-hydroxycholest-
5-en-7-one (10-18) and the isomeric 5,6-epoxides and 5α-chole-
stane-3β,5,6β-triol (13,16,18-22).

II. EXPERIMENTAL APPROACHES

We have tried to use cholesterol as substrate, with attendant product analysis, as means of recognition of the nature of oxidizing agents and oxidation processes in several
experimental systems, both chemical (23-27) and enzymic, including lipid peroxidation systems (10-12). As shown in
TABLE I, different product distributions characterize the
several dioxygen species and related active oxidants poten-
tially encountered in biological and chemical systems. Other
active oxidants such as the dioxygen cation O_2^+ not a likely
biochemical species yield some of these same oxidation products (28). The triatomic species ozone gives totally different oxidation products with cholesterol.

TABLE I. Products of Cholesterol Oxidations

Oxidant	Initial Products	Subsequent Products
$^3O_2$ (autoxidation)	Cholesterol 7-hydroperoxides	Cholest-5-ene-3β,7-diols 3β-Hydroxycholest-5-en-7-one Cholesterol 5,6-epoxides Cholesta-3,5-dien-7-one 5α-Cholestane-3β,5,6β-triol
$^1O_2$	3β-Hydroxy-5α-cholest-6-ene 5-hydroperoxide	5α-Cholest-6-ene-3β,5-diol Cholesta-4,6-dien-3-one 7-Oxygenated sterols[a]
O_2^-	No detectable products	Possibly autoxidation and $O_2^=$ products[b]
$O_2^=$	Cholesterol 5,6-epoxides	5α-Cholestane-3β,5,6β-triol 3β,6-Dihydroxy-5α-cholestan-6- one
HO·	Cholest-5-ene-3β,7-diols 3β-Hydroxycholest-5-en-7-one Cholesterol 5,6-epoxides	5α-Cholestane-3β,5,6β-triol

a From isomerization of the 5α-hydroperoxide

b From O_2^- dismutation

Accordingly, it is conceptually possible to distinguish among the several oxidants by analysis of the oxidized cholesterol derivatives formed. Thin-layer chromatography in conjunction with detection of ultraviolet light absorbing steroids (the 7-ketones, etc), of sterol hydroperoxides with N,N-dimethyl-p-phenylenediamine, and of all sterols with 50% sulfuric acid, warming to full color display or to charring, provides rapid and satisfactory means of recognition of these products (29,30). In TABLE II a systematic treatment of such data leads to product patterns that can be used to deduce the nature of the oxidizing species implicated.

Thin-layer chromatography obviously should be conducted with appropriate authentic reference sterols. Resolution of all sterols involved may not be available, as cholesterol 7α-hydroperoxide and 3β-hydroxy-5α-cholest-6-ene-5-hydroperoxide are not resolved. However, their sodium borohydride reduction products cholest-5-ene-3β,7α-diol and 5α-cholest-6-ene-3β,5-diol respectively, are readily resolved.

Similarly, the isomeric cholesterol 5,6-epoxides are not resolved in simple systems, and the 5,6-epoxides and 3β-hydroxycholest-5-en-7-one may present separation problems. Borohydride reduction of such mixtures yields the epimeric cholest-5-ene-3β,7-diols from the 7-ketone and unaltered 5,6-epoxides, this mixture being readily resolved. Subsequent lithium aluminum hydride reduction of the 5,6-epoxides yields alcohols that may be resolved (8). Other methods are also available for analysis of the isomeric 5,6-epoxides, the resolution of which is important to further interpretations of which epoxidation process be implicated. The 5β,6β-epoxide predominates in oxidations by $O_2^=$ and sterol hydroperoxides (8,23,25); the 5α,6α-epoxide predominates in hydroxyl radical (HO·) oxidations (27).

Confirmation of product identity is essential to sound deductions about associated oxidizing processes. Gas chromatography and high performance liquid chromatography aid in the matter. The three sterol hydroperoxides involved give individual pyrolysis patterns recognized by gas chromatography (7), and all are resolved on microparticulate high performance liquid chromatography (31). The isomeric cholesterol 5,6-epoxides are resolved by high performance liquid chromatography also (22,31,32).

Independent confirmation of product identities by spectral means is also advisable, particularly where recovery from tissues and metabolizing systems be the case. We have increasingly turned to the use of chemical ionization (CI) mass spectrometry for confirmation of identities made by chromatography. In a limited series such as the monohydroxylated cholesterols

TABLE II. Differentiation Among Cholesterol Oxidation Products

UV	Peroxide Text	H$_2$SO$_4$ Color	Product Identified	Oxidation Implicated
-	+	Blue	Cholesterol 7-hydroperoxides	$^3O_2$
-	-	Blue	Cholest-5-ene-3β,7-diols	$^3O_2$,[a] HO·
+	-	_b	3β-Hydroxycholest-5-en-7-one	$^3O_2$,[a] HO·
+	-	Tan	Cholesta-3,5-dien-7-one[c]	$^3O_2$
-	-	Tan	Cholesterol 5,6-epoxides	$^3O_2$,[a] $O_2^=$, HO·
-	-	Tan	5α-Cholestane-3β,5,6β-triol[d]	$^3O_2$, $O_2^=$, HO·
-	+	Green-blue	3β-Hydroxy-5α-cholest-6-ene-5-hydroperoxide	$^1O_2$
-	-	Blue	5α-Cholest-6-ene-3β,5-diol	$^1O_2$
+	-	Tan	Cholesta-4,6-dien-3-one	$^1O_2$

a Inference of $^3O_2$ confirmed by the presence of the cholesterol hydroperoxides
b No color developed; detected after charring
c Derived by dehydration of 3β-hydroxycholest-5-en-7-one
d Derived by hydration of the 5,6-epoxides

or established cholesterol autoxidation products, CI mass spectra are suitable for the matter. The class of monohydroxy-cholesterols are characterized by four ions m/z $420 [M+NH_4]^+$, $402 [M+NH_3-OH]^+$, $385 [M-OH]^+$, and $367 [M-H_2O-OH]^+$ using NH_3 as reagent gas, by five ions m/z $403 [M+H]^+$, $402 [M]^+$, $401 [M-H]^+$, $385 [M-H_2O+H]^+$, and $367 [M-2H_2O+H]^+$ using CH_4 (33-35). Monitoring these ions is about as close as one may come to routine screening of complex tissue lipids for $C_{27}H_{46}O_2$ stenediols as a class. Confirmation of structure assignments necessarily may also be made with conventional methods, including derivatization, electron impact mass spectrometry, etc. We have used CI mass spectrometry to obtain results discussed in this report.

These approaches using product analyses and cholesterol as probe have been used in several chemical and biological systems under controlled conditions, some of which are summarized in TABLE III. Each was conducted in aqueous media, thereby to approach conditions like those in living systems. From results in TABLE III, it appears that several oxidizing species $^3O_2$, $^1O_2$, $O_2^=$, and HO· may be recognized under favorable circumstances.

These results leave much to be desired, and the use of cholesterol as probe must be very carefully controlled. Besides the technical problems of analysis and the dynamics of the systems, several other matters may limit success. The short lifetime of $^1O_2$ in water, the sluggishness of $O_2^=$ reaction, and lack of reaction with O_2^- are factors that presently cannot be countered satisfactorily.

We have also examined other unsaturated sterols as potential probes, including cholest-4-en-3β-ol, 5α-cholest-7-en-3β-ol, and 5α-lanost-8-en-3β-ol, and all three give different products in autoxidations than formed in $^1O_2$ oxidations. However, the use of these sterols as probes is not as far developed as is the case of cholesterol.

III. MULTIPLE ORIGINS OF OXIDIZED STEROLS

In tissues and in systems involving enzymic metabolism, cholesterol metabolites as well as oxidation products from the action of discrete oxygen species may be encountered. Accordingly, it is necessary to be able to recognize acknowledged metabolites among all oxidation products. Six specific monohydroxylated cholesterol metabolites have been established: cholest-5-ene-3β,7α-diol, cholest-5-ene-3β,25-diol, and (25R)-cholest-5-ene-3β,26-diol associated with liver bile acids biosynthesis, (20S)-cholest-5-ene-3β,20-diol and (22R)-cholest-5-ene-3β,22-diol associated with endocrine steroid hormone biosynthesis, and (24S)-cholest-5-ene-3β,24-diol found in brain. All but the (22R)-3β,22-diol and (24S)-3β,24-diol are also cholesterol autoxidation products. Cholesterol 5α,6α-epoxide is also regarded as a genuine cholesterol metabolite (19-22).

Moreover, products of *in vitro* cholesterol lipid peroxidation (the epimeric 7-hydroperoxides, cholest-5-ene-3β,7-diols, 3β-hydroxycholest-5-en-7-one, and isomeric 5,6-epoxides) may be regarded broadly as enzyme products although not metabolites from specific enzyme-substrate interactions.

Cholest-5-ene-3β,7α-diol has multiple origins, being a cholesterol metabolite, autoxidation product, and product of HO· attack and of lipid peroxidation. Cholesterol 5α,6α-epoxide is all of these and also a product of $O_2^=$ oxidation of cholesterol. Clearly the presence of these two oxidized sterols in a biological system cannot *per se* infer origins.

However, cholest-5-ene-3β,7β-diol has no demonstrated genuine metabolite status but is formed by autoxidation, lipid peroxidation, HO· attack, or the epimerization of cholest-5-ene-3β,7α-diol (3). Cholesterol 5β,6β-epoxide derives as product of autoxidation, lipid peroxidation, and reaction with $O_2^=$ or HO· but has no demonstrated genuine metabolite status. Finally, 3β-hydroxycholest-5-en-7-one is an autoxidation and lipid peroxidation product as well as product of HO· oxidation of cholesterol. Cholest-5-ene-3β,7β-diol and 3β-hydroxycholest-5-en-7-one are interconverted *in vitro* by liver systems (43-45).

Because of these problems a rather careful, complete analysis of all implicated cholesterol oxidation products must be made in most cases before reliable assessment of the nature of the contributing oxidants can be made.

At the parts-per-million level at which these several cholesterol metabolites and related oxidation products have been found in tissues, technical difficulties of differentiation among monooxygenase action, autoxidation, lipid peroxidation, and other oxidative action greatly limit progress in applying these approaches to tissues. As an alternative to tissue analyses seeking the free sterols, with the attendant uncertainty as to potential origins but also with the extreme limitation that concurrent unrecognized autoxidation could compromise results in any experiment, we have turned to analyses of fatty acyl esters of these oxidized cholesterol derivatives.

IV. HUMAN TISSUE STEROL ESTERS ANALYSES

The presence of oxidized cholesterol derivatives in tissues being compromised by the uncertainties of insidious autoxidation, it is of interest to examine the presence of esterified

TABLE III. Product Analyses in Aqueous Systems

System	Products Identified	Oxidation Implicated
Chemical Systems:		
Aeration ($^3O_2$) (29)	Cholesterol 7-hydroperoxides Cholest-5-ene-3β,7-diols 3β-Hydroxycholest-5-en-7-one Cholesterol 5,6-epoxides 5α-Cholestane-3β,5,6β-triol	$^3O_2$
H_2O_2 (23,25)	Cholesterol 5,6-epoxides 5α-Cholestane-3β,5,6β-triol	$O_2^=$
H_2O_2 dismutation (23,25)	Cholesterol 7-hydroperoxides Cholest-5-ene-3β,7-diols 3β-Hydroxycholest-5-en-7-one 5α-Cholest-6-ene-3β,5-diol Cholesta-4,6-dien-3-one	$^3O_2$ $^1O_2$
Photosensitized oxygenation (36-38)	3β-Hydroxy-5α-cholest-6-ene-5-hydroperoxide Cholest-5-ene-3β,7-diols 3β-Hydroxycholest-5-en-7-one Cholesterol 5,6-epoxides 5α-Cholestane-3β,5,6β-triol	$^1O_2$
Radiolysis of water (27)		HO·

Biological Systems:

System	Products	Oxidant
Soybean lipoxygenase (10,12,13,15,17,18)	Cholesterol 7-hydroperoxides Cholest-5-ene-3β,7-diols 3β-Hydroxycholest-5-en-7-one Cholesterol 5,6-epoxides 5α-Cholestane-3β,5,6β-triol	$^3O_2$
Horseradish peroxidase (10,12)	Cholesterol 7-hydroperoxides Cholest-5-ene-3β,7-diols 3β-Hydroxycholest-5-en-7-one	$^3O_2$
Xanthine oxidase (24)	No products detected	$-$[a]
Rat liver microsomes (11-14,16-18)	Cholesterol 7-hydroperoxides Cholest-5-ene-3β,7-diols 3β-Hydroxycholest-5-en-7-one Cholesterol 5,6-epoxides 5α-Cholestane-3β,5,6β-triol	$^3O_2$
Bovine liver microsomes (22)	Cholesterol 5,6-epoxides 5α-Cholestane-3β,5,6β-triol	$-$[b]
Bovine adrenal cortex microsomes (20)	Cholesterol 5,6-epoxides 5α-Cholestane-3β,5,6β-triol	$-$[b]
Sheep vesicular gland microsomes (39)	No products detected	$-$[a]
Human erythrocyte photooxygenation (40,41)	3β-Hydroxy-5α-cholest-6-ene-5-hydroperoxide	$^1O_2$
Rat lung macrophages (42)	7-Oxygenated products	$^3O_2$ or HO·

[a] Consistent with O_2^-, specific enzymic, or no oxidation

[b] Other oxidation products not analyzed. Consistent with $^3O_2$, $O_2^=$, of HO· oxidation

oxidized sterols in tissues. It is recognized that cholesterol esters are less sensitive to air oxidations than cholesterol, and the enzymic monohydroxylation of cholesterol esters is not indicated (46-49). However, cholesterol fatty
acyl esters are autoxidized under conditions of heating (3) or
irradiation (50,51) in air.

   Esterified oxidized cholesterol derivatives have been
found in tissues by others, as summarized in TABLE IV. To
these discoveries we now add our own recent results, summarized in TABLE V.

A.   Results with Human Aorta

   Total lipids of human intimal and medial tissue were
chromatographed repeatedly, using authentic stenediol esters
as guides in later phases to recover sterol esters whose identities were established by combinations of chromatography and
mass spectrometry, including saponifications and identifica-
tions of the free sterols and fatty acids. Individual stenediol mono- and di-esters of cholesterol metabolites cholest-
5-ene-25-diol and cholest-5-ene-3β,26-diol were identified at
levels ranging from 24-100 µg/g tissue, cf. TABLE V.

   In addition to these esters, the presence of both mono-
and di-esters of the epimeric cholest-5-ene-3β,7-diols was
demonstrated. In this case individual sterol esters were not
resolved, but stenediol esters involving the common fatty
acids palmitic, stearic, oleic, linoleic, and linolenic were
recognized. Cholesterol esters of these and of arachidonic
acid were also recognized.

   Our data establishing cholest-5-ene-3β,7-diol and cholest-
5-ene-3β,26-diol esters in aortal tissue confirms prior note of
these same esters (58-60). Esters of cholest-5-ene-3β,25-diol
are reported here for the first time. Esters of cholest-5-ene-
3β,24-diol previously noted in human aorta (58,60) were not
encountered in the present study, but the search protocol may
have excluded these esters.

B.   Results with Human Plasma

   Total lipids of fresh human plasma were examined in comparison with human plasma stored for over six months in blood
bank plasma bottles. In this study only the sterols present
following saponification of resolved sterol ester fractions
were identified. As shown in TABLE V, six sterols were identified, the isomeric cholest-5-ene-3β,24-, 3β,25- and 3β,26-
diols being considered as cholesterol metabolites in all pro-
bability.

   However, the epimeric cholest-5-ene-3β,7-diols and 3β-hy-
droxycholest-5-en-7-one comprise a trio of oxidized chole-

TABLE IV. Esterified Oxidized Cholesterol Products in Tissues

Tissue	Fatty Acyl Esters	Sulfate Esters
Human serum	Cholest-5-ene-3β,7-diols(50-55) Cholest-5-ene-3β,25-diol(50,51)	Cholest-5-ene-3β,22-diol(62)
Human plasma	—	Cholest-5-ene-3β,25-diol(62)
		Cholest-5-ene-3β,24-diol(63)
Human urine	—	Cholest-5-ene-3β,26-diol(63)
Human meconium and feces	—	Cholest-5-ene-3β,26-diol(63) (22R)-Cholest-5-ene-3β,22-diol(64)
		Cholest-5-ene-3β,23-diol(65)
Human liver	Cholest-5-ene-3β,7-diols(56) 3β-Hydroxycholest-5-en-7-one(56) Cholesterol 5,6-epoxides(56)	Cholest-5-ene-3β,24-diol(66) Cholest-5-ene-3β,26-diol(66) —
Human aorta	Cholest-5-ene-3β,7-diols(42,57,58) Cholest-5-ene-3β,24-diol(58,60) Cholest-5-ene-3β,26-diol(42,58-60)	—
Rat liver, serum, skin	Cholest-5-ene-3β,7α-diol(53,54)	—
Bovine adrenal cortex	(20S)-Cholest-5-ene-3β,20-diol(61)	—

TABLE V. Oxidized Sterol Esters in Human Tissues

Parent Sterol	Fatty Acyl Ester	Level
Human <u>Aorta</u>		
Cholest-5-ene-	Diesters	–
3β,7-diols	Monoesters[a]	–
Cholest-5-ene-	3β,25-Dipalmitate	32 μg/g
3β,25-diol	3β,25-Dioleate	24 μg/g
	3β-Octadecadienoate	–
	Monoesters[a]	–
Cholest-5-ene-	Myristate palmitate	–
3β,26-diol	Palmitate Octadecatrienoate	–
	3β,26-Dipalmitate	100 μg/g
	3β,26-Distearate	31 μg/g
	3β,26-Dioleate	60 μg/g
	3β,26-Dioctadeca- trienoate	–
	Myristate	–
	Octadecenoate	–
	3β-Palmitate	–
	Monoesters[a]	–
Cholesterol	Esters[b]	–
Human-<u>Liver</u>		
3β-Hydroxycholest- 5-en-7-one	3β-Palmitate	70 ng/g
Cholest-5-ene- 3β,25-diol	3β-Palmitate	100 ng/g
Human <u>Plasma</u>		
Cholest-5-ene- 3β,7α-diol	Esters	19 ng/mL[c]
Cholest-5-ene- 3β,7β-diol	Esters	11 ng/mL[c]
3β-Hydroxycholest- 5-en-7-one	Esters	6 ng/mL[c]
Cholest-5-ene- 3β,24-diol	Esters	7 ng/mL[c]
Cholest-5-ene- 3β,25-diol	Esters	5 ng/mL[c]
Cholest-5-ene- 3β,26-diol	Esters	106 ng/mL[c]

[a] Fatty acids implicated: 16:0, 18:0, 18:1, 18:2, 18:3
[b] Fatty acids implicated: 16:0, 18:0, 18:1, 18:2, 18:3, 20:4
[c] Sterol levels in saponified ester fraction

sterol derivatives that may be regarded as evidence of autoxidation, lipid peroxidation, or HO· oxidation of cholesterol.
Adventitious autoxidation may be ruled out, as fresh plasma
was used directly and the amounts of esterified sterols found
were at the same levels as found for the esterified cholesterol metabolites. Analyses of aged human plasma failed to
detect esterified cholest-5-ene-3β,24-diol and cholest-5-ene-
3β,26-diol, but cholest-5-ene-3β,25-diol was found at 20 ng/mL
level esterified. Moreover, the epimeric cholest-5-ene-3β,7-
diols and 3β-hydroxycholest-5-en-7-one were found esterified
at much higher levels (500-1000 ng/mL) than in fresh plasma.

We tentatively hold that the increased levels of esterified 7-oxygenated sterols in aged human plasma represent an
insidious autoxidation of cholesterol esters and/or esterifi-
cation of cholesterol autoxidation products formed during
storage frozen. The lower levels found in fresh plasma then
represent sterol esters present in human blood at the time of
blood donation.

C.   Results with Human Liver

Only preliminary results have been obtained with human
liver, but the detection of at least two specific esters has
been made. The fractionation protocol was modified to utilize
high performance liquid chromatography throughout, starting
with preparative equipment and proceeding through intermediate-
size semipreparative columns to analytical columns for ultimate work, using synthesized authentic sterol esters as guides.

Two components with retention properties exactly those of
cholest-5-ene-3β,25-diol 3β-palmitate and 3β-hydroxycholest-
5-en-7-one 3β-palmitate were recovered in sufficuent purity
for confirmation of identities by CI mass spectra, ions m/z
658 $[M+NH_4]^+$, 367 $[M-C_{16}H_{32}O_2-OH]^+$, and 274 $[C_{16}H_{32}O_2+NH_4]^+$
(with NH_3) being observed for the cholest-5-ene-3β,25-diol
3β-palmitate, ions m/z 383 $[M-C_{16}H_{32}O_2+H]^+$ and 257 $[C_{16}H_{32}O_2+H]^+$
(with CH_4) recorded for the 3β-hydroxycholest-5-en-7-one 3β-
palmitate.

Human liver contains many other lipid esters, including
sterol esters, but our abbreviated screening techniques in
this study did not lead us to other identifiable oxidized
sterol esters. The work is much in need of expansion and
definitive treatment.

V.   CONCLUSIONS

The present results demonstrate that discrete esterified
oxidized cholesterol derivatives exist in human tissues at low

levels. Whether the sterol esters serve a physiological role
is not addressed in our study, but the sterol esters may be
regarded as metabolites derived from *in vivo* processes.

In these studies the autoxidation of cholesterol was mini-
mized by exercise of care in tissue collection and sterol analysis, and the autoxidation of cholesterol esters should be
even less conspicuous. The sterol esters found here accordingly do not appear to arise via autoxidation of cholesterol
esters. The autoxidation of cholesterol esters obviously cannot account for the presence of stenediol diesters anyway.
The lipid peroxidation of cholesterol esters has not been described.

Rather than posit an insidious cholesterol autoxidation
coupled with *post mortem* transesterification, the *in vivo* es-
terification of stenediols may be posed. This is almost sure-
ly the case for the cholest-5-ene-3β,26-diol esters encountered in the aorta, where an accumulation of (25R)-cholest-5-
ene-3β,26-diol and cholest-5-en-3β,26-diol esters has been
repeatedly demonstrated. Diesters of metabolites cholest-5-
ene-3β,7α-diol and cholest-5-ene-3β,25-diol are also reasonably accepted as being formed *in vivo*.

The same argument for cholest-5-ene-3β,7β-diol and 3β-
hydroxycholest-5-en-7-one esters implies that these two
sterols also be present *in vivo* before tissue specimens were
taken. An *in vivo* origin of cholest-5-ene-3β,7β-diol could be
via epimerization of metabolite cholest-5-ene-3β,7α-diol (3),
and enzymic dehydrogenation of cholest-5-ene-3β,7β-diol could
account for *in vivo* origins of 3β-hydroxycholest-5-en-7-one
(43-45).

However, another equally probable origin is suggested,
that *in vivo* lipid peroxidation of cholesterol, yielding cho-
lest-5-ene-3β,7α-diol, cholest-5-ene-3β,7β-diol, and 3β-hy-
droxycholest-5-en-7-one in analogy to the *in vitro* case.
These putative *in vivo* lipid peroxidation products then undergo acylations in subsequent metabolic disposition, providing
the fatty acyl esters found here and in prior studies.

Our present studies must be regarded as incomplete, and
conclusions associated therewith as tentative. However, the
work with plasma does offer an approach to the detection of
putative *in vivo* lipid peroxidations in which initial or
early products rather than terminal products such as the
breath gases ethane or pentane, malondialdehyde, etc. are
analyzed.

REFERENCES

1. Smith, L.L., in "Autoxidation in Food and Biological Systems", (M.G. Simic and M. Karel, eds.), Plenum Press, New York (1980), In Press.
2. Smith, L.L., "Cholesterol Autoxidation", Plenum Press, New York (1981), monograph in preparation.
3. Teng, J.I., Kulig, M.J., Smith, L.L., Kan, G., and van Lier, J.E., J. Org. Chem. 38, 119 (1973).
4. Smith, L.L., Teng, J.I., Kulig, M.J., and Hill, F.L., J. Org. Chem. 38, 1763 (1973).
5. van Lier, J.E., and Smith, L.L., Steroids 15, 485 (1970).
6. Smith, L.L., Kulig, M.J., and Teng, J.I., Steroids 22, 627 (1973).
7. Teng, J.I., Kulig, M.J., and Smith, L.L., J. Chromatog. 75, 108 (1973).
8. Smith, L.L., and Kulig, M.J., Cancer Biochem. Biophys. 1, 79 (1975).
9. van Lier, J.E., and Smith, L.L., J. Org. Chem. 35, 2627 (1970).
10. Teng, J.I., and Smith, L.L., J. Am. Chem. Soc. 95, 4060 (1973).
11. Smith, L.L., and Teng, J.I., J. Am. Chem. Soc., 96, 2640 (1974).
12. Teng, J.I., and Smith, L.L., Bioorg. Chem. 5, 99 (1976).
13. Aringer, L., Lipids 15, 563 (1980).
14. Mendlesohn, D., Mendelsohn, L., and Staple, E., Biochim. Biophys. Acta 97, 379 (1965).
15. Björkhem, I., Eur. J. Biochem. 18, 299 (1971).
16. Mitton, J.R., Scholan, N.A., and Boyd, G.C., Eur. J. Biochem. 20, 569 (1971).
17. Johansson, G., Eur. J. Biochem. 21, 68 (1971).
18. Aringer, L., and Eneroth, P., J. Lipid Res. 15, 389 (1974).
19. Watabe, T., and Sawahata, T., Biochem. Biophys. Res. Commun. 83, 1396 (1978).
20. Watabe, T., and Sawahata, T., J. Biol. Chem. 254, 3854 (1979).
21. Watabe, T., Kanai, M., Isobe, M., and Ozawa, N., Biochem. Biophys. Res. Commun. 92, 977 (1980).
22. Watabe, T., Kanai, M., Isobe, M., and Ozawa, N., Biochim. Biophys. Acta 619, 414 (1980).
23. Smith, L.L., and Kulig, M.J., J. Am. Chem. Soc. 98, 1027 (1976).
24. Smith, L.L., Teng, J.I., and Kulig, M.J., Chem. Phys. Lipids 20, 211 (1977).
25. Smith, L.L., Kulig, M.J., Miiller, D., and Ansari, G.A.S., J. Am. Chem. Soc. 100, 6206 (1978).

26. Smith, L.L., and Stroud, J.P., Photochem. Photobiol. 28, 479 (1978).
27. Ansari, G.A.S., and Smith, L.L., Photochem. Photobiol. 30, 147 (1979).
28. Sanche, L., and van Lier, J.E., Chem. Phys. Lipids 16, 225 (1976).
29. Smith, L.L., Matthews, W.S., Price, J.C., Bachmann, R.C., and Reynolds, B., J. Chromatog. 27, 187 (1967).
30. Smith, L.L., and Hill, F.L., J. Chromatog. 66, 101 (1972).
31. Ansari, G.A.S., and Smith, L.L., J. Chromatog. 175, 307 (1979).
32. Tsai, L.S., Hudson, C.A., and Meehan, J.J., Lipids 15, 124 (1980).
33. Lin, Y.Y., and Smith, L.L., Biomed. Mass Spectrom. 5, 604 (1978).
34. Lin, Y.Y., and Smith, L.L., Biomed. Mass Spectrom. 6, 15 (1979).
35. Lin, Y.Y., Lipids 15, 756 (1980).
36. Lightner, D.A., and Norris, R.D., New Eng. J. Med. 290, 1260 (1974).
37. Suwa, K., Kimura, T., and Schaap, A.P., Biochem. Biophys. Res. Commun. 75, 785 (1977).
38. Foote, C.S., Shook, F.C., and Abakerli, R.A., J. Am. Chem. Soc. 102, 2503 (1980).
39. Marnett, L.J., Wlodawer, P., and Samuelsson, B., J. Biol. Chem. 250, 8510 (1975).
40. Lamola, A.A., Yamane, T., and Trozzolo, A.M., Science 179, 1131 (1973).
41. De Goeij, A.F.P.M., and Van Steveninck, J., Clin. Chim. Acta 68, 115 (1976).
42. Foote, C.S., in "Biochemical and Clinical Aspects of Oxygen", (W.S. Caughey, ed.), p. 603. Academic Press, New York (1979).
43. Björkhem, I., Einarsson, K., and Johansson, G., Acta Chem. Scand. 22, 1595 (1968).
44. Van Cantfort, J., Life Sci. 11, Part II, 773 (1972).
45. Aringer, L., and Eneroth, P., J. Lipid Res. 14, 563 (1973).
46. Katayama, K., and Yamasaki, K., Yonago Acta Med. 12, 103 (1968).
47. Ogura, M., Shiga, J., and Yamasaki, K., J. Biochem. 70, 967 (1970).
48. Mayer, D., Koss, F.-W., and Glasenapp, A., Z. Physiol. Chem., 353, 921 (1972).
49. Balasubramaniam, S., Mitropoulos, K.A., and Myant, N.B., Biochim. Biophys. Acta 398, 172 (1975).
50. Claude, J.R., and Beaumont, J.L., Compt. Rend. 260, 3204 (1965).
51. Claude, J.R., and Beaumont, J.L., J. Chromatog. 21, 189 (1966).

52. Crastes de Paulet, A., and Crates de Paulet, P., in "The Enzymes of Lipid Metabolism" (P. Desnuelle, ed.), p. 109. Pergamon Press, New York, (1961).

53. Boyd, G.S., and Mawer, E.B., Biochem. J. 81, 11P (1961).

54. Boyd, G.S., Federation Proc. 21, Suppl. II, 86 (1962).

55. Claude, J.-R., and Beaumont, J.-L., Ann. Biol. Clin. 22, 815 (1964).

56. Assmann, G., Fredrickson, D.S., Sloan, H.R., Fales, H.M., and Highet, R.J., J. Lipid Res. 16, 28 (1975).

57. Claude, J.R., Clin. Chim. Acta 17, 371 (1967).

58. Brooks, C.J.W., Steele, G., Gilbert, J.D., and Harland, W.A., Atherosclerosis 13, 223 (1971).

59. Gilbert, J.D., Brooks, C.J.W., and Harland, W.A., Biochim. Biophys. Acta 270, 149 (1972).

60. Teng, J.I., and Smith, L.L., Texas Rpts. Biol. Med. 33, 293 (1975).

61. Roberts, K.D., Bandy, L., and Lieberman, S., Biochemistry 8, 1259 (1969).

62. Eberlein, W.R., and Patti, A.A., J. Clin. Endocrinol. Metab. 25, 1101 (1965).

63. Summerfield, J.A., Billing, B.A., and Shackleton, C.H.L., Biochem. J. 154, 507 (1976).

64. Gustaffson, J.-Å., and Sjövall, J., Eur. J. Biochem. 8, 467 (1969).

65. Eneroth, P., and Gustaffson, J.-Å., FEBS Lett. 3, 129 (1969).

66. Lavy, U., Burstein, S., and Javitt, N.E., J. Lipid Res. 18, 232 (1977).

INTERACTION BETWEEN AN ORGANIC HYDROPEROXIDE AND AN UNSATURATED PHOSPHOLIPID AND α-TOCOPHEROL IN MODEL MEMBRANES

Minoru Nakano

Gunma University
College of Medical Care and Technology
Maebashi

I. INTRODUCTION

It has been well documented that oxygen, iron, and an enzyme, NADPH-dependent cytochrome P-450 reductase, are involved in microsomal lipid peroxidation (1,2). NADPH-dependent cytochrome P-450 reductase catalyzes the reduction of iron, probably in a chelated form, in the presence of NADPH (3). The iron in ferrous state can easily be conjugated with oxygen and the resulting iron-oxygen complex (Fe^{3+}----O_2^-) may abstract hydrogen atom from unsaturated fatty acid in phospholipid, yielding lipid radicals as an initiator of the lipid peroxidation (4,5).

Even though the existence of lipid hydroperoxides in intact biological membrane has not yet been clearly demonstrated, these hydroperoxides are also considered to be precursors of iron-induced microsomal lipid peroxidation (3,6,7). Glutathione peroxidase in cytosol is an important enzyme in catabolizing lipid hydroperoxide and preventing the agents from propagating lipid peroxidation (8-10).

On the other hand, α-tocopherol is believed to scavenge lipid radicals generated in biological membranes and also to facilitate the molecular packing with biological membrane (11).

Little is known, however, of the effect of organic hydroperoxides on unsaturated phospholipids and of the interaction of hydroperoxides with α-tocopherol in biological membrane. The present work was undertaken to determine the behavior of an organic hydroperoxide in model membrane, using cholesterol-5α-hydroperoxide as an organic hydroperoxide.

LIPID PEROXIDES IN BIOLOGY AND MEDICINE

II. MATERIALS

Microsomal phospholipid, from rat liver microsomes, was usually prepared by the methods of Folch et al. (12). In some cases, cholesterol, its esters and free fatty acid contaminants were completely removed by silicic acid column chromatography, using chloroform (washing agent) and methanol (eluting agent). Egg lecithin was prepared by the method of Brandt and Lands (13).

Microsomal NADPH-dependent cytochrome P-450 reductase (specific activity, 18-25 μmol ferricytochrome c reduced/min/ mg protein) was prepared from rat liver microsomes by an established method (14).

[4-^{14}C]Cholesterol-5α-hydroperoxide was prepared by the dye-sensitized photooxygenation of [4-^{14}C] cholesterol (15). Other cholesterol analogues were prepared by the methods described previously (16).

Unless stated otherwise, α-tocopherol and its analogues (Eisai Co. Ltd., Tokyo) were dl-configurations and their purities were more than 98.9%. Specific activity of ^{14}C-labeled α-tocopherol was 0.1 Ci/mol and purity was 95%.

III. INCUBATION EXPERIMENTS AND ANALYSIS

Micelle system constituted of liposomes (10.1 μmol/lipid phosphorous) containing cholesterol-5α-hydroperoxide (1 μmol), or both cholesterol-5α-hydroperoxide (1 μmol) and one of the other lipid-soluble compounds (1 μmol), and 0.1 M Tris-HCl buffer (pH 6.8), in a total volume of 12 ml. The incubation of the media was continued for the time stated in the results.

Iron-containing system was composed of liposomes (10.1 μmol/lipid phosphorous) with or without lipid-soluble compound (1 μmol), 100 μM $Fe(NO_3)_3$, 1.67 mM ADP, 50 μM EDTA, NADPH-dependent cytochrome P-450 reductase (2.7 units), 0.14 mM NADPH, and 0.1 M Tris-HCl buffer (pH 6.8) in a total volume of 12 ml. Controls, which are described in the results, were included in these experiments. The reaction was initiated by the addition of NADPH and continued at 25°C in a water bath with continuous shaking. At the time cited, an aliquot of the reaction mixture or total reaction mixture was extracted with 2 vols of $CHCl_3/CH_3OH$ (2 : 1, by volume) and the extract was evaporated to dryness in vacuo. The residue was dissolved in 1.0 ml of $CHCl_3/CH_3OH$ (1 : 1, by volume) for analysis of radioactive cholesterol and its analogues on silica gel H plates (20 × 20 cm), which was developed in diethyl ether/cyclohexane (9 : 1, by volume) (17). In some cases, the lipid sample was

treated with NaBH₄ before thin layer chromatography. Samples
containing radioactive α-tocopherol and its analogues were
chromatographed on silica gel G plates in cyclohexane/diethyl
ether (7 : 3, by volume). Malondialdehyde was measured by the
thiobarbiturate method (18). NADPH was measured by fluorometry at 450 nm with excitation wavelength at 360 nm (19).
Oxygen uptake was determined by means of the Yanako oxygeno-
meter (Modle PO-100A) and calculated using 258 nmol/ml of
oxygen concentration in the initial incubation mixture. Fatty
acid composition of phospholipid was measured by a modification of the methods described by May and McCay (20), using
Shimadzu gas chromatograph (Model GC-4CM).

IV. RESULTS AND DISCUSSION

1. An allylic rearrangement of cholesterol-5α-hydroperoxide
in phospholipid micelles

     When [4-^{14}C]cholesterol-5α-hydroperoxide, emulsified with
a saturated phospholipid, was incubated at 25°C, it was
rapidly converted to cholesterol-7α-hydroperoxide (allylic
rearrangement). Such a rearrangement increased linearly up
to the time at which one-half of the 5α-hydroperoxide was isomerized. However, the allylic rearrangement was strikingly
inhibited by the replacement of the saturated phospholipid

TABLE I. Isomerization of Cholesterol-5α-hydroperoxide in
Phospholipid Micelles[a]

Phospholipid	Product Yield (%)		
	5α-OH	7α-OOH	Others
Dipalmitoyl phosphatidylcholine	1.5	53.5	3.2
Egg yolk lecithin	1.9	11.7	4.5
Microsomal phospholipid	1.5	4.3	1.5

[a] The system described in the text (micelle system) was incubated for 1 hr at 25°C with continuous shaking. Radioactivity
incorporated was compared with control (unincubated chole-
sterol-5α-hydroperoxide, purity 95%). 5α-OH, cholesterol-5α-
hydroxyde; 7α-OOH, cholesterol 7α-hydroperoxide. Results are
expressed as the mean of two experiments.

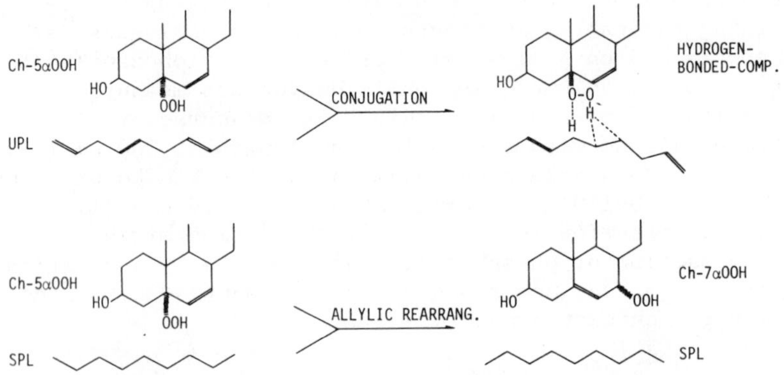

Figure 1. Formation of a hydrogen-bonded complex between cholesterol-5α-hydroperoxide and unsaturated fatty acid in phospholipid micelle. UPL, unsaturated phospholipids; SPL, saturated phospholipid.

with unsaturated phospholipid, microsomal phospholipid, or egg yolk lecithin (Table I). Under these conditions, further degradation of cholesterol-5α-hydroperoxide and of cholesterol-7α-hydroperoxide, as well as the formation of malondialdehyde from unsaturated phospholipids was negligible. These results may indicate that double bond in unsaturated fatty acid in phospholipid makes a hydrogen bonded complex with OOH group in cholesterol-5α-hydroperoxide thereby protecting the allylic rearrangement (Fig. 1). It has been reported that the compounds which have unshared electrons or π-electron clouds can produce hydrogen bonded complexes with hydroperoxides (21-23).

2. Behavior of cholesterol-5α-hydroperoxide produced from cholesterol in phospholipid micelles

To study this, [4-^{14}C]cholesterol, emulsified either with dipalmitoyl phosphatidylcholine (saturated phospholipid) or microsomal phospholipid (unsaturated phospholipid), was exposed to singlet oxygen according to the method described by Suwa et al. (24) and the radioactive compounds were separated by TLC (Fig. 2).

In these experiments cholesterol-5α-hydroperoxide and cholesterol-6β-hydroperoxide, which are known to be the singlet oxygen derived products (15), were not well distinguished from the 7β-hydroperoxide and the 7α-hydroperoxide, respectively, on the chromatoplate developed in diethyl ether/cyclohexane. However, the treatment of these hydroperoxides

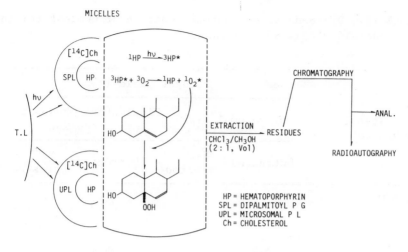

Figure 2. Processes for dye-sensitized photooxygenation of cholesterol in phospholipid micelles and analysis of the products. The liposomal suspension, which consisted of 0.5 mg of hematoporphyrin, 10.1 μmol of phospholipid and 1 μmol of [4-^{14}C]cholesterol in 12 ml of 0.1 M Tris-HCl buffer (pH 6.7), was dialyzed at 4°C against 1 l of 0.1 M Tris-HCl buffer (pH 6.8) for 18 hr to remove hematoporphyrin that had not been incorporated in the liposomes. Nitrogen gas was passed through the dialysis medium continuously. The dialyzed liposomal suspension was then transferred to a Pyrex tube. Photooxygenation was achieved by bubbling oxygen through the liposomal suspension in a water-cooled immersion irradiation apparatus using a 500W tungsten lamp at a distance of 7 cm. The irradiation was continued for 20 min.

with an excess of NaBH₄ gave their corresponding alcohols, which can be clearly separated from each other on the plate developed with the same solvent. Since further degradation of the hydroperoxides to their corresponding alcohols or keto analogues was negligible, the alcohols obtained by the treatment of the products with NaBH₄ should be derived mainly from their corresponding hydroperoxides.

The results obtained under these conditions (Table II) indicate that cholesterol-5α-hydroperoxide produced can be isomerized in saturated phospholipid micelles, but not in unsaturated phospholipid micelles even though lipid peroxidation could be detected (30.1 nmol malondialdehyde formed/ml). This can be explained from the assumption that the 5α-hydroperoxide produced by singlet oxygen can easily make a hydrogen bonded complex with a double bond which could not be attacked by singlet oxygen. Singlet oxygen-derived conjugated lipid

TABLE II. Dye-Sensitized Photooxygenation of Cholesterol in Phospholipid Micelles[a]

Cholesterol analogues	Product yield (%)			
	Microsomal lipid		Dipalmitoyl phosphatidyl-choline	
	Untreated	NaBH₄-treated	Untreated	NaBH₄-treated
5α-OOH	11.0		20.8	
7β-OOH		0.2		0.0
7α-OOH	4.9		34.8	
6β-OOH				
5,6-Oxide	0.0	0.4	0.6	0.3
7-Keto	0.1	0.3	1.1	0.2
5α-OH	0.1	10.9	0.7	17.5
7β-OH	0.1	0.1	0.3	2.7
7α-OH	0.1	1.2	1.3	17.6
6β-OH	0.1	3.6	0.7	19.4
Others	0.5	2.7	0.1	6.3

[a] 5α-OOH, cholesterol-5α-hydroperoxide; 7β-OOH, cholesterol-7β-hydroperoxide; 7α-OOH, cholesterol-7α-hydroperoxide; 6β-OOH, cholesterol-6β-hydroperoxide; 5,6-oxides, cholesterol-5,6-oxide (a mixture of α- and β-oxides); 7-keto, 7-ketocholesterol; 5α-OH, cholesterol-5α-hydroxide; 7β-OH, cholesterol-7β-hydroxide; 7α-OH, cholesterol-7α-hydroxide; 6β-OH, cholesterol-6β-hydroxide. Radioactivity incorporated was compared with controls. Results are expressed as the mean of two experiments.

peroxide (25) would not be easily decomposed into free radicals which cause chain cleavage.

On the other hand, cholesterol in the unsaturated phospholipid micelles excluding hematoporphyrin, which had been aerated and irradiated identically to the dye-sensitized photooxygenation procedure, was oxidized to a lesser extent with a little lipid peroxidation.

3. Effects of α-tocopherol analogues on the allylic rearrangement of the 5α-hydroperoxide in saturated phospholipid
micelles

    Some of the lipid-soluble compounds, α-tocopherol, α-
tocopheryl nicotinate, α-tocopheryl acetate, 2,5-di-tert-
butylhydroquinone, and butyrated riboflavin, were tested as
possible inhibitors on the allylic rearrangement. The results
are summarized in Table III. Of the compounds tested, α-tocopherol and 2,5-di-tert-butylhydroquinone were the most potent
inhibitors and possessed weak O-O bond splitting activities.
α-Tocopheryl acetate and nicotinate, which have an esterified
OH group on their benzene ring, were less effective and non-
effective on the allylic rearrangement, respectively. Butyrated riboflavin had no demonstrable effect.
    These results may indicate that the OH group in the α-
tocopherol molecule as well as that in 2,5-di-tert-butyl-
hydroquinone also acts as to produce a hydrogen-bonded complex
with the 5α-hydroperoxide in saturated phospholipid micelles
(Fig. 3). Under the conditions listed in Table III, 5.8% of

TABLE III. Effect of Lipid-Soluble Compound on the Isomerization of Cholesterol-5α-hydroperoxide in Dipalmitoyl Phosphatidylcholine Micelles[a]

Added substances	Product Yield (%)		
	5α-OH	7α-OOH	Others
None	1.5	53.5	2.8
Tocopherol	9.0	0.0	6.4
Tocopheryl acetate	1.1	13.6	3.2
Tocopheryl nicotinate	2.4	51.1	4.1
2,5-Di-tert-butylhydroquinone	9.5	0.0	0.0
Riboflavin tetrabutyrate	1.8	56.4	3.2

[a] The system described in the text was incubated for 1 hr at
25°C with continuous shaking. Radioactivity incorporated was
compared with control (unincubated cholesterol-5α-hydroper-
oxide, purity 96%). The abbreviations were the same as those
described in the legend for Tables I and II. The results are
expressed as the mean of two experiments.

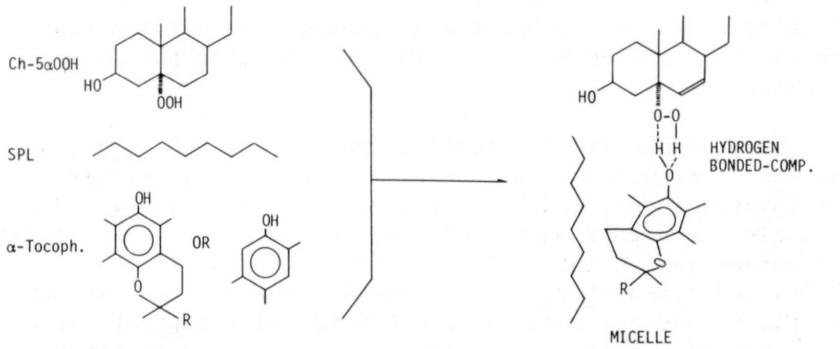

Figure 3. Formation of a hydrogen-bonded complex between cholesterol-5α-hydroperoxide and α-tocopherol (or, 2,5-di-tert-butylhydroquinone) in saturated phospholipid micelle.

TABLE IV. Effect of α-Tocopherol Analogues and 2,5-Di-tert-butylhydroquinone on Iron-Induced Cholesterol-5α-hydroperoxide Decomposition in Dipalmitoyl Phosphatidylcholine Micelles[a]

Cholesterol analogues	Product Yield (%)			
	None	Toco-pherol	Toco-trienol	2,5-Di-tert-butylhydro-quinone
5α-OH	1.0	15.9	16.9	10.4
7α-OOH	53.9	0.0	0.0	3.6
7α-OH	11.6	2.0	1.1	1.2
7-Keto	29.9	0.5	0.0	0.0
Unknown	2.7	0.9	0.4	0.3
Origin[b]	2.0	0.1	0.7	0.0
Others	0.0	8.4	8.4	1.5

[a] The system described in the text (iron-containing system) was incubated for 1 hr at 25°C with continuous shaking. The abbreviations were the same as those described in the legend for Tables I and II.
[b] Those retained at the origin of the chromatoplate. Radio-activity incorporated was compared with control (unincubated cholesterol-5α-hydroperoxide, purity 95%). Results are expressed as the mean of two experiments.

radioactive α-tocopherol present in the saturated phospholipid micelles containing the 5α-hydroperoxide was converted to tocopherylquinone (2.7%), and a mixture of tocopherol dimer and trimer (3.1%) during 1 hr incubation. Thus, it can be calculated that approximately 77% of the 5α-hydroperoxide reduced is α-tocopherol-dependent, assuming that 1 mol of α-tocopherol is consumed for the reduction of 1 mol of the 5α-hydroperoxide with the formation of tocopherylquinone and the other products (26.27).

4. Fe^{2+}-Mediated degradation of the 5α-hydroperoxide in saturated phospholipid micelles with or without α-tocopherol analogues

When the 5α-hydroperoxide in saturated phospholipid micelles was exposed to the NADPH-dependent iron-containing system, the 7α-hydroperoxide, which had been produced by the allylic rearrangement (Table I), was decomposed to 7α-hydroxy- and 7-ketocholesterol, with a lower product ratio of 7α-hydroxy- to 7-ketocholesterol of approximately 0.4 (Table IV). Both the inhibition of allylic rearrangement and reduction of the 5α-hydroperoxide by α-tocopherol and 2,5-di-tert-butyl-hydroquinone (Table III) were not significantly affected.

5. Fe^{2+}-Mediated allylic rearrangement and degradation of the 5α-hydroperoxide in unsaturated phospholipid micelles with or without α-tocopherol analogues

The exposure of the 5α-hydroperoxide in microsomal phospholipid to the NADPH-dependent iron reduction system (iron-containing system) caused a partial reduction of the 5α-hydroperoxide to the corresponding alcohol and a significant isomerization of the 5α-hydroperoxide to the 7α-hydroperoxide, which could be further decomposed to 7-ketocholesterol and 7α-hydroxycholesterol (Table V). As shown in the same tabel, α-tocopherol, α-tocotrienol, or 2,5-di-tert-butylhydroquinone present in micelles inhibited both the allylic rearrangement and the further degradation, with the production of the 5α-alcohol in an appreciable amount. Such protections of the allylic rearrangement and the further degradation were not observed with α-tocopheryl nicotinate, instead of α-tocopherol.
In some cases of iron-containing system, three parameters, such as NADPH consumption, oxygen consumption, and malondialdehyde formation, were measured and depicted in Figure 4. Furthermore, the change in fatty acid composition in phospholipid during 1 hr-incubation was analyzed by gas chromatography (Fig. 5). During iron-induced lipid

TABLE V. Effect of α-Tocopherol and 2,5-Di-tert-butylhydroquinone on Iron-Induced Cholesterol-5α-hydroperoxide Decomposition in Microsomal Phospholipid Micelles

Cholesterol analogues	Product yield (%)				
	None	Tocopherol	Tocotrienol	Tocopheryl nicotinate	2,5-Di-tert-butyl-hydroquinone
5α-OH	5.3	58.4	46.8	6.1	35.5
7α-OOH	28.4	1.1	3.5	28.6	15.4
7α-OH	23.2	10.6	8.0	22.8	10.5
7-Keto	13.4	-0.9	0.0	13.4	0.0
Unknown	4.8	7.3	6.3	4.9	11.6
Origin[b]	9.8	2.7	3.2	9.2	4.8
Others	0.0	-2.7	3.2	0.8	3.7

[a] The system described in the text (iron-induced system) was incubated for 1 hr at 25°C with continuous shaking. The abbreviations were the same as those described in the legend for Tables I and II.
[b] Those retained at the origin of the chromatoplate. Radioactivity incorporated was compared with control (NADPH omitted from the experimental system). Results are expressed as the mean of two experiments.

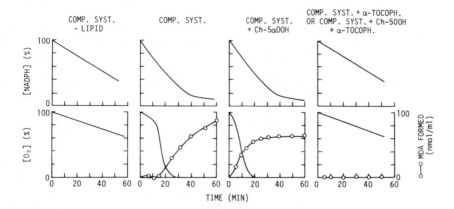

Figure 4. Malondialdehyde formation, oxygen consumption and NADPH consumpsion in various systems. Complete system (COMPT. SYST.) consisted of liposomes prepared with microsomal phospholipid and iron-containing system (NADPH, Fe^{3+}-ADP-EDTA and cytochrome P-450 reductase), described in the text. Incubation was carried out at 25°C. Concentrations of oxygen and NADPH were represented as percent to their initial concentrations. Ch-5α-OOH, cholesterol-5α-hydroperoxide.

peroxidation, the 5α-hydroperoxide in the micelles greatly enhanced malondialdehyde formation and oxygen uptake in an early stage of the reaction over the complete system (reconstructed microsomal lipid peroxidation system), suggesting that 5α- and 7α-alkoxy radical act as initiators for the lipid peroxidation. Under the same conditions, changes in the unsaturated fatty acids in micelles, especially in linoleic, arachidonic and docosahexaenoic acids, were observed both in the presence and absence of the 5α-hydroperoxide in complete system, but the former caused more significant effect on the decomposition of these unsaturated fatty acids.

α-Tocopherol (or α-tocotrienol) added to the above two systems inhibited malondialdehyde formation completely, protected the change in fatty acid composition, and did not elevate NADPH consumption and oxygen consumption over a control (complete system minus phospholipid). This indicates that little or no NADPH was used for the reduction of the 5α-hydroperoxide even in the presence of α-tocopherol. Unexpectedly, only 13.3% of radioactive α-tocopherol present in the micelles was converted to tocopherylquinone, and a mixture of dimer and trimer during Fe^{2+}-mediated rearrangement and degradation of the 5α-hydroperoxide. This value is obviously low, assuming that only α-tocopherol is an electron donor for the reduction of the 5α-hydroperoxide and the 7α-hydroperoxide to

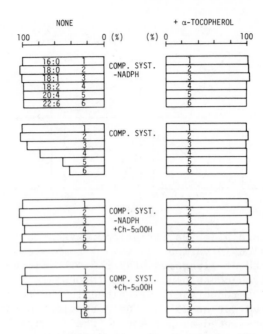

Figure 5. Change in the fatty acid composition of microsomal phospholipid during lipid peroxidation and its protection with α-tocopherol. Purified microsomal phospholipid was used in this experiment. The incubation was carried out at 25°C for 1 hr and the fatty acid composition in the microsomal phospholipid was analyzed as described in the text. Values (mean of three experiments) represent percentage of each fatty acid in unincubated microsomal phospholipid. 16 : 0, palmitic; 18 : 0, stearic; 18 : 1, oleic; 18 : 2, linoleic; 20 : 4, arachidonic; 22 : 6, docosahexaenoic acids. Other abbreviations were the same as those described in the legend for Figure 4.

their corresponding alcohols, corresponding to 70% of the 5α-hydroperoxide added. Therefore, the hydrogen atom required for the reduction of the 5α-hydroperoxide in the iron-induced lipid peroxidation could be mainly supplied from polyunsaturated fatty acids in the phospholipid. The resulting fatty acid radicals may be protected from destructive peroxidation either by radical chain termination or accepting an electron from Fe^{2+} in the compact structure, fortified by the molecular packing effect of the side chain in vitamin E within polyunsaturated phospholipid membrane (11).

From these results obtained with the 5α-hydroperoxide-microsomal phospholipid micelles with or without α-tocopherol, an iron-atom in a partially chelated form, which had been

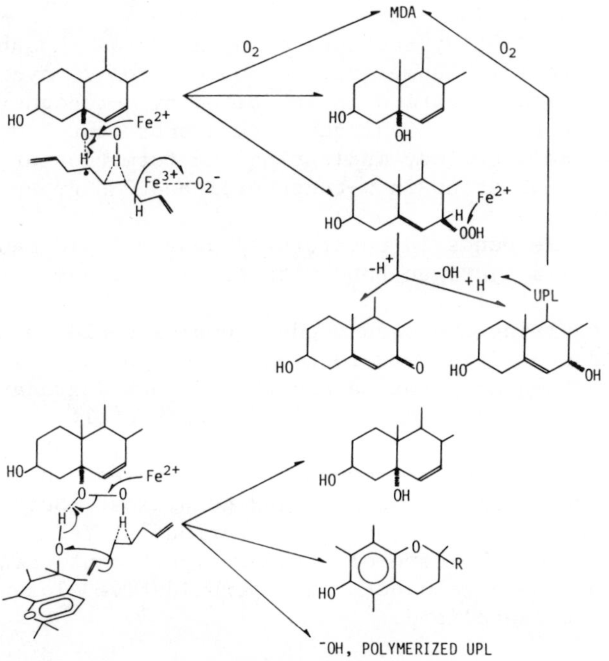

Figure 6.   Possible formation of hydrogen-bonded complexes and their degradation in the presence of Fe^{2+}.   MDA, malondialdehyde; UPL, unsaturated phospholipid.

reduced to the ferrous state by NADPH in an electron transfer reduction by cytochrome P-450 reductase (3), is considered to act as an electron donor for splitting the O-O bond in the 5α-hydroperoxide, yielding the 5α-alkoxy radical.   In the absence of α-tocopherol, such a reaction also derived both the keto and alcohol analogues from the 7α-hydroperoxide, probably through an alkoxy radical, in keeping with the mechanism of the cumylhydroperoxide decomposition by Fe^{2+} (28).   Hydrogen atom in unsaturated phospholipid could be required for the predominant formation of the 7α-alcohol in the reaction (Table IV and V).   An outline of the mechanism explained here is shown in Figure 6.

V.   CONCLUSION

The behavior of an organic hydroperoxide in the presence of lipid and/or α-tocopherol in model membranes has been studied using ^{14}C-labeled cholesterol-5α-hydroperoxide as an organic hydroperoxide.

Cholesterol-5α-hydroperoxide in saturated phospholipid micelles is rapidly isomerized to cholesterol-7α-hydroperoxide. Such an isomerization is inhibited by α-tocopherol, but not by an α-tocopheryl analogue with a substituted OH group, present in the micelles, indicating the formation of hydrogen bonds between OH group in α-tocopherol and OOH group in the 5α-hydroperoxide.

The double bonds in unsaturated phospholipid can also serve to form a hydrogen bond with OOH in the 5α-hydroperoxide moiety in the micelles.

The resulting hydrogen-bonded complex could be decomposed by iron-induced lipid peroxidation, accompanied by isomerization of the 5α-hydroperoxide and the further degradation to the 7-ketocholesterol and 7α-hydroxycholesterol.

When three components, such as unsaturated phospholipid, 5α-hydroperoxide, and α-tocopherol, are present in the same micelles, they form hydrogen bonded complexes. Such complexes could be decomposed by iron in the ferrous state, yielding mainly 5α-hydroxycholesterol without significant change in the structure of the α-tocopherol and peroxidative cleavage of unsaturated phospholipid.

ACKNOWLEDGMENTS

The author wishes to acknowledge the assistance in experimental studies of Mr. Katsuaki Sugioka, Mr. Takehito Oki, and Dr. Tetsuya Nakamura.

REFERENCES

1. Hochstein, P., and Ernster, L. (1963) Biochem. Biophys. Res. Commun. 12, 388.
2. Poyer, J. L., and McCay, P. B. (1971) J. Biol. Chem. 246, 263.
3. Noguchi, T., and Nakano, M. (1974) Biochim. Biophys. Acta 368, 446.
4. Nakano, M., and Noguchi, T. (1979) in "Biochemical and Medical Aspects of Active Oxygen" (O. Hayaishi, and K. Asada eds.), p. 29. University of Tokyo Press, Tokyo.
5. Svingen, B. A., Bugge, J. A., O'Neal, F. O., and Aust, S. D. (1979) J. Biol. Chem. 254, 5829.
6. Wills, E. D. (1966) Biochem. J. 99, 667.
7. Tyler, D. D. (1975) FEBS Lett. 51, 180.
8. Rotruck, J. T., Pope, A. L., Ganther, H. E., Swanson,

A. B., Hoffman, D. C., and Hockstera, W. G. (1973)
Science 179, 588.

9. Little, D., and O'Brien, P. J. (1968) Biochem. Biophys.
Res. Commun. 31, 145.

10. Chow, C. K., Reddy, K., and Tappel, A. L. (1973) J. Nutr.
103, 618.

11. Lucy, J. A. (1978) in "Tocopherol, Oxygen, and Biomembranes" (C. DeDuve, and O. Hayaishi eds.), p. 109.
Elsevier/North Holland, Amsterdam.

12. Folch, J., Lee, M. M., and Sloane-Staney, G. H. (1957) J.
Biol. Chem. 234, 2295.

13. Brandt, A. E., and Lands, W. E. M. (1967) Biochim. Bio-
Phys. Acta 144, 605.

14. Omura, T., and Takesue, S. (1970) J. Biochem. 67, 249.

15. Kulig, M. J., and Smith, L. L. (1973) J. Org. Chem. 38,
3639.

16. Nakano, M., Sugioka, K., Nakamura, T., and Oki, T. (1980)
Biochim. Biophys. Acta 619, 274.

17. Chioe, E., Powrie, M. D., and Fennema, O. (1968) Lipids
3. 551.

18. Johanson, G. (1971) Eur. J. Biochem. 21, 68.

19. Ernster, L., and Nordenbrand, K. (1967) in "Method in
Enzymology" (S. P. Colowick, and N. O. Kaplan, eds.),
Vol. 10, p. 574. Academic Press, New York.

20. May, H. E., and McCay, P. B. (1968) J. Biol. Chem. 243,
2288.

21. Barnard, D., Hargrave, K. R., and Higgins, G. M. C.
(1956) J. Chem. Soc. 2845.

22. Oswald, A. A. Jr., Noel, F., and Stephenson, A. J. (1961)
J. Org. Chem. 26, 3969.

23. Denison, E. T. (1964) Chem. Abstr. 61, 13163.

24. Suwa, K., Kimura, T., and Schapp, P. B. (1977) Biochem.
Biophys. Res. Commun. 75, 785.

25. Rawls, H. R., and Van Santen, P. J. (1970) Ann. N. Y.
Acad. Sci. 171, 135.

26. Svanholin, U., Beckgarrs, K., and Parken, V. D. (1974) J.
Am. Chem. Soc. 96, 2409.

27. Skinner, W. A., and Alaupovic, P. (1963) J. Org. Chem.
28, 2852.

28. Karash, M. S., Fono, A., and Nudenberg, W. (1950) J. Org.
Chem. 15, 763.

A SUPEROXIDE-ACTIVATED LIPID-ALBUMIN
CHEMOTACTIC FACTOR FOR NEUTROPHILS[1]

Joe M. McCord
William F. Petrone

Department of Biochemistry
University of South Alabama
Mobile, Alabama

INTRODUCTION

Superoxide dismutase [EC 1.15.1.1] is an ubiquitously
distributed intracellular enzyme which efficiently catalyzes
the dismutation of superoxide free radical anions (1,2):

$$O_2^- + O_2^- + 2H^+ \rightarrow O_2 + H_2O_2$$

When native or chemically modified superoxide dismutase is
administered by injection to laboratory animals it displays
anti-inflammatory activity in a number of models of induced
inflammation (3,4). Because phagocytosing neutrophils produce
substantial quantities of superoxide radical for bactericidal
purposes (5,6), we originally assumed that the primary
mechanism of the enzyme's anti-inflammatory action was the
protection of host tissue from the cytotoxic effects of the
phagocyte-produced superoxide. Systematic study of the
phenomenon, however, especially by histological methods,
suggested that superoxide was somehow involved in the process
whereby inflammatory cells congregate at the site of the
developing inflammation (3,4). That is, in animals which were

[1]Supported in part by Research Grant AM-20527 from the
National Institutes of Health. J.M.McC. is the recipient of a
Research Career Development Award from the National Institutes
of Health.

treated with superoxide dismutase, neutrophils failed to
accumulate at the site of the inflammatory stimulus, whereas
untreated control animals showed the expected heavy, diffuse
infiltration by inflammatory cells. To explain these observations, we proposed the existence in extracellular fluids of a
superoxide-activated chemotactic factor. Such a factor would
allow resting neutrophils to recognize and respond to a nearby
neutrophil which had been activated by an immune stimulus.
This mechanism would allow for amplification and perpetuation
of the chemotactic response for as long as the stimulus
remained. We describe below evidence which supports the
existence of a superoxide-activated chemotactic factor in
plasma, as well as the release of such a factor from activated
neutrophils.

MATERIALS AND METHODS

     Ferricytochrome c, sodium xanthine, xanthine oxidase
(grade III), catalase, zymosan, human serum albumin (fatty
acid free), bovine serum albumin (Cohn fraction V), and ovalbumin were obtained from Sigma. Superoxide generation and
superoxide dismutase activity were assayed by the cytochrome
reduction assay (1). Hanks' balanced salt solution was from
GIBCO. Lucite blind well chemotaxis chambers were obtained
from Nucleopore (Pleasantown, CA). Three-μm pore cellulose
acetate filters for chemotaxis were from Millipore.
     Human leukocytes and plasma were from blood of healthy
volunteers. Leukocytes were obtained by mixing blood with
6% dextran in saline (1:4). After the red cells sedimented,
the cells of the upper layer were collected, washed once with
Hanks' balanced salt solution, and resuspended in this
solution at 3-5 x 10^6 neutrophils per ml. Plasma was prepared
by centrifugation of heparinized blood.
     Chemotactic activity was assessed by the Boyden method
(7). Filters were stained with hematoxylin and the lower
surfaces examined microscopically. Ten fields were counted
and averaged.
     Lipids were extracted from plasma by mixing 20 ml plasma
and 60 ml of chloroform. The chloroform layer was evaporated
under nitrogen and the residue dissolved in 100 μl of dimethyl
sulfoxide. This solution was dissolved in 2 ml Hanks'
balanced salt solution with and without 0.4% fatty acid-free
human serum albumin or ovalbumin.
     Zymosan was opsonized by mixing 10 mg zymosan and 1 ml
plasma and incubating at 37° for 30 min. The opsonized
zymosan was washed twice and resuspended in Hanks' balanced
salt solution containing 1% bovine serum albumin.

RESULTS

Human plasma was exposed *in vitro* to superoxide generated enzymatically by xanthine oxidase and xanthine. The relative roles of superoxide and hydrogen peroxide were assessed by including superoxide dismutase or catalase in the reaction mixtures. The chemotactic activity of the solution was measured by the Boyden chamber technique, and the results are shown in Table I. Because experiments were performed on different days with cells and plasma from different donors, each experiment contains positive (10% superoxide-treated plasma) and negative (10% untreated plasma) controls for internal comparison. Exposure of plasma to superoxide resulted in the generation of a potent chemotactic activity for neutrophils. Omission of either part of the superoxide generating system prevented the formation of the factor, as did superoxide dismutase when added before, but not after the

TABLE I. *Superoxide-Dependent Chemotactic Activity in Plasma*

Exp.	Description[a]	Cells/ Field[b]	% Inhibition[c]
1	Normal plasma	4.3 ± 3.0	
	Superoxide-treated plasma	16.2 ± 4.2	
	minus xanthine	6.4 ± 2.5	87
	minus xanthine oxidase	4.3 ± 1.6	100
2	Normal plasma	1.3 ± 0.9	
	Superoxide-treated plasma	20.7 ± 3.9	
	plus SOD at t = 0	1.5 ± 1.4	99
	plus SOD at t = 60	19.0 ± 3.5	n.s.
3	Normal plasma	2.3 ± 1.4	
	Superoxide-treated plasma	16.3 ± 3.4	
	plus catalase at t = 0	14.2 ± 3.8	n.s.
	plus catalase at t = 60	12.3 ± 5.5	n.s.

[a]*Plasma was used at a concentration of 10% in Hanks' balanced salt solution. Superoxide-treated plasma was incubated for 1 hr at 37° with 0.3 mM xanthine and 0.01 unit xanthine oxidase per ml. Superoxide dismutase (SOD) was added at 50 µg/ml at 0 min or after 60 min. Catalase was added at 83 µg/ml.*
[b]*Mean ± SEM*
[c]*n.s. = not significant*

incubation period during which exposure to the radical occurred. Catalase did not cause significant inhibition of the production of the factor, whether added before or after exposure to superoxide had occurred. Heating the superoxide-activated plasma to 56° for 30 min resulted in loss of 53% of the chemotactic activity. The activity was stable at 4° for 24 hr, as well as to lyophilization. It was not dialyzable.

The possibility that the superoxide-dependent chemotactic factor might stimulate superoxide production by neutrophils was examined by incubating the cells in normal or superoxide-exposed 10% plasma in the presence of 30 µM ferricytochrome c. At 10 and 20 min, aliquots of supernatant were spectrally examined to determine the rates of cytochrome reduction. No stimulation of superoxide production was observed.

Chromatography of normal plasma on Sephadex G-100 resolved two major protein peaks. When fractions were assayed by the Boyden technique, weak chemotactic activity was associated with the high molecular weight globulin fraction as others have reported (8). When superoxide-exposed plasma was chromatographed, a strong new peak of chemotactic activity appeared. It chromatographed with the albumin fraction. The chemotactically active fractions were pooled, diluted, and rechromatographed on a Whatman DE-52 anion exchange column equilibrated with 0.1 M NaCl and 5 mM phosphate, pH 7.2. The column was eluted with a linear saline gradient and yielded four distinct protein peaks. Chemotactic activity was associated only with the last protein eluted. By polyacrylamide

TABLE II. *Reconstitution of Superoxide-Dependent Chemotactic Activity*

Test solution[a]	Cells/Field
Extract in serum albumin	15.6
+ superoxide-generating system	54.8
+ superoxide-generating system + SOD[b]	14.6
Extract in ovalbumin	3.5
+ superoxide-generating system	2.0
Albumin alone	1.2
+ superoxide-generating system	2.6

[a]*Chloroform extract prepared as described in text. The superoxide-generating system is described in Table I.*

[b]*Superoxide dismutase (SOD) was added at 50 µg/ml prior to the addition of xanthine oxidase.*

gel electrophoresis, this protein migrated identically to commercial human serum albumin.

Albumin purified by the above procedure from fresh, untreated human plasma became weakly chemotactic when exposed to superoxide. Commercially obtained bovine serum albumin or fatty acid-free human serum albumin showed no superoxide-dependent chemotactic activity.

Based on these results, we speculated that the superoxide-dependent chemotactic factor might be a lipid-albumin complex. To test this, a chloroform extract of untreated plasma was prepared. The chloroform was evaporated under nitrogen and the residue was dissolved in dimethyl sulfoxide. When this solution was added to Hanks' balanced salt solution the opalescent product had no chemotactic activity for neutrophils, before or after exposure to superoxide. When the dimethyl sulfoxide extract was added to Hanks' containing 0.2% fatty acid-free human serum albumin or ovalbumin, clear solutions resulted. The ovalbumin-solubilized extract showed no significant chemotactic activity before or after exposure to superoxide (see Table II). The serum albumin-solubilized extract, however, was chemotactic, and furthermore showed a substantial increase following exposure to superoxide.

Because metabolically activated neutrophils release a number of lipid products and have been reported to release a chemotactic activity as well, we examined the medium in which neutrophils had phagocytosed zymosan particles to determine whether the neutrophils themselves might be a source of the superoxide-dependent chemotactic factor. Table III shows that

TABLE III. *Release of Superoxide-Dependent Chemotactic Activity from Phagocytosing Neutrophils*

Description[a]	Cells/Field
Phagocytosing neutrophils	
in 1% serum albumin	20.6
in 1% serum albumin + SOD	< 1
in 1% ovalbumin	< 1

[a]*Each incubation contains 2.7 x 10^7 neutrophils and 1 mg of opsonized and washed zymosan in 1 ml Hanks' balanced salt solution, with additions as indicated. The mixtures were incubated at 37° for 40 min, then centrifuged. Supernatants were assayed by the Boyden technique. Superoxide dismutase (SOD) was present at 120 µg/ml, as indicated.*

when neutrophils phagocytosed zymosan in Hanks' balanced salt
solution containing 1% bovine serum albumin, the supernate
contained a chemoattractant when assayed by the Boyden
technique. Superoxide dismutase prevented the appearance of
this activity, and if serum albumin were replaced with ovalbumin no chemotactic activity appeared. The release of the
chemotactically inactive precursor lipid from neutrophils was
demonstrated by incubating cells and opsonized zymosan in 10%
serum albumin in the presence of superoxide dismutase.
Following the 40 min incubation, the supernatant was chromatographed on Sephadex G-100 in order to separate superoxide
dismutase from the albumin-lipid complex. The albumin content
was diluted to 1% and a superoxide-generating system was
added. The result was a significant increase in chemotactic
activity.

Finally, the chloroform extractability of the neutrophil-
released lipid was demonstrated, both before and after its
activation by superoxide. When supernatant from phagocytosing
neutrophils was extracted and reconstituted as described
above, the solution was chemotactic (28 cells/field). When
superoxide dismutase was present during phagocytosis, the
extracted and reconstituted supernatant was not chemotactic
(8.4 cells/field). That the precursor lipid was present was
confirmed by exposing this reconstituted solution to the
superoxide-generating system, resulting in the development of
chemotactic activity (36 cells/field).

DISCUSSION

Superoxide reacts with an albumin-bound lipid present in
plasma or released from phagocytosing neutrophils to produce
a powerful chemoattractant for neutrophils. Serum albumin is
required for expression of chemotactic activity and cannot be
replaced by ovalbumin. Superoxide dismutase, but not
catalase, prevents the activation of the factor, indicating
that the lipid reacts with O_2^- per se and not with H_2O_2 nor
with radicals derived secondarily via a Haber-Weiss mechanism.
The identity of the lipid is not yet known, but its extractability into chloroform at neutral pH suggests that it may
not be a free fatty acid.

Based on experimental observations of the nature of the
anti-inflammatory activity of superoxide dismutase in vivo
(3,4), we believe the superoxide-dependent chemotactic factor
plays an important role in the development of inflammation.
Neutrophils may initially be drawn to the site of an immune
stimulus by complement factors or by chemotactic factors of

bacterial origin. Once the first neutrophils arrive at the
site, they are stimulated to produce superoxide. The radical
reacts with the precursor lipid present in plasma (and
presumably in interstitial fluid) or being released from the
neutrophils themselves, producing a powerful chemotactic
gradient to summon greater numbers of neutrophils to the site.
Thus, the initial signals are amplified and sustained for as
long as immune stimulus remains at the site. When the
stimulus has been removed, the last wave of cells to arrive
produces no superoxide and thus the production of the chemotactic factor ceases as well. It is important to point out
that this factor, unlike others, appears to be purely chemotactic without simulating superoxide production by the
neutrophils. If this were not the case, the factor would
catalyze its own production and the inflammatory episode could
never resolve.

REFERENCES

1. McCord, J.M., and Fridovich, I., *J. Biol. Chem. 244,* 6049
    (1969).
2. McCord, J.M., *Science 185,* 529 (1974).
3. Petrone, W.F., English, D.K., Wong, K., and McCord, J.M.,
    Proc. Natl. Acad. Sci. USA 77, 1159 (1980).
4. McCord, J.M., Stokes, S.H., and Wong, K., *in* "Advances in
    Inflammation Research" (G. Weissmann, B. Samuelsson, and
    R. Paoletti, eds.), Vol. 1, p. 273. Raven Press, New
    York (1979).
5. Babior, B., Kipnes, R., and Curnutte, *J. Clin. Invest. 52,*
    741 (1973).
6. Johnston, R.B., Jr., Keele, B.B., Jr., Misra, H.P.,
    Lehmeyer, J.E., Webb, L.S., Baehner, R.L., and Rajago-
    palan, K.V., *J. Clin. Invest. 55,* 1357 (1975).
7. Boyden, S., *J. Exp. Med. 115,* 453 (1962).
8. Snyderman, R. and Mergenhagen, S.E., *in* "Biological
    Activities of Complement" (D.G. Ingram, ed.), p. 116.
    Karger, Basel (1972).

THE ROLE OF OXYGEN AND OXYGEN RADICALS IN THE BIOSYNTHESIS
AND REGULATION OF MITOCHONDRIAL AND CYTOPLASMIC
SUPEROXIDE DISMUTASE IN EUKARYOTES

Anne P. Autor

Department of Pharmacology
University of Iowa
Iowa City, Iowa

BIOLOGY OF OXYGEN

The interactions of oxygen with biological systems and the
consequences, thereof, are many and varied. In air living
plants and animals, reactions which use oxygen as a substrate
are, in general, very tightly controlled. Such reactions
involve the reduction of oxygen for metabolic purposes and,
because of the electronic characteristics of molecular oxygen,
are constituted to ensure the complete reduction of oxygen
to water. Many examples are now known, however, where biological systems are exposed to the reactive products of the
partial reduction of oxygen. This can occur through the
intracellular production of partially reduced oxygen moieties,
including superoxide radical (O_2^-), hydrogen peroxide (H_2O_2)
and other hydroperoxides (1-7), through the extracellular
release of partially reduced oxygen for specific purposes such
as phagocytosis (8-12), and via exposure to autoxidizing
xenobiotics, drugs and toxic agents (13-18). In all cases,
the consequences to biological systems exposed to uncontrolled
oxygen radical flux are functional damage, structural damage,
and ultimately cell death (15,16,19-21).
Aerobic cells are equipped with a variety of means of
chemically quenching or catalytically removing oxygen and
partially reduced oxygen. Some cell components simply undergo
non-catalytic chemical reactions with oxygen-derived free
radicals. These include ascorbic acid and α-tocopherol

LIPID PEROXIDES IN BIOLOGY AND MEDICINE

131

(22,23) and under certain circumstances polyunsaturated fatty
acids (24,25). Several enzymes are present in subcellular
compartments such as mitochondria, peroxisomes and cytosol
whose sole and vital function is the catalytic conversion of
oxygen radicals or partially reduced oxygen to oxygen-
containing but biologically inocuous products. The most
important of these enzymes appear to be the superoxide
dismutases (26), catalase (27) and glutathione peroxidase
(28). Compelling evidence has accumulated demonstrating the
protection against oxygen or oxygen radical toxicity afforded
biological systems by these enzymes. Experimental evidence
includes not only studies which showed that the endogenous
enzymes are protective (15,29-32) but also that exogenously
supplied enzymes provide unequivocal protection both in
whole animals (33-35) and in isolated cell systems (15,16,21,
36) exposed to oxygen-derived radicals.

PULMONARY MACROPHAGES

Model System

     Because of a continued interest in pulmonary oxygen toxic-
ity in this laboratory, the biochemical characteristics, the
developmental biochemical changes and the toxic response to
oxygen and oxygen radicals of several of the many pulmonary
cells have been studied. The most thoroughly studied has been
the phagocytic cell of the lung, the pulmonary macrophage.
This cell type proved to have a number of advantages for our
studies, including ease of recovery, an easily measurable cell
function, a complex plasma membrane structure and the capacity
to induce both catalase and mitochondrial Mn superoxide dismutase in hyperoxia in an age defined manner. A series of
studies from this laboratory have shown two unique and experimentally exploitable characteristics of rat pulmonary macrophages: 1) the basal level of both catalase and Mn superoxide
dismutase is four- to five-fold higher in the cells of rats
immediately after birth compared with cells from adults and 2)
both enzymes are inducible by hyperoxia whether the cells are
incubated in vitro or the animals are exposed in vivo (37,38).
The first characteristic appears to explain, at least partially, why neonatal cells are much more resistant to oxygen
radicals than adult cells (15,39,40). The second characteristic has been exploited to investigate oxygen control of the
biosynthesis of enzymes which metabolize oxygen and oxygen
metabolites (37).

Regulation by Oxygen

   The oxygen induction of catalase and Mn superoxide dismutase in pulmonary macrophages occurs when isolated, inducible cells are incubated in an atmosphere of 95%+ oxygen.
Inducibility is a direct function of the age of the animal
from which the cells are taken. The enzymes are not inducible
immediately after birth but are gradually inducible after 2 to
3 days, reaching a peak of induction capacity at 10 days of
age. Thereafter inducibility diminishes until it disappears
at approximately two weeks of age. Under the conditions of
in vitro oxygen exposure (i.e. five hours incubation), adult
cells do not induce either enzyme.
   The age pattern of inducibility is superimposed upon the
age-dependent changes in the basal activity of both catalase
and Mn superoxide dismutase described above. Because the
coordinated effect of these enzyme alterations acts to maintain the two enzymes at a high level immediately after birth,
we interpreted this to suggest that an endogenous inducer may
be present which causes the high basal levels of enzyme in the
cells immediately after birth. Mitochondria produce a low
level of H_2O_2 which originates from O_2^- production from one (or
more) components of the electron transport chain (3,4).
Therefore, because it is the mitochondrial form of superoxide
dismutase which is induced, H_2O_2 and O_2^- production by mitochondria and washed submitochondrial particles, respectively,
was studied in pulmonary macrophages. An interesting
relationship was seen in which mitochondria from cells of neo-
nates produced high levels of H_2O_2 (and O_2^-) concomitant with
the high intrinsic level of mitochondrial superoxide dismutase. Both of these characteristics diminished in concert in
cells from increasingly older animals (38).
   The production of H_2O_2 and O_2^- by mitochondria under conditions of increased oxygen tension was investigated to determine whether the system was responsive to hyperoxia. Both
H_2O_2 and O_2^- did indeed increase under conditions where
measurement was conducted at increased oxygen tension (Table
I). It is important to note that the responsiveness of H_2O_2
and O_2^- generating process to increased oxygen tension occurs
only in the inducible cells of the neonatal animals. At
twenty-five days there is no response to oxygen tensions
higher than 21%. Indeed, in cells from adult animals mitochondrial generation of H_2O_2 (and O_2^-) is actually inhibited
by oxygen concentrations above 21%. We have interpreted
these results to mean that an endogenous inducer of mitochondrial superoxide dismutase (and catalase) may exist specifically in neonatal animals which is sensitive to increased
oxygen tension and which may be related to the increased
generation of oxygen radicals from the mitochondrial electron

TABLE I. Production of H_2O_2 and O_2^- by Mitochondria
or Submitochondrial Particles from Rat
Pulmonary Macrophages

Animal Age	Oxygen Concentration[a] (percent)	H_2O_2[b] nmole/min/mg protein	O_2^-[c] nmole/min/mg protein
4 days	21	1.4	48
	95+	3.9	117
10 days	21	1.0	45
	95+	2.6	121
25 days	21	0.8	35
	95+	0	35
Adult	21	0.5	28
	95+	0	10

[a]*Measurement of H_2O_2 and O_2^- production was conducted using an Aminco DW2A spectrophotometer with the cuvette chamber covered by a gassing tent containing oxygen to the indicated percentage.*

[b]*Measured by using cytochrome c peroxidase (7,41).*

[c]*Measured by assessing the rate of superoxide dismutase inhibitable reduction of nitroblue tetrazolium to blue formazan.*

transport chain. Supporting this line of reasoning is preliminary evidence that the presence of butylated hydroxytoluene
during incubation of inducible cells in hyperoxia prevents
oxygen induction of catalase and mitochondrial superoxide
dismutase (42).

SACCHAROMYCES CEREVISIAE

Superoxide Dismutase and Oxygen Toxicity

Most oxygen-metabolizing enzymes of Saccharomyces
cerevisiae appear in the active form only when the yeast is
grown aerobically. Detailed studies which have investigated

oxygen control of heme proteins, including catalase (43) and
mitochondrial cytochromes (44,45), point to a dependence on
the presence of heme for assembly of the active protein. In
the absence of oxygen, the synthesis of heme a is blocked
(46). Under anaerobic conditions the accumulation of apo-
protein or unassembled sub-units has been demonstrated.

     Growth of S. cerevisiae aerobically was reported to yield
cells which contained 6.5 times more superoxide dismutase and
2.3 times more catalase when compared with cells grown
anaerobically (29). Equal increases of both the mitochondrial
Mn superoxide dismutase and the cytosolic Cu/Zn superoxide
dismutase were recorded. The oxygen-induced increase of
enzyme activity was associated with resistance to the toxicity
of hyperbaric oxygen. This recalls the same association
between elevated Mn superoxide dismutase and catalase activities and resistance to the toxic effects of oxygen radicals in
pulmonary macrophages described above. Both the mitochondrial
and the cytosolic superoxide dismutases of S. cerevisiae have
been purified and characterized (47,48) and neither form of
superoxide dismutase contains heme.

     Although the controlling effect of oxygen on the synthesis
of active superoxide dismutase has been known for some time,
little is understood about the nature of that control. S.
cerevisiae presents a good experimental system for the study
of various aspects of the biosynthesis of superoxide dismutase. Studies were undertaken to investigate two aspects of
the biosynthesis of superoxide dismutase 1) post-translational
modification and 2) control exerted by oxygen. Attention was
directed primarily to the mitochondrial superoxide dismutase
in order to address the question of the nature of the initial
translation product, as well as the means by which the polypeptides of superoxide dismutase gain entrance into the mitochondrial matrix. Earlier work had demonstrated that Mn
superoxide dismutase is in the mitochondrial matrix and
synthesized from the nuclear genome (49,50). Yeast Mn superoxide dismutase has a molecular weight of 96,000 and is
composed of four identical sub-units, each with one atom of
manganese.

Preparation of Antibodies

     The experiments to be described depended for the most
part on the use of antibodies specific to either Mn superoxide
dismutase or Cu/Zn superoxide dismutase. Purified Cu/Zn
superoxide dismutase was prepared according to a published
method (48) and Mn superoxide dismutase according to a method
(47) slightly modified in this laboratory. Rabbit antisera to

both enzymes were prepared (51,52). The specificity of the
antisera to both enzymes was tested with the immunoelectrophoretic transfer technique using [^{125}I]-protein A which
reacts specifically with the immune complex immobilized on
nitrocellulose sheets (53). Each antisera was specific for
the appropriate antigen (Figs. 1-3). Although cross reactivity among enzymes of each type from varied species occurred
(Fig. 3), no cross reactivity occurred between the Cu/Zn and
the Mn superoxide dismutases. Interestingly, the Rf of the
Cu/Zn superoxide dismutase sub-units was slightly different
for each of the species tested (Fig. 3).

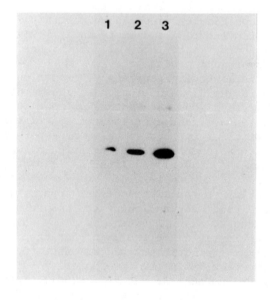

FIGURE 1. Autoradiograph of purified yeast Mn superoxide
dismutase. Track 1, 10 µg protein; Track 2, 50 µg protein;
Track 3, 100 µg protein. Samples were dissociated with
sodium dodecyl sulfate in the presence of mercaptoethanol
and electrophoresed on a 12.5% polyacrylamide gel (59). The
gel was subjected to immunoelectrotransfer to nitrocellulose
(53). The nitrocellulose sheet was treated with rabbit
antisera to yeast Mn superoxide dismutase - [^{125}I]-protein A,
dried and autoradiographed (53,59). By permission (Ref. 51).

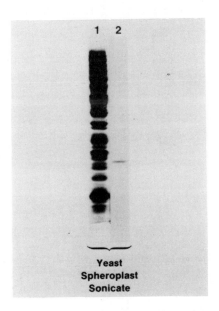

FIGURE 2. Electrophoresis and autoradiography of yeast
spheroplast sonicate treated with antisera to yeast Mn
superoxide dismutase. Track 1, 100 µg of yeast spheroplast
protein was electrophoresed on a 12.5% polyacrylamide gel
and stained with Coomassie blue. Track 2, 100 µg of yeast
spheroplast protein was electrophoresed, electrotransferred
and autoradiographed as described in Fig. 1. By permission
(Ref. 51).

Precursor of Mn Superoxide Dismutase

Extensive studies of the events crucial to the biogenesis
of mitochondria have identified several mitochondrial enzymes,
most of which are incorporated into the mitochondrial
membrane, which are synthesized in the cytoplasm and thence
transported to the mitochondria (54). The initial polypeptide
with few exceptions is synthesized as a larger molecular
weight precursor to the final mature polypeptide. The transport of the precursor polypeptide to the mitochondria and its
conversion by proteolytic cleavage to the mature form has been
termed vectorial processing (54). It has been established

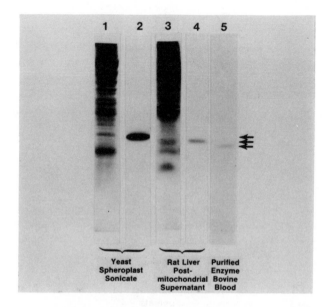

FIGURE 3.   Immunoreactivity of Cu/Zn superoxide dismutase.
Electrophoresis employed 15% polyacrylamide gels for all
samples.   Track 1, Coomassie blue stained gel with 100 µg of
yeast spheroplast sonicate protein, Track 3, Coomassie blue
stained gel with 100 µg of rat liver post-mitochondrial
supernatant protein.   Track 2, autoradiograph of Track 1;
Track 4, autoradiograph of Track 3; Track 5, autoradiograph
of 100 ng of purified bovine blood Cu/Zn superoxide dismutase.

that passage into the mitochondrial membranes is an energy-
dependent process (64).

   Like many other enzymes of the mitochondria which are
synthesized in the cytoplasm, Mn superoxide dismutase is
synthesized initially as a higher molecular weight precursor.
Using the uncoupler carbonyl cyanide m-chlorophenylhydrazone
(CCCP) to block the transport of the precursor (55), a polypeptide immunoreactive with antisera to Mn superoxide dismutase but of larger molecular weight than seen in normally
grown cells accumulated in yeast spheroplasts labelled with
[35S] sulfate (Fig. 4).   A similar higher molecular weight
polypeptide also accumulated in a cell-free protein synthe-
sizing system of reticulocyte lysate (56) programmed with
yeast RNA and containing [35S] methionine.

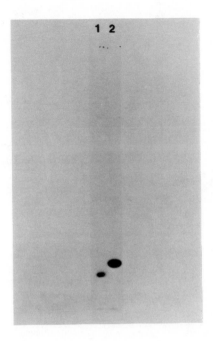

FIGURE 4. Mature and precursor Mn superoxide dismutase.
Track 1, sonicate from yeast spheroplasts labelled with [^{35}S]
sulfate and immunoprecipitated with antisera to Mn superoxide
dismutase plus fixed Staphylococcus aureus. Track 2 sonicate
from yeast spheroplasts labelled with [^{35}S] sulfate in the
presence of CCCP and immunoprecipitated as described above
(55). Samples were heated to 95° in 3% SDS and mercaptoethanol and electrophoresed on a 10% polyacrylamide gel.
After electrophoresis, the immunoreactive enzyme was located
on the gel by fluorography (60). By permission (Ref. 51).

Analysis by peptide mapping was employed to compare the
two polypeptides of different molecular weights but similar
immunoreactivity with antisera to yeast Mn superoxide dismutase. Peptides were prelabelled with [^{35}S] sulfate and subjected to limited proteolysis according to an established
method (57). The peptide patterns obtained were virtually
identical thus supporting the identification of the large
molecular weight polypeptide as a precursor to the mature
form. This interpretation was further supported by
pulse chase experiments in which spheroplasts labelled with
[35S] methionine and chased with cold methionine according to
established procedures (58) (Fig. 5), contained the higher

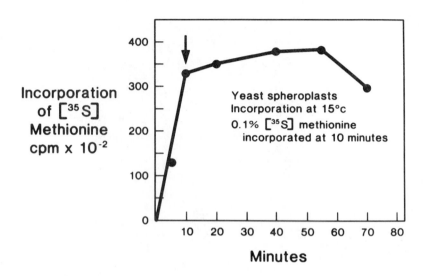

FIGURE 5. Yeast spheroplasts were prepared according to published methods (61) and incubated at 15° with [35S] methionine. Samples were removed for radioactive counting at the indicated times. Cold methionine was added after 10 minutes. Samples for analysis by gel electrophoresis, electrotransfer and autoradiography were taken at 12, 20, 30 and 45 minutes.

molecular weight immunoreactive polypeptide at the time of addition of cold methionine but lost the radiolabelled precursor upon incubation with cold methionine (51).

With the accumulated evidence that Mn superoxide dismutase is initially synthesized as a higher molecular weight precursor polypeptide before final processing to the mature form occurring presumably in the mitochondria, the effect of oxygen on the accumulation and/or processing of the precursor was next studied. Yeast cells were grown under strict anaerobic conditions and harvested according to established procedures (44). After polyacrylamide gel electrophoresis, electrophoretic transfer and autoradiography of sonicated cell protein were conducted. A comparison of the amount of mature Mn superoxide dismutase from aerobically and anaerobically

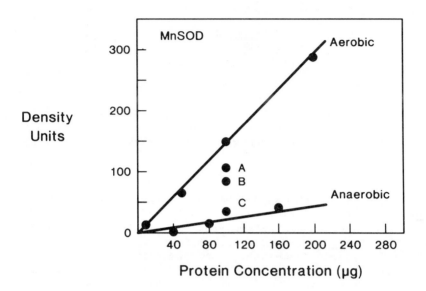

FIGURE 6. Mature Mn superoxide dismutase in sonicated
S. cerevisiae grown aerobically and anaerobically and in a
heme deficient mutant, GLI. Mature Mn superoxide dismutase
from the specified strains of S. cerevisiae was detected by
electrophoretic transfer (53) and autoradiography of cell
sonicates as a function of protein concentration. Density
units (ordinate) are an absorbance measurement of immunoreactive bands at the established position of mature Mn
superoxide dismutase on the autoradiograph film and were
obtained with an RFT II scanning densitometer. Mature Mn
superoxide dismutase from 100 µg protein samples of GLI
sonicate are designated A (mutant grown in δ-aminolevulinic
acid [δ-ala]) (mutant grown in δ-ala, ergosterol and Tween-
80) and C (mutant grown in ergosterol and Tween-80).

grown yeast sonicates was made by a densitometric scan of
the autoradiograph film. A measurement of the density of
positive bands on the film as a function of protein concentration revealed a linear relationship (Fig. 6) thus
validating the use of the density scan to provide comparative
data. At an equivalent protein concentration of the yeast
cell sonicate (i.e. 100 mg), anaerobically grown cells

contained approximately 15% of the mature polypeptide of Mn superoxide dismutase found in the aerobically grown cells. No measurable superoxide dismutase activity was detected in the anaerobically grown cells which, furthermore, contained less than 5% of the normal cytochrome oxidase activity of aerobically grown cells (51).

GL1 heme-deficient S. cerevisiae (62) whether supplemented with δ-aminolevulinic acid alone or in combination with Tween-80 (oleic acid) and ergosterol required for anaerobic growth in the absence of heme contained approximately one-fourth less mature Mn superoxide dismutase polypeptide chains. Cells grown only with ergosterol and Tween-80 contained approximately 35-40% of mature superoxide dismutase of the fully supplemented cells (Fig. 6). No evidence of immunoreactive material at an equivalent position to the precursor peptide was observed.

Because no superoxide dismutase activity, either mitochondrial or cytosolic, could be detected in the cell sonicates of anaerobically grown yeast, a similar immunoelectrophoretic transfer analysis was employed for the analysis of Cu/Zn superoxide dismutase. Anaerobically grown cells contained approximately 7% of the content of Cu/Zn superoxide dismutase found in aerobically grown cells (Table II).

Fully supplemented hemeless mutant yeast contained the same or less immunoreactive Cu/Zn superoxide dismutase when compared with an equivalent cell protein content of wild type yeast (Table III). Mutant yeast grown without heme contained 30-40% of the Cu/Zn superoxide dismutase of fully supplemented cells (Table III).

Although the results reported here represent only a first step in understanding oxygen control of the biosynthesis of superoxide dismutase a few tentative conclusions can be made. Firstly, it is clear that mitochondrial, Mn superoxide dismutase is synthesized in the cytosol as a larger molecular weight precursor polypeptide. Although not all mitochondrial proteins are synthesized in this manner, two other enzymes of the mitochondrial matrix are also known to be first synthesized as larger molecular weight precursors; carbamyl phosphate synthetase from rat liver (63) and citrate synthetase of Neurospora crassa (65). Work is in progress to determine whether Mn superoxide dismutase is processed to the mature form in the same way as other precursor mitochondrial proteins (54). One of the known steps in this process, energy-dependent transport of the precursor protein across the mitochondrial membranes, appears to be necessary for the processing of Mn superoxide dismutase, as judged from the results obtained with CCCP (Fig. 4).

TABLE II. Immunoreactive Cu/Zn Superoxide Dismutase in Aerobically and Anaerobically Grown S. cerevisiae[a]

Conditions of Growth	Protein Concentration μg[b]	Density Units[c]
Aerobic	200	370
	100	220
	50	100
Anaerobic	160	30
	80	10
	40	0

[a]After electrophoresis of sonicated yeast samples on polyacrylamide slab gels, gels were subjected to electrotransfer, treated with antisera to Cu/Zn superoxide dismutase plus ^{125}I-protein A and autoradiographed, as described in the legend to Fig. 1.

[b]Total protein of the yeast sonicate applied to each gel track.

[c]Arbitrary units obtained by scanning the autoradiograph film with an RFT II scanning densitometer.

Although no active superoxide dismutase is measurable in anaerobically grown S. cerevisiae, the cells appear to accumulate small amounts of the polypeptide of mature Mn superoxide dismutase. Cu/Zn superoxide dismutase polypeptides are also accumulated to a small extent in anaerobically grown cells. Like other cytosolic enzymes studied, the sub-units are synthesized directly with no formation of a larger molecular weight precursor.

Although oxygen regulation of protein biosynthesis is expressed through the oxygen requirement for heme biosynthesis in the case of proteins containing heme prosthetic groups, regulation by heme is a complex process (43-45). The analysis of the two types of eukaryotic superoxide dismutase in hemeless mutants was an attempt to separate heme/oxygen control of biosynthesis from the requirement for heme as a prosthetic group. The results thus far obtained do not give a clear answer. Immunoreactive superoxide dismutase of both

TABLE III. Immunoreactive Cu/Zn Superoxide Dismutase
in Wild-Type and a Heme-Deficient Mutant of
S. cerevisiae[a]

	Protein Concentration μg	Density Units
Wild-type yeast	100	395
Hemeless mutant yeast		
grown in δ-ala[b]	100	150
grown in δ-ala, ergosterol and Tween-80	100	195
grown in ergosterol and Tween-80	100	60

[a]*See Table II and Fig. 3 for experimental conditions.*

[b]*Strain GLI lacking δ-ala synthetase.*

type is present in supplemented and unsupplemented hemeless
mutants. The latter cells contain about 50% of the content
of supplemented cells. It is not clear as yet whether this
is a reflection of direct heme control or an indication of
the general state of growth of the unsupplemented cells out
recently published work suggests that oxygen independent of
heme is required for some aspect of post-translational
processing (51).

ACKNOWLEDGMENTS

The author acknowledges with gratitude the collaboration
of Professor Gottfried Schatz (Biozentrum, University of
Basel, Switzerland) in whose laboratory the work with
Saccharomyces cerevisiae was conducted. The author thanks
Drs. E.G. Gollub and D.B. Sprinson (Columbia University) who
provided the GLI mutant S. cerevisiae for the analysis
reported in this paper.

REFERENCES

1. Fridovich, I., Science 201, 875 (1978).
2. Hemler, M.E. and Lands, W.E.M., J. Biol. Chem. 255, 6253 (1980).
3. Loschen, G., Azzi, A., Richter, C., and Flohé, L., FEBS Lett. 42, 68 (1974).
4. Cadenas, E., and Boveris, A., Biochem. J. 188, 31 (1980).
5. Autor, A.P., and Stevens, J.B., in "Chemical and Biochemical Aspects of Superoxide and Superoxide Dismutase" (J.V. Bannister and H.A.O. Hill, eds.), p. 104. Elsevier, Amsterdam (1980). Vol. 11A.
6. Cohen, G., and Cederbaum, A.I., in "Microsomes, Drug Oxidations, and Chemical Carcinogenesis" (M.J. Coon, A.H. Conney, R.W. Estabrook, H.V. Gelboin, J.R. Gillette and P.J. O'Brien, eds.), p. 307. Academic Press, New York (1980).
7. Boveris, A., Oshino, N., and Chance, B., Biochem. J. 128, 617 (1972).
8. Babior, B.M., New Engl. J. Med. 298, 659 (1978).
9. Tauber, A.I., and Goetzl, E.J., Biochem. J. 18, 5576 (1979).
10. Autor, A.P., and Hoffman, M., Bull. europ. Physiopath. resp. 17 (Suppl.), 153 (1981).
11. Hirai, K-I., Ueno, S., and Ogawa, K., Acta Histochem. Cytochem. 13, 113 (1980).
12. Petrone, W.F., English, D.K., Wong, K., and McCord, J.M., Proc. Natl. Acad. Sci. (USA) 77, 1159 (1980).
13. Bachur, N.R., Gordon, S.L., Gee, M.V., and Kon, H., Proc. Natl. Acad. Sci. (USA) 76, 954 (1979).
14. Myers, C.E., McGuire, W.P., Liss, R.H., Ifrim, I., Grotzinger, K., and Young, R.C., Science 197, 165 (1977).
15. Autor, A.P., McLennan, G., and Fox, A.W., in "Molecular Basis of Environmental Toxicity" (R.J. Bhatnagar, ed.), p. 51. Ann Arbor Press (1979).
16. McLennan, G., Oberley, L.W., and Autor, A.P., Radiat. Res. 84, 122 (1980).
17. Farrington, J.A., Ebert, M., Land, E.J., and Fletcher, K., Biochim. Biophys. Acta 314, 372 (1973).
18. Fischer, L.J., and Hamburger, S.A., Life Sciences 26, 1405 (1980).
19. Marklund, S.L., and Westman, G., in "Chemical and Biochemical Aspects of Superoxide and Superoxide Dismutase" (J.F. Bannister and H.A.O. Hill, eds.), p. 318. Elsevier, Amsterdam (1980). Vol. 11B.
20. Lavelle, F., Michelson, A.M., and Dimitrijevic, L.D., Biochem. Biophys. Res. Commun. 55, 350 (1973).

21. Fischer, L.J., and Hamburger, S.A., Endocrinology 108, 2331 (1981).

22. Perkins, R.C., Beth, A.H., Wilkerson, L.S., Serafin, W., Dalton, L.R., Park, C.R., and Park, J.H., Proc. Natl. Acad. Sci. (USA) 77, 790 (1980).

23. Tappel, A.L., and Zalkin, H., Nature 35, 4705 (1960).

24. Kehrer, J.P., and Autor, A.P., Arch. Biochem. Biophys. 181, 73 (1977).

25. Kehrer, J.P., and Autor, A.P., Tox. Appl. Pharmacol. 44, 423 (1978).

26. Yost, F.J., and Fridovich, I., Arch. Biochem. Biophys. 176, 514 (1976).

27. Halliwell, B., New Phytol. 73, 1075 (1974).

28. Flohé, L., Gunzler, W.A., and Ladenstein, R., in "Glutathione: Metabolism and Function" (I.M. Arias and W.B. Jakoby, eds.), p. 115. Raven Press, New York (1976).

29. Gregory, E.M., Goscin, S.A., and Fridovich, I., J. Bacteriol. 117, 456 (1974).

30. Gregory, E.M., Yost, F.J., and Fridovich, I., J. Bacteriol. 115, 987 (1973).

31. Yam, J., Frank, L., and Roberts, R.J., Proc. Soc. Exptl. Biol. Med. 157, 293 (1978).

32. Stevens, J.B., and Autor, A.P., Fed. Proc. 39, 3138 (1980).

33. Edsmyr, F., Huber, W., and Menander, K.B., Curr. Therap. Res, 19, 198 (1976).

34. Petkau, A., Kelly, K., Chelack, W.S., Pleshack, S.D., Barefoot, C., and Meeker, B.E., Biophys. Biochem. Res. Commun. 67, 1167 (1975).

35. McLennan, G., and Autor, A.P., in "Pathology of Oxygen" (A.P. Autor, ed.), Academic Press (In Press).

36. Grankvist, K., Marklund, S.L., Sehlin, J., and Täljedal, I.-B., Biochem. J. 182, 17 (1979).

37. Autor, A.P., and Stevens, J.B., in "Chemical and Bio-chemical Aspects of Superoxide and Superoxide Dismutase" (J.V. Bannister and H.A.O. Hill, eds.), p. 104. Elsevier, Amsterdam (1980). Vol. 11A.

38. Stevens, J.B., and Autor, A.P., Fed. Proc. 39, 3138 (1980).

39. Megivern, K.A., Wallace, K.B., and Autor, A.P., Fed. Proc. 39, 690 (1980).

40. Autor, A.P., Fox, A.W., and Stevens, J.B., in "Biochemi-cal and Clinical Aspects of Oxygen" (W.S. Caughey, ed.), p. 767. Academic Press, New York (1979).

41. Yonetani, T., J. Biol. Chem. 240, 4509 (1965).

42. Autor, A.P., and Stevens, J.B., Europ. J. Cell Biol. 22, 283 (1980).
43. Zimniak, P., Hartter, E., Woloszczuk, W., and Ruis, H., Eur. J. Biochem. 71, 393 (1976).
44. Djavadi-Ohaniance, L., Rudin, Y., and Schatz, G., J. Biol. Chem. 253, 4402 (1978).
45. Woodrow, G., and Schatz, G., J. Biol. Chem. 254, 6088 (1979).
46. Falk, J.E., Dresel, E.B., and Rimington, C., Nature 172, 292 (1953).
47. Ravindranath, S.D., and Fridovich, I., J. Biol. Chem. 250, 6107 (1975).
48. Goscin, S.A., and Fridovich, I., Biochim. Biophys. Acta 289, 276 (1972).
49. Weisiger, R.A., and Fridovich, I., J. Biol. Chem. 248, 3582 (1973).
50. Weisiger, R.A., and Fridovich, I., J. Biol. Chem. 248, 4793 (1973).
51. Autor, A.P., J. Biol. Chem. 257, 2713 (1982).
52. Poyton, R.D., and Schatz, G., J. Biol. Chem. 250, 762 (1975).
53. Towbin, H., Staehelin, T., and Gordon, J., Proc. Natl. Acad. Sci. (USA) 73, 1083 (1979).
54. Schatz, G., FEBS Lett. 103, 203 (1979).
55. Nelson, N., and Schatz, G., Proc. Natl. Acad. Sci. (USA) 76, 4365 (1979).
56. Pelham, H.R.B., and Jackson, R.M., Europ. J. Biochem. 67, 247 (1976).
57. Cleveland, D.W., Fischer, S.G., Kirschner, M.W., and Laemmli, U.K., J. Biol. Chem. 252, 1102 (1977).
58. Côté, C., Solioz, M., and Schatz, G., J. Biol. Chem. 254, 1437 (1979).
59. Cabral, F., and Schatz, G., Meth. Enzymol. 56, 602 (1979).
60. Bonner, W.M., and Lasky, R.A., Europ. J. Biochem. 46, 83 (1974).
61. Hutchison, H.T., and Hartwell, L.H., J. Bacteriol. 94, 1697 (1967).
62. Gollub, E.G., Liu, K., Dayan, J., Aldersberg, M., and Sprinson, D.B., J. Biol. Chem. 252, 2846 (1977).
63. Shore, G.C., Carignan, P., and Raymond, Y., J. Biol. Chem. 254, 3141 (1979).
64. Gasser, S.M., Ohashi, A., Daum, G., Böhni, P.C., Gibson, J., Reid, G.A., Yonetani, T., and Schatz, G., Proc. Natl. Acad. Sci. (USA), 79 267 (1982).
65. Harmey, M.A., and Neupert, W., FEBS Lett. 108, 385 (1979).

ROLE OF GSH PEROXIDASE IN LIPID PEROXIDE METABOLISM

Leopold Flohé

Center of Research
Grünenthal GmbH
Aachen, FRG

I. INTRODUCTION

GSH peroxidase was discovered by Mills in 1957 (1) as an enzyme protecting hemoglobin from oxidative destruction by H_2O_2. In 1962 the enzyme was rediscovered by Neubert et al. (2) as a "contraction factor" of mitochondria, i.e. as a compound preventing loss of contractibility of mitochondria under special conditions. A few years later Little and O'Brien (3) found that GSH peroxidase not only catalyzes the reduction of H_2O_2 but also of organic hydroperoxides including those derived from unsaturated lipids. Since then, the interest of physiologists, cell biologists, and clinicians has centered on the role of GSH peroxidase in lipid peroxide metabolism. It could soon be demonstrated that the strange contraction factor activity of the enzyme was not a property of its own but intimately related to the prevention of oxidative lipid destruction in mitochondrial membranes (4). The discovery of deficiencies of GSH peroxidase further supported the view that the enzyme protects biomembranes from oxidative attack (5). After it had been established that selenium is a stoichiometric constituent of GSH peroxidase (6), the scope of hypothetical claims as to the physiological role of the enzyme was broadened again due to the knowledge gained from the trace element research and correspondingly the number of related publications grew logarithmically.

As time elapsed, other enzymes acting on lipid hydroperoxides have been discovered, and some apparently established roles of GSH peroxidase have been questioned. It thus appears time to review the present state of knowledge in order to

clarify which questions may be considered to be satisfactorily
answered and which ones would benefit from further experiments.

II.   ENZYMOLOGICAL DATA AND PHYSIOLOGICAL FUNCTION

While only less than ten publications on GSH peroxidase
appeared in the sixties, our knowledge of the structural and
functional features of the enzyme by now seems to be almost
completed.   The data are summarized in tables I and II, and
only the functional parameters relevant to physiology shall be
discussed here.

From kinetic studies and structural investigations, we
know that the selenocysteine residues of GSH peroxidase shuttle between different redox states during catalysis, i.e. the
enzyme itself is present in either a reduced, oxidized or partially oxidized state depending on the steady state concentration of the substrates.   At physiological levels of GSH,
however, the enzyme is largely reduced and therefore the rate
equation for reasonable limits of hydroperoxide concentrations
(e.g. below 10^{-6} M) can be simplified as follows:

$$- \frac{d \ (ROOH)}{dt} = k_{+1} \cdot (E_o) \cdot (ROOH)$$

This equation allows at least a rough calculation of removal rates of hydroperoxides in tissues, if the molarity of
GSH peroxidase and the rate constant k_{+1} for a given hydroperoxide is known.   The simplified rate equation further reveals
a basic principle of the GSH peroxidase reaction: Unless the
levels or the regeneration of GSH drop considerably, the reaction rate in a specified tissue only depends on the nature of
ROOH, which affects k_{+1}, and on the concentration of the hydroperoxide.   The biological and medical implications of the
kinetic behaviour of GSH peroxidase have recently been discussed in detail (19).

Unfortunately, the seemingly complete kinetic analysis of
bovine GSH peroxidase is of limited value to clarify the role
of GSH peroxidase in lipid peroxide metabolism.   This is particularly true, if species other than cows are considered, because it cannot reasonably be assumed that the kinetic parameters of GSH peroxidase do not differ between species.   Further, the "lipid peroxides" frequently quoted as key mediators
of tissue damage are ill-defined chemical entities.   We know
that the highly reactive selenocysteine residues of GSH peroxidase are exposed at the surface of the protein and correspondingly react with a large variety of hydroperoxides, but
the pertinent rate constants for naturally occurring hydroperoxides (apart from H_2O_2) remain to be worked out.

In summary, before a reliable conclusion as to the role
of GSH peroxidase in e.g. human tissue can be theoretically

TABLE I. Structural Characteristics of GSH Peroxidase

Molecular weight: 76000 - 85000 (7,8,9)

Molecular weight of subunits: 17000 - 22000 (8-10)

Prosthetic groups: 4 selenium atoms per molecule (6,9-11)

Primary structure: Preliminarily derived from x-ray data (12). The proposed amino acid sequence of the subunit is:

```
        10          20          30          40          50
RKLFGNGAAEAQGGGTRNLKNSPKPLLLRLVAPASeCPLTSLWQHFNSLIGNG
        60          70          80          90          100
GPKGFAVHAFVVSLRMHLNPPYRRLENRVQRVGSGGGFQPNFQNYRQLKTQGR
        110         120         130         140         150
QAFPLFSNLRTAKGVVNQQNGIFTSPTQLTWGPIFQSKSGQNWYQFLVGPNAA
        160         170
PNYKFMRLQGNRIQNIAQRFK
```

Tertiary structure: Known from x-ray analysis at 2.8 Å resolution (12)

Quaternary structure: Assembly of 4 identical subunits (12,13)

TABLE II. Functional Characteristics of GSH Peroxidase

Reaction catalyzed: $2GSH + ROOH \longrightarrow GSSG + ROH + H_2O$ (1,3)

Specificity for donor (GSH): high (14)

Specificity for acceptor (ROOH): very low (for review see 15)

Initial velocity equation:

$$-\frac{d\,(ROOH)}{dt} = v = (E_o)\left(\frac{1}{k_{+1}\,(ROOH)} + \frac{1}{k_{+2}\,(GSH)} + \frac{1}{k_{+3}\,(GSH)}\right)^{-1}$$

(16)

k_{+1} value for H_2O_2: 2×10^8 M^{-1} sec^{-1} (16)

k_{+1} values for ROOH: between 5 and 0.7×10^7 M^{-1} sec^{-1} (17)

Assumed catalytic mechanism: redox shuttle of selenocysteine residues (18,19)

derived from enzymological data, some requirements would have
to be fulfilled. Specifically, the minimum prerequisites
would be to determine
  - the essential kinetic parameters of human GSH peroxi-
    dase;
  - the molarity of GSH peroxidase in the tissue and cell
    compartment under consideration;
  - the chemical nature and the concentration range of in-
    dividual lipid peroxides;
  - the k_{+1} values for individual lipid peroxides;
  - corresponding data on competing enzymes.
Since this brief catalogue of unresolved problems could
keep several outstanding laboratories busy for years, it appears not only justified, but attractive to consider alternative approaches to pinpoint the biological role of GSH peroxidase.

III. IN VITRO MODEL SYSTEMS REVEALING A ROLE OF GSH PEROXI-
      DASE IN LIPID PEROXIDE METABOLISM

    As already mentioned, GSH peroxidase reduces almost every
hydroperoxide to the corresponding alcohol. The known substrates comprise hydroperoxides of fatty acids, of fatty acid
esters and of steroids. However, the significance of this
catalytic potential under in vivo conditions is still debated.
It therefore appears justified to recall some antique though
not antiquated experiments in order to clarify the point(s) of
disagreement.
    In 1959 Lehninger and Schneider (20) described that iso-
lated mitochondria when exposed to GSH start swelling irrever-
sibly. This phenomenon can be suppressed by GSH peroxidase
("contraction factor II") and to some degree by catalase
("contraction factor I")(2). Hunter et al. (21) showed that
the irreversible high-amplitude swelling of mitochondria is
associated with oxidative destruction of phospholipids in the
mitochondrial membranes. When it had become evident that GSH
peroxidase could reduce hydroperoxides of fatty acids (3), we
reinvestigated the phenomena observed by Lehninger's group to
clarify whether the so-called contraction factor activity of
GSH peroxidase was linked to its peroxidatic function or an
independent activity. Our findings may be briefly summarized
as follows:
    1. The GSH peroxidase level of rat liver mitochondria
depends on age and sex. The tendency of mitochondria to swell
and to form malonaldehyde is inversely correlated with their
content of GSH peroxidase (22).
    2. When the activity of GSH peroxidase in mitochondria

is experimentally manipulated by change of the pH, again lipid peroxidation and high-amplitude swelling decrease with increasing peroxidase activity (22).

3. Similarly, isolated bovine GSH peroxidase added to isolated rat mitochondria in the presence of GSH inhibits both swelling (2) and lipid peroxidation (4).

4. Lipid peroxidation also occurs in purified inner membranes of rat liver mitochondria when exposed to GSH. Again GSH peroxidase is inhibitory (4).

5. In these experiments the lipid peroxidation revealed from formation of malonaldehyde and a substantial loss of intact phospholipids, particularly of phosphatidylethanolamine. Concomitantly the formation of new types of phospholipids, preliminarily identified as lysophosphatides, was observed. All effects were antagonized by GSH peroxidase. The putative products of the GSH peroxidase reaction, i.e. hydroxylipids, however, were not detected (23).

6. Catalase was inactive or much less active than GSH peroxidase in preventing oxidative lipid destruction in isolated mitochondria or purified mitochondrial membranes (4).

Almost identical results were obtained by McCay et al. (24) who investigated lipid peroxidation in microsomes exposed to ADP, trivalent iron, and a NADPH-generating system.

1. Lipid peroxidation is detectable by malonaldehyde formation and by a substantial loss of extractable polyunsaturated lipids, though the time course of these events is much faster than in swelling mitochondria.

2. GSH plus GSH peroxidase effectively inhibited lipid peroxidation.

3. Catalase was inactive in preventing lipid destruction.

4. The putative products of the GSH peroxidase reaction could not be detected despite the availability of sensitive analytical methods.

The striking analogy of the quoted experiments unfortunately did not provoke an identical interpretation concerning the precise catalytic role of GSH peroxidase.

We concluded (so far in complete agreement with the McCay's group) that GSH peroxidase prevents lipid peroxidation by scavenging hydrogen peroxide and hereby slowing down H_2O_2-dependent free radical attack of the lipids (4). This conclusion was clearly supported by the protective activity of GSH peroxidase, but also of catalase and superoxide dismutase in some of our experiments. Since, however, GSH peroxidase protects the unsaturated lipid much better than catalase, it was and is our conviction that scavenging of H_2O_2 can only be part of the story and therefore we invoked a direct reduction of lipid hydroperoxides by GSH peroxidase. McCay et al. (24) exclude this latter hypothesis, because they failed to recover

the assumed GSH peroxidase products. In consequence, an H_2O_2-
dependent free radical chain was considered the most likely
primary cause of lipid peroxidation. This is the only point
of disagreement.

Our reasoning is as follows: Both catalase and GSH peroxidase, if present in equimolar concentrations, metabolize
H_2O_2 at comparable rates. This can be readily deduced from
the rate equations and relevant rate constants of the enzymes
(see chapter II and Ref. 25). If catalase in contrast to GSH
peroxidase does not affect a model system, unless it is added
in large excess, such findings clearly rule out a crucial
function of H_2O_2, but point to the involvement of some key
intermediates specifically metabolized by GSH peroxidase.
From the accumulated knowledge of the substrate specificity of
GSH peroxidase some organic hydroperoxide has to be considered,
which, however, remains to be identified. I therefore conclude this chapter with the perhaps acceptable generaliza-
tions:

    1. GSH peroxidase effectively inhibits or delays lipid
peroxidation in a variety of model systems.

    2. The relative contribution of H_2O_2 and other hydroper-
oxides in triggering lipid peroxidation can be qualitatively
estimated by investigating the protective potency of catalase
and GSH peroxidase, respectively. Lack of efficacy of catalase in preventing lipid peroxidation in vitro rules out an
important mediator role of H_2O_2.

    3. Lipid peroxidation in isolated mitochondria exposed
to autoxidizing GSH is sustained in part by H_2O_2 and $\cdot O_2^-$, but
in addition by hydroperoxides only metabolized by GSH peroxidase.

    4. Unfortunately, neither these hypothesized hydroper-
oxides nor the putative products thereof have been identified
so far.

IV. DOES GSH PEROXIDASE ANTAGONIZE LIPID PEROXIDATION IN
     VIVO?

The identification of the in vivo function of GSH peroxidase has to consider interference of various competing or sup-
plementing enzymatic systems, species differences, tissue distribution and subcellular compartmentation of relevant enzymes
and co-substrates. Despite the complexity of the problem and
the uncertainties outlined in the previous chapters, at least
some definite statements as to the physiological role of GSH
peroxidase should be possible, if the multiple approaches of
enzymologists, cell physiologists, physicians and nutrition-
ists are compiled. For obvious reasons such an analytical

review can only reveal whether or not functional integrity of
the GSH peroxidase systems is required in a specified biological system.

A. Competing Enzymes

Enzymatic activities overlapping with GSH peroxidase have
been established for GSH-S-transferases B and AA (the so-called non-Se-GSH peroxidase), prostaglandin hydroperoxidase and
catalase. At present it is difficult to generally predict to
what extent this overlap in catalytic potential results in
real competition for the common substrates of the enzymes, but
it is possible in special cases.
    1. GSH peroxidase and catalase effectively compete for
one common substrate of physiological relevance only, i.e.
H_2O_2, which may be indirectly linked to lipid peroxidations.
In highly organized cells such as the hepatocytes, the two
enzymes, however, are localized in different compartments.
While catalase is restricted to the peroxisomes, GSH peroxidase resides in the cytosol and the mitochondrial matrix (26).
Thus, a functional competition between these enzymes will be
of subordinate importance, unless the subcellular organization
is degenerated as in erythrocytes. This prognosis was veri-
fied by perfusion studies with rat liver clearly indicating
that the fate of H_2O_2 depends on the subcellular site of its
formation (27-29).
    2. Theoretically GSH-S-transferase can compete with GSH
peroxidase for organic hydroperoxides, but not for H_2O_2 (30,
31). Rate constants of GSH-S-transferase for relevant hydroperoxides are not available. However, perfusion studies indicate that this enzyme can substitute for GSH peroxidase in
metabolizing artificial substrates such as t-butyl hydroperoxide in animals made deficient in GSH peroxidase by means of
a selenium-deprived diet (32). Also the observation that in
guinea pigs, in contrast to all other vertebrates investigated, GSH peroxidase is replaced by GSH-S-transferase suggests
that in this species at least lipid peroxidation may be balanced by the latter enzyme. Similarly, induction of GSH-S-transferase in response to selenium deficiency in rats indicates
that GSH-S-transferase can take over the job usually done by
GSH peroxidase. However, it should be stressed that the non-
selenium-dependent GSH peroxidase activity appears to be restricted to the cytosol (33) and thus will hardly be able to
substitute for the real GSH peroxidase fraction of mitochondria.
    3. Prostaglandin hydroperoxidase and GSH peroxidase
share the property to reduce hydroperoxy intermediates of the
arachidonic acid cascade (34,35). Unfortunately, the enzymology of prostaglandin hydroperoxidase is not yet very advanced:

The physiological hydrogen donor of the prostaglandin peroxidase reaction has not yet been identified. The stoichiometry
of the reaction varies with the (artificial) cosubstrates employed and the enzymatic activity undergoes rapid self-de-
struction (35). These observations cast doubt on the assumption that the prostaglandin hydroperoxidase activity, which is
associated with the cyclo-oxygenase complex, functions in an
analogous way in vivo. Any prognosis as to the physiological
relevance of this enzyme therefore appears to be premature.
However, the alternative possibility, i.e. that GSH peroxidase
takes part in the arachidonic acid cascade as the key enzyme
reducing the hydroperoxy-intermediates, seems not to be
favoured at present (34).

B.   Strategies to Define the Role of GSH Peroxidase in Vivo

      In principle, the following methods are required to solve
our problems: a) a measure of the enzyme in action, b) non-
invasive or at least non-destructive indicators of ongoing
lipid peroxidation and c) a method to selectively manipulate
the activity of GSH peroxidase. Such methods are available
and it depends on the genius of the experimenter   to combine
them in a way that they yield reliable conclusions.
      Most of our present insight into GSH-dependent hydroper-
oxide metabolism is derived from studies with perfused organs,
particularly liver, which can be observed under conditions
closely resembling the in vivo situation (36). Two direct
indicators of GSH peroxidase action in liver were introduced
by Sies et al. (37): (1) If GSH is oxidized, it is released
from the liver cells and can be detected in the effluent, more
specifically in the bile (28). (2) Simultaneously NADPH oxidation by glutathione reductase as a consequence of GSSG formation can be monitored by means of surface fluorescence.
      The only non-invasive method to measure lipid peroxida-
tion at present seems to be the detection of exhaled alkanes
(38). Measurement of surface luminescence of organs as indicator of ongoing lipid peroxidation could also prove useful
but has not yet been widely applied (39,40).
      The best way to manipulate GSH peroxidase activity still
is to restrict selenium intake (41), since selective inhibitors of the enzyme are not available. In case of human
studies we depend on the limited access to patients with
inborn errors (5) or dietary deficiencies (42).

C. Facts

      Despite the complexity of the hydroperoxide metabolism
skillful combinations of the methodology outlined above have
elucidated some aspects of in vivo function of GSH peroxidase

beyond any reasonable doubt.

In human red blood cells GSH peroxidase protects the cell membrane against oxidative attack. This statement is based on the most convincing experiment we can think of, i.e. an experiment of nature: Patients with a substantial deficiency in GSH peroxidase suffer from hemolytic episodes, if exposed to prooxidant drugs (5). Similarly, erythrocytes of rats made deficient in GSH peroxidase by restriction of selenium intake became susceptible to peroxide-induced hemolysis (41). This experiment, incidentally, bridged the gap between selenium biochemistry and glutathione research.

In rat liver extraperoxisomal hydrogen peroxide, particularly that of mitochondrial origin is clearly metabolized by GSH peroxidase (37). Correspondingly, metabolism of substrates of monoamine oxidase giving rise to H_2O_2 formation at the outer mitochondrial membrane is balanced by GSH peroxidase (28,29).

Microsomal metabolism of drugs (e.g. aminopyrine, ethylmorphine and hexobarbital) associated with hydroperoxide formation also results in responses typical for functioning GSH peroxidase in rat liver (GSSG release; NADPH decline). In this case, however, it is not clear at present to what extent GSH-S-transferase contributes to GSH-dependent hydroperoxide removal, since similar responses are observed in selenium-deficient rats. This finding, however, proves that the hydroperoxides formed during microsomal drug metabolism cannot exclusively represent H_2O_2, because GSH-S-transferase cannot substitute for GSH peroxidase in H_2O_2 metabolism (28,29).

General protection against lipid peroxidation by GSH peroxidase in the intact animal is evident from the data of Hafeman and Hoekstra (43): Combined deficiency of selenium and vitamin E resulted in a prefatal exponential rise in ethane evolution and finally in death. Both selenium and vitamin E significantly reduced ethane evolution and prevented mortality completely. The protective effect of selenium and vitamin E was also observed, when ethane evolution resulted from poisoning by carbon tetrachloride (44).

REFERENCES

1. Mills, G. C. J. Biol. Chem. 229, 189 (1957).
2. Neubert, D., Wojtczak, A. B. and Lehninger, A. L. Proc. Natl. Acad. Sci. USA 48, 1651 (1962).
3. Little, C. and O'Brien, P. J. Biochem. Biophys. Res. Commun. 31, 145 (1968).
4. Flohé, L. and Zimmermann, R. In "Glutathione" (Flohé, L., Benöhr, H. Ch., Sies, H., Waller, H. D. and Wendel, A.

eds.) p. 245, Georg Thieme, Stuttgart (1974).

5. Necheles, T. F. In "Glutathione" (Flohé, L., Benöhr, H. Ch., Sies, H., Waller, H. D. and Wendel, A. eds.) p. 173, Georg, Thieme, Stuttgart (1974).

6. Flohé, L., Günzler, W. A. and Schock, H. H. FEBS Lett. 32, 132 (1973).

7. Schneider, F. and Flohé, L. Hoppe Seylers Z. Physiol. Chem. 348, 540 (1967).

8. Flohé, L., Eisele, B. and Wendel, A. Hoppe Seylers Z. Physiol. Chem. 352, 151 (1971).

9. Nakamura, W., Hosoda, S. and Hayashi, K. Biochim. Biophys. Acta 358, 251 (1974).

10. Oh, S. H., Ganther, H. E. and Hoekstra, W. G. Biochemistry 13, 1825 (1974).

11. Awasthi, Y. C., Beutler, E. and Srivastava, S. K. J. Biol. Chem. 250, 5144 (1975).

12. Ladenstein, R., Epp, O., Bartels, K., Jones, A., Huber, R. and Wendel, A. J. Mol. Biol. 134, 199 (1979).

13. Flohé, L., Günzler, W. A. and Ladenstein, R. In Glutathione (Arias, I. M. and Jakoby, W. B. eds.) p. 115, Raven Press, New York (1976).

14. Flohé, L., Günzler, W. A., Jung, G., Schaich, E. and Schneider, F. Hoppe Seylers Z. Physiol. Chem. 352, 159 (1971).

15. Flohé, L., Günzler, W. A. and Loschen, G. In "Trace Metals in Health and Disease" (Kharasch, N. ed.) p. 263, Raven Press, New York (1979).

16. Flohé, L., Loschen, G., Günzler, W. A. and Eichele, E. Hoppe Seylers Z. Physiol. Chem. 353, 987 (1972).

17. Günzler, W. A., Vergin, H., Müller, I. and Flohé, L. Hoppe Seylers Z. Physiol. Chem. 353, 1001 (1972).

18. Wendel, A., Pilz, W., Ladenstein, R., Sawatzki, G. and Weser, U. Biochim. Biophys. Acta 377, 211 (1975).

19. Flohé, L. In "Oxygen Free Radicals and Tissue Damage". Ciba Foundation Symposium 65 (new series), p. 95, (1979).

20. Lehninger, A. L. and Schneider, M. J. Biophys. Biochem. Cytol. 5, 109 (1959).

21. Hunter, F. E., Jr., Scott, A., Weinstein, J. and Schneider, A. J. Biol. Chem. 239, 622 (1964).

22. Flohé, L. and Zimmermann, R. Biochim. Biophys. Acta 223, 210 (1970).

23. Zimmermann, R. Lipidperoxidation in Mitochondrien. Thesis, University of Tübingen (1973).

24. McCay P. B., Gibson, D. D., Fong, K.-L. and Hornbrook, K. R. Biochim. Biophys. Acta 431, 459 (1976).

25. Schonbaum, G. R. and Chance, B. In "The Enzymes" (Boyer, P. D. ed.) Vol. XIII, p. 363, Academic Press, New York (1976).

26. Flohé, L. and Schlegel, W. Hoppe Seylers Z. Physiol. Chem.

352, 1401 (1971).

27. Sies, H., Gerstenecker, C., Summer, K. H., Menzel, H. and Flohé, L. In "Glutathione" (Flohé, L., Benöhr, H. Ch., Sies, H., Waller, H. D. and Wendel, A. eds.) p. 261, Georg Thieme, Stuttgart (1974).

28. Sies, H., Wahlländer, A. and Waydhas, Ch. In "Functions of Glutathione in Liver and Kidney" (Sies, H. and Wendel, A. eds.) p. 120, Springer-Verlag, Berlin (1978).

29. Sies, H., Bartoli, G. M., Burk, R. F. and Waydhas, C. Eur. J. Biochem. 89, 113 (1978).

30. Burk, R. F. and Lawrence, R. A. In "Functions of Gluta-thione in Liver and Kidney" (Sies, H. and Wendel, A. eds.) p. 114, Springer-Verlag, Berlin (1978).

31. Prohaska, J. R. Biochim. Biophys. Acta 611, 87 (1980).

32. Burk, R. F., Nishiki, K., Lawrence, R. A. and Chance, B. J. Biol. Chem. 253, 43 (1978).

33. Lawrence, R. A. and Burk, R. F. J. Nutr. 108, 211 (1978).

34. Christ-Hazelhof, E. and Nugteren, D. H. In "Functions of Glutathione in Liver and Kidney" (Sies, H. and Wendel, A. eds.) p. 201, Springer-Verlag, Berlin (1978).

35. Ohki, S., Ogino, N., Yamamoto, S. and Hayaishi, O. J. Biol. Chem. 254, 829 (1979).

36. Sies, H. Trends Biochem. Sci. 5, 182 (1980).

37. Sies, H., Gerstenecker, C., Menzel, H. and Flohé, L. FEBS Lett. 27, 171 (1972).

38. Riely, C. A., Cohen, G. and Lieberman, M. Science 183, 208 (1974).

39. Chance, B., Boveris, A., Nakase, Y. and Sies, H. In "Functions of Glutathione in Liver and Kidney" (Sies, H. and Wendel, A. eds.) p. 95, Springer-Verlag, Berlin (1978).

40. Boveris, A., Cadenas, E., Reiter, R., Filipkowski, M., Nakase, Y. and Chance, B. Proc. Natl. Acad. Sci. USA 77, 347 (1980).

41. Rotruck, J. T., Hoekstra, W. G., Pope, A. L., Ganther, H., Swanson, A. and Hafeman, D. Fed. Proc. 31, 691 (1972).

42. Lombeck, I., Kasperek, K., Harbisch, H. D., Becker, K., Schumann, E., Schröter, W., Feinendegen, L. E. and Bremer, H. J. Eur. J. Pediatr. 128, 213 (1978).

43. Hafeman, D. G. and Hoekstra, W. G. J. Nutr. 107, 656 (1977).

44. Hafeman, D. G. and Hoekstra, W. G. J. Nutr. 107, 666 (1977).

MECHANISM OF PROTECTION AGAINST MEMBRANE PEROXIDATION[1]

James F. Mead

Laboratory of Nuclear Medicine and Radiation Biology
and Department of Biological Chemistry
University of California
Los Angeles, California

Guey-Shuang Wu
Robert A. Stein
David Gelmont
Alex Sevanian
Erick Sohlberg

Laboratory of Nuclear Medicine and Radiation Biology
University of California
Los Angeles, California

Ronald N. McElhaney

Department of Biochemistry
University of Alberta
Edmonton, Canada

It is now generally acknowledged that the major site of oxidative damage to tissues is in the membranous systems of the cell but the mechanism of damage, the species immediately involved and the means by which they are derived are still obscure. For this reason, any preventative measures are still

[1]Research supported in part by U. S. Public Health Service Research Career Award No. 5-K06-GM19177-18 from the Division of General Medical Sciences, National Institutes of Health and by Contract DE-AMO3-76-SF00012 between the Department of Energy and the University of California.

empirical, although it is recognized that antioxidants of various sorts are effective.

In several laboratories, including ours, attempts to clarify the mechanisms involved are being made with various model systems which isolate and simplify certain aspects of membrane properties, permitting them to be studied separately and easily. In our laboratory, a series of models of increasing complexity has been used to afford considerable insight into the nature of membrane peroxidation and its prevention.

The simplest of these is a monolayer of polyunsaturated fatty acid (usually linoleic acid) adsorbed on silica gel (1-5). In this model, fatty acids are positioned with chains more or less parallel and with unsaturated centers more or less in a plane, an arrangement that should promote a radical-propagated chain reaction most efficiently. The fatty acid chains are about 3 Å apart, a distance not too different from those in the membrane phospholipids. Moreover, other membrane constituents can be incorporated into the system, such as saturated fatty acids, sterols and antioxidants (6,7). A schematic representation of this model showing a linoleic acid monolayer and including a saturated fatty acid, cholesterol and α-tocopherol is depicted in Fig. 1.

Studies with this system have revealed that in such an ordered arrangement of unsaturated fatty acids, the principal products are not the predicted fatty acid hydroperoxides but are the epoxides. A suggested mechanism for the transfer of oxygen from the initially formed hydroperoxide to an adjacent molecule is depicted in Fig. 2, but since hydroperoxides are generally poor reagents for epoxidation, the actual process may be more complex. This adjacent molecule can be linoleic acid or another olefin such as oleic acid or cholesterol, and it should be pointed out that this mechanism represents a means by which the epoxides (including several mutagenic and carcinogenic compounds) may be formed in the membrane.

Introducing increased space between the linoleic acid chains by lowering the linoleic acid:silica ratio below 20 mg/g, or by introducing saturated fatty acids into the monolayer, decreases the proportion of unsubstituted fatty acid epoxides and leads to increasing amounts of 11-hydroxy-9,10-epoxy-12-octadecenoic acid and the isomeric 11-hydroxy-12,13-epoxy-9-octadecenoic acid (6) formed presumably by intramolecular oxygen transfer because of the increasing difficulty of intermolecular transfer, as depicted in Fig. 3. It should be noted that the hydroxyepoxides are the most potent carcinogens derived from the aromatic hydrocarbons (8). Fig. 4 lists the formulas of the types of compounds formed in this and succeeding models during peroxidation.

Once the products and mechanisms of peroxidation in the

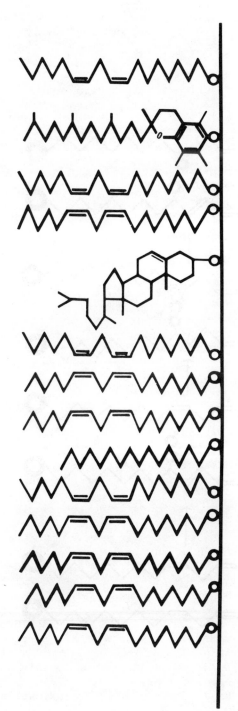

Fig. 1. Schematic representation of a linoleic acid monolayer on silica gel. Included is a molecule of palmitic acid and one each of cholesterol and α-tocopherol.

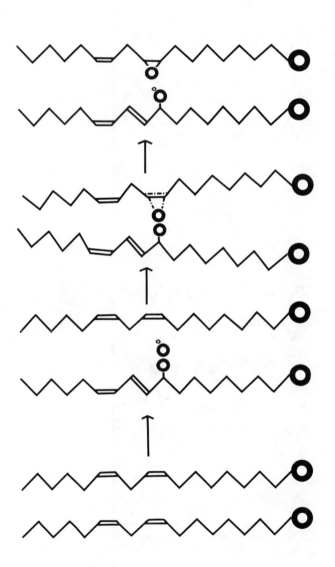

Fig. 2. Schematic representation of proposed mechanism of oxygen transfer from peroxy radical to an adjacent unsaturated fatty acid in the monolayer.

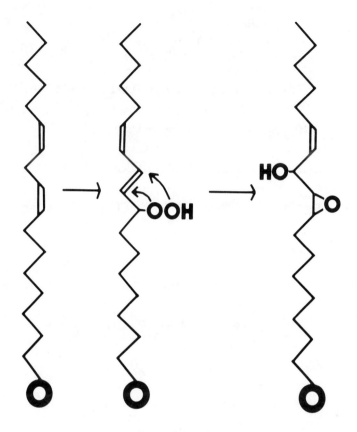

Fig. 3. Schematic representation of proposed mechanism of intramolecular transfer of oxygen and hydroxyl in 9-hydroperoxy-10,12-octadecadienoic acid.

simple monolayer system were recognized, it became important to ascertain whether they would also be found in model systems more closely related to biomembranes. For this purpose, use was made of the bilayer phosphatidylcholine liposome system, which may show many of the features of simple membranes (9).

When pure soybean phosphatidylcholine liposomes were in-cubated at 40°C in 0.1M KCl in 0.01M Tris buffer, pH > 7 the rates of disappearance of the linoleic acid (70% of total) and linolenic acid (6% of total) followed typical autocatalytic kinetics (Fig. 5) and the products were typical of bulk phase autoxidation. Thus, from linoleic acid the main products (fol-lowing reduction of the hydroperoxides, I and II, during the isolation procedure) were 9- and 13-hydroxyoctadecadienoates, III and IV, as has been reported by Porter et al. (10).

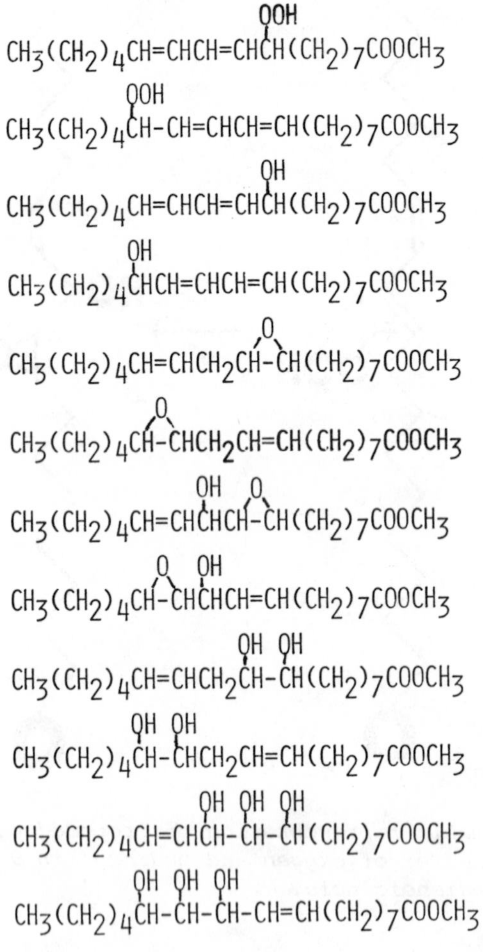

$$\overset{\displaystyle OOH}{CH_3(CH_2)_4CH=CHCH=CHCH(CH_2)_7COOCH_3} \qquad I$$

$$\overset{\displaystyle OOH}{CH_3(CH_2)_4CH-CH=CHCH=CH(CH_2)_7COOCH_3} \qquad II$$

$$\overset{\displaystyle OH}{CH_3(CH_2)_4CH=CHCH=CHCH(CH_2)_7COOCH_3} \qquad III$$

$$\overset{\displaystyle OH}{CH_3(CH_2)_4CHCH=CHCH=CH(CH_2)_7COOCH_3} \qquad IV$$

$$CH_3(CH_2)_4CH=CHCH_2CH\overset{\displaystyle O}{-}CH(CH_2)_7COOCH_3 \qquad V$$

$$CH_3(CH_2)_4CH\overset{\displaystyle O}{-}CHCH_2CH=CH(CH_2)_7COOCH_3 \qquad VI$$

$$\overset{\displaystyle OH\quad O}{CH_3(CH_2)_4CH=CHCHCH-CH(CH_2)_7COOCH_3} \qquad VII$$

$$\overset{\displaystyle O\quad OH}{CH_3(CH_2)_4CH-CHCHCH=CH(CH_2)_7COOCH_3} \qquad VIII$$

$$\overset{\displaystyle OH\quad OH}{CH_3(CH_2)_4CH=CHCH_2CH-CH(CH_2)_7COOCH_3} \qquad IX$$

$$\overset{\displaystyle OH\quad OH}{CH_3(CH_2)_4CH-CHCH_2CH=CH(CH_2)_7COOCH_3} \qquad X$$

$$\overset{\displaystyle OH\quad OH\quad OH}{CH_3(CH_2)_4CH=CHCH-CH-CH(CH_2)_7COOCH_3} \qquad XI$$

$$\overset{\displaystyle OH\quad OH\quad OH}{CH_3(CH_2)_4CH-CH-CH-CH=CH(CH_2)_7COOCH_3} \qquad XII$$

Fig. 4. Methyl esters of products of peroxidation of linoleic acid in monolayer on silica gel, as an acyl moiety of PC in soybean PC and in soybean PC-DPPC liposome systems and in acholeplasma laidlawii membranes.

I, II	*Isomeric 9- and 13-Hydroperoxyoctadecadienoates*
III, IV	*Isomeric 9- and 13-Hydroxyoctadecadienoates Reduction Products of I and II.*
V, VI	*Isomeric 9,10- and 12,13-Epoxyoctadecenoates*
VII, VIII	*Isomeric 11-Hydroxy-9,10- and 12,13-epoxyoctadecenoates*
IX, X	*Isomeric 9,10-Dihydroxy-12-octadecenoate and 12,13-Dihydroxy-9-octadecenoate, Hydrolysis*

However, with increasing incorporation of the saturated dipal-
mitoylphosphatidylcholine into the liposomes, the rates of disappearance changed until at a ratio of DPPC/SPC of 4:1, apparent first order kinetics were obtained (Fig. 6). At this ratio, the products were also quite different in that the simple
hydroperoxides had decreased considerably and there were large
increases in the epoxides, V and VI, and particularly the hydroxyepoxides, VII and VIII and their hydrolysis products, the
di- and trihydroxyoctadecenoates (IX, X, XI and XII). The extent of formation of these products, with time in the DPPC-SPC
liposomes is shown in Fig. 7.

It seems evident that in the bilayer liposomes, the ordered
arrangement of the unsaturated hydrocarbon chains and particularly, the inclusion of saturated chains in the arrangement,
have the effect of promoting an intermolecular or intramolecular transfer of oxygen from the hydroperoxides or peroxyradi-
cals initially formed to neighboring susceptible molecules or
to reactive carbons in the same molecules. The question then
had to be asked whether, in a naturally occurring biomembrane
with its more complex structure, the same reactions yielding
the same products would occur.

To answer this question, membranes of Acholeplasma laidlawii were prepared by Ronald McElhaney with two different ratios of palmitate to linoleate in the membrane phospholipids -
90:10 and 40:60 (11). These membranes were incubated at 37°C
in the presence of ferrous ion and ascorbate as initiating
agents. Although the 90:10 ratio membranes understandably oxidized very slowly and yielded only small amounts of products,
the 40:60 ratio membranes oxidized rapidly and extensively
yielding as major products fatty acid epoxides and, particularly hydroxyepoxides and their hydrolysis products, trihydroxy-
octadecenoic acids. Thus, the products of peroxidation of this
simple membrane are almost identical to those derived from liposome preparations of similar composition.

The final question, of course, must be whether the same reactions that occur in model systems and in simple membranes occur in the membranes of living animals. The answer to this
question came from experiments with rats breathing filtered air
or air containing 6 ppm NO_2 (12). In both cases, extraction
and analysis of the lipid fractions of the rat lungs revealed
the presence of epoxides of fatty acids and cholesterol, with
the relative amounts of the different fatty acid epoxides

	Products of V and VI
XI, XII	Isomeric 9,10,11-Trihydroxy-12-octadecenoate and 11,12,13-Trihydroxy-9-octadecenoate, Hydrolysis Products of VII and VIII.

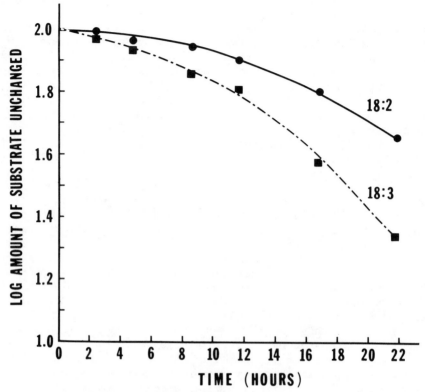

Fig. 5. *Plot of log of disappearances with time of the major polyunsaturated fatty acids during incubation of liposomes of pure soybean phosphatidylcholine at 40°C in 0.1M KCl, 0.01M tris buffer.*

roughly proportional to those of the fatty acids from which they were presumably derived. The most interesting aspect of these findings stemmed from the location of the fatty acid epoxides in the different lipid fractions and the effect of NO_2 on their amounts. In the case of the lung tissue lipids, a 24 hr exposure to NO_2 increased the epoxide content of the phospholipid fatty acids and the cholesterol but not of the triglyceride fatty acids (Fig. 8). In the lung lavage, however, NO_2 did not bring about any increase in epoxide content of the phospholipid fatty acids but markedly increased that of the triglyceride fatty acids as well as that of the cholesterol (Fig. 9).

There are several possible explanations for these findings. First it follows logically from the previous experiments that epoxide formation should be greatest in the ordered arrangement of the tissue phospholipids and that the epoxidation of cholesterol as a membrane constituent would follow. One of the major

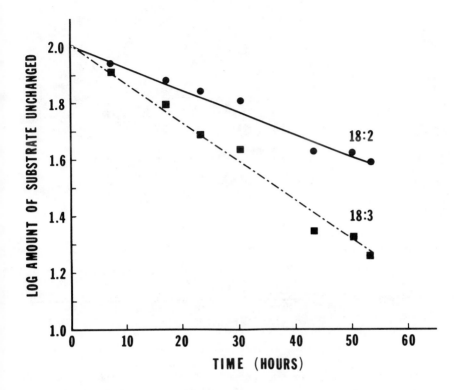

Fig. 6. Plot of log of disappearance with time of the major polyunsaturated fatty acids during incubation of liposomes containing soybean phosphatidylcholine: dipalmitoyl phosphatidylcholine (1:4) at 40°C in 0.1M KCl, 0.01M tris buffer.

protective devices against retention of a damaged fatty acid in a membrane phospholipid would logically be its removal. Experiments with phospholipase A_2, either the soluble enzyme from viper venom or the particulate enzyme from the microsomes, revealed that hydrolysis of epoxystearate from the 2-position of phosphatidylcholine is about twice as rapid as is hydrolysis of oleic acid from which it was derived (13) (Table I). Moreover, once removed from the phospholipid, the fatty acid epoxide is a much better substrate for epoxide hydrolase than if it remained bound (14) (Table II). Thus a protective mechanism against accumulation of oxidized fatty acids in membrane lipids is their rapid removal and hydrolysis.

    An additional protective device, might also operate here. If phospholipase C were also more active against a phospholipid

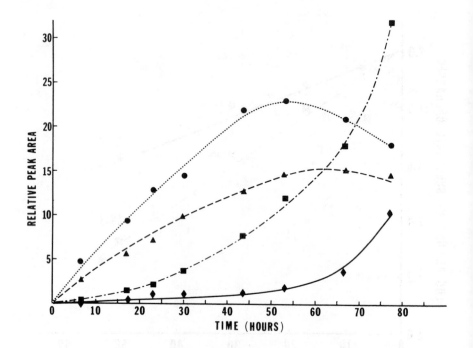

Fig. 7. Change in the concentrations of the oxygenated products during incubation of soybean PC: dipalmitoyl PC (1:4) liposomes at 40°C in 0.1M KCl, 0.01M tris buffer ·· ● ··· ● ···, III; — ▲ - - ▲ - -, IV; — ■ ·· ■ ···, XI & XII; ——◆——◆——, VII & VIII.

containing an oxidized fatty acid, then diacylglycerol thus formed could be removed from the membrane and acylated to a triacylglycerol, which would appear in the lung washings. This possibility, however, has yet to be proven. Considering the occurrence and damaging effects of these products of peroxidation in the membranes it is not surprising that protective devices abound. The hydrolytic reactions for removal and detoxification of epoxides have already been mentioned. Prevention of the initial oxidation reactions by segregation of reactive substrates from initiators is well known. Enzymes, such as glutathione peroxidase, superoxide dismutase and catalase, destroy their respective substrates. However, we have concerned ourselves with two main means of protection. First, as has been found in all the model systems and in the simple membranes, the inclusion of saturated fatty acids in the membrane phospholipids decreases the rate of peroxidation and changes the products. It is not yet known, however, whether the effect

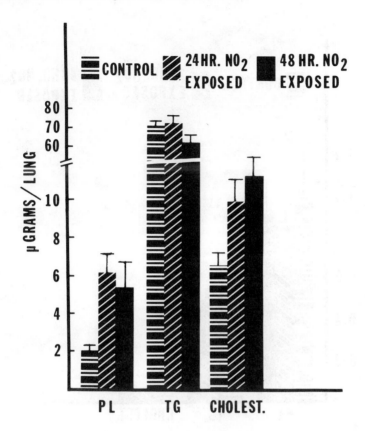

Fig. 8. Amounts of fatty acid and cholesterol epoxides in lipid fractions of lung tissues following 24 or 48 hr exposure of rats to filtered air or 6 ppm NO_2.

of saturated fatty acids in the same molecule as the unsaturated fatty acids is the same as when in admixture as separate molecules and this is the subject of further experiment.

The second protective system under investigation is the inclusion of antioxidants in the models and membranes. In the linoleic acid monolayer, as in bulk phase lipids, α-tocopherol introduces an induction period that is proportional to its concentration relative to that of linoleate. During this induction period little oxidation of linoleate occurs, but when the tocopherol content has been reduced to about 10% of its original value, linoleate oxidation proceeds as if no antioxidant had been present (7). Gamma-tocopherol at the same concentration

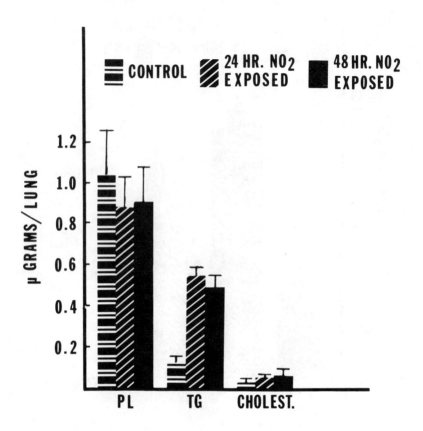

Fig. 9. Amounts of fatty acid and cholesterol epoxides in lipid fractions of lung lavage following 24 or 48 hr exposure of rats to filtered air or 6 ppm NO₂.

(0.05 mole %) gives a somewhat longer induction period while a synthetic antioxidant, 3-(ω-carboxynonyl)-4-methoxy-5-pentyl-phenol, gives no induction period at all but results in a very much lower rate of oxidation for a very long time (Fig. 10). A combination of these two types of antioxidants resulted in a very long induction period.

Turning to the bilayer liposome system we find that protection by α-tocopherol takes a course almost identical to that with the monolayer system. At 0.05 mole percent, α-tocopherol incorporated into the liposomes produced an induction period of about 7 hours under the conditions of incubation used (Fig. 11 a & b). It is interesting to note that at the end of the

TABLE I.   Phospholipase A_2 Activity (C. adamanteus) Against
Phosphatidylcholine and Phosphatidylcholine
Epoxide

PC - nmoles/mg lipase/min	PC epoxide - nmoles/mg lipase/min
Enzyme 51.0 ± 6.9	95.0 ± 10.4
+ EDTA 18.2 ± 0.74	11.8 ± 1.42
+ 1mM pBPB 23.1 ± 3.15	35.0 ± 5.09

induction period, the initial rates of decrease of linoleic and
linolenic moieties are more rapid than if no antioxidant had
been present (compare Figs. 11a and 11b) even though the concentration of tocopherol is well below that reported to have a
prooxidant effect (15).

In the living animal, the effect of antioxidants has been
studied in many laboratories. In this Laboratory, these studies have taken two directions. First, it has been shown that
in NO_2-exposed rats, the concentrations of epoxides in the
lung lipids are inversely related to the tissue tocopherol concentration (16) (Table III). Second, an in vivo measure of
tissue damage has followed the studies of Tappel and his co-
workers (17,18) and others (19) showing that pentane expiration
is a sensitive measure of lipid peroxidation in tissues. When

TABLE II.   Epoxide Hydratase Activity in Rat Lung and
Liver Microsomes (expressed as pmoles/mg
protein/min)

Substrate	Lung	Liver
Methylepoxystearate (20 µM)	159 ± 16.7	1301 ± 30.2
Epoxystearic acid (20 µM)	180 ± 12.9	---
Cholesterol epoxide (20 µM)	10 ± 2.3	183 ± 17.4
Styrene oxide (50 µM)	200 ± 7.6	1700 ± 19.9
Phosphatidylcholine epoxide (20 µM)	4 ± 0.75	16 ± 2.1
Phosphatidylcholine epoxide + pBPB (1 mM)	15 ± 2.4	28 ± 4.2

Table III. Phospholipid and Phospholipid Epoxide Content
in Rat Lungs

	Phospholipids mg/g lung	Fatty acid epoxides of phospholipids µg/g lung	Phospholipid epoxides as % of total phospholipids
Vit. E Supplemented: Control	23.1 ± 2.08	31.7 ± 5.16	0.14 ± 0.013
NO$_2$-exposed	20.9 ± 2.11	39.1 ± 5.80	0.19 ± 0.020
Vit. E Deficient: Control	19.7 ± 1.91	32.1 ± 4.66	0.17 ± 0.016
NO$_2$-exposed	23.6 ± 3.80	48.7 ± 8.43	0.21 ± 0.033

Long-Evans rats were made tocopherol deficient by maintaining
nursing mothers on a tocopherol-deficient diet and then weaning
the pups to the same diet, their breath was found to contain a
ratio of pentane/isopentane of about 100. Following one 200 mg
oral dose of α-tocopherol, within 24 hr the ratio had changed
to 1-2. This ratio is used since isopentane is a very unlikely
metabolic product and probably measures atmospheric hydrocarbon
contamination. This would seem to indicate a very rapid onset
of, and recovery from the deficiency in this strain of rats.
It is possible that expired pentane could be correlated very
closely with tissue (i.e., liver) tocopherol, thus simplifying
the analysis of the deficiency state.

In conclusion, it appears that the danger of a peroxidative
reaction occurring in the membranous systems of the cell is
sufficiently great that a number of protective measures have
developed to counter it. In our present state of knowledge
there is only one that we can manipulate safely by dietary
means. For this reason, it would seem to be of considerable
importance to correlate the antioxidant state of the tissues,
for which there are a number of measures, with both the dietary
and tissue content of tocopherol. As our knowledge grows we
can apply similar correlations with the other protective devices in an effort to reverse the present trend of steadily increasing exposure of man to his own industrial pollutants.

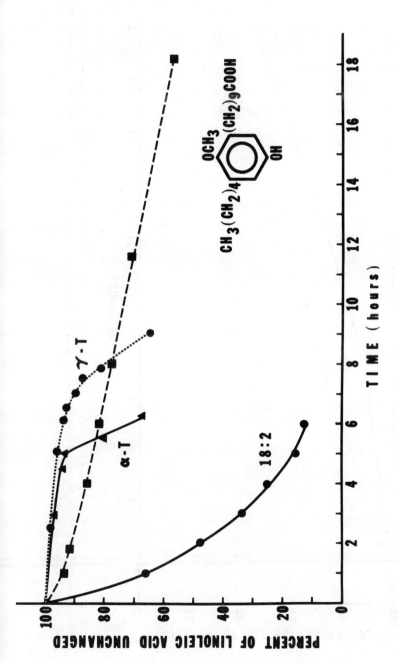

Fig. 10. Effect of antioxidants on rate of decrease of linoleic acid in monolayers exposed to air at 60°C. ── no antioxidant; ····· 0.05 mole % of α-tocopherol; ── 0.05 mole % of γ-tocopherol; ──▲── 0.05 mole % of α-tocopherol; ──■── 0.4 mole % of FAHQ.

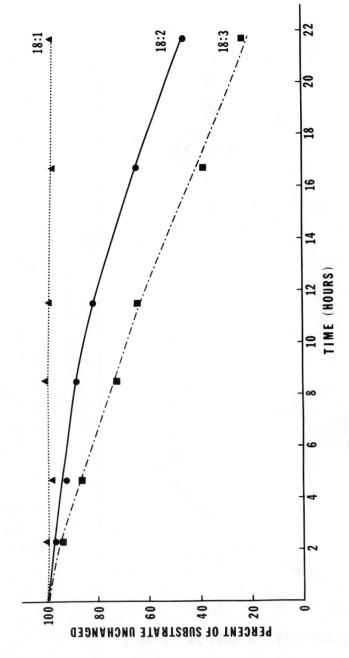

Fig. 11a. Plot of disappearance with time of the major unsaturated fatty acids during incubation of pure soybean phosphatidylcholine liposomes in 0.1M KCl, 0.01M tris buffer at 40°C. Absence of antioxidant.

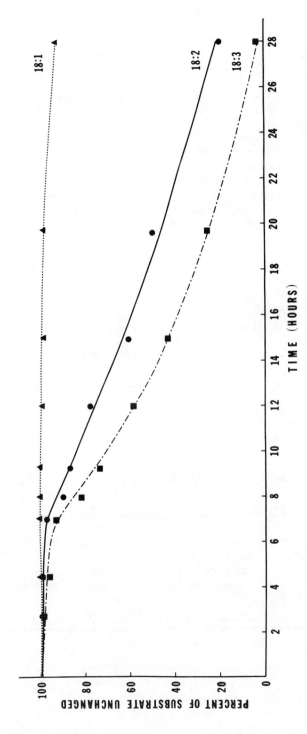

Fig. 11b. Presence of 0.05 mole % α-tocopherol.

REFERENCES

1. Porter, W. L., Levasseur, L. A. and Henick, A. S., *Lipids*
    6, 1 (1971).
2. Porter, W. L., Levasseur, L. A. and Henick, A. S., *Lipids*
    7, 699 (1972).
3. Porter, W. L., Henick, A. S., and Levasseur, L. A., *Lipids*
    8, 31 (1973).
4. Wu, G.-S., and Mead, J. F., *Lipids 12*, 965 (1977).
5. Wu, G.-S., Stein, R. A., and Mead, J. F., *Lipids 12*, 971
    (1977).
6. Wu, G.-S., Stein, R. A., and Mead, J. F., *Lipids 13*, 517
    (1978).
7. Wu, G.-S., Stein, R. A., and Mead, J. F., *Lipids 14*, 644
    (1979).
8. Wislocki, P. G., Wood, A. W., Chang, R. L., Levin, W.,
    Yagi, H., Hernandez, O., Jerina, D. M., and Conney, A. H.,
    Biochem. Biophys. Res. Comm. 68, 1006 (1976).
9. Wu, G.-S., Stein, R. A., and Mead, J. F., *J. Am. Oil Chem.
    Soc. 57*, 145A (1980).
10. Porter, N. A., Wolf, R. A. and Weenen, H., *Lipids 15*, 163
    (1980).
11. Silvius, J. R., and McElhaney, R. N., *Canad. J. Biochem.
    56*, 462 (1978).
12. Sevanian, A., Mead, J. F., and Stein, R. A., *Lipids 14*,
    634 (1979).
13. Sevanian, A., Stein, R. A., and Mead, J. F., *J. Am. Oil
    Chem. Soc. 57*, 168A (1980).
14. Sevanian, A., Personal Communication.
15. Cillard, J., Cillard, P., Cormier, M., and Girre, L.,
    J. Am. Oil Chem. Soc. 57, *252 (1980).*
16. *Sevanian, A., El Sayed, N. M., and Hacker, A. D., Lipids*
    (in press).
17. Dillard, C. J., Dumelin, E. E., and Tappel, A. L., *Lipids
    12*, 109 (1977).
18. Dumelin, E. E., Dillard, C. J., Purdy, R. E., and Tappel,
    A. L., *Fed. Proc. 36*, 1160 (1977).
19. Cohen, G., in "Oxygen Free Radicals and Tissue Damage",
    Ciba Foundation Symposium 65, p. 177. Excerpta Medica,
    Amsterdam (1979).

CONTROL OF LIPID PEROXIDATION
BY A HEAT-LABILE CYTOSOLIC FACTOR

Paul B. McCay
Donald D. Gibson

Biomembrane Research Laboratory
Oklahoma Medical Research Foundation
Oklahoma City, Oklahoma

I.   INTRODUCTION

Biological control of extraneous oxidative reactions in living cells is important in terms of prevention of disruption of cellular structures containing easily oxidizable substances such as highly unsaturated lipids. Since membranes of organelles in animal tissues contain relatively large amounts of polyunsaturated fatty acid moieties, they are particularly vulnerable to oxidative attack. Such attacks can be initiated by the enzymic activity of several endogenous oxidoreductases in the microsome (such as gulonolactone oxidase (1) and NADPH-cytochrome P-450 reductase (2,3)), and in mitochondria (NADPH oxidase (4)). The mechanism of the oxidative attack on membrane lipids in these systems is still uncertain. In addition, a number of other enzyme systems, both soluble and membrane-bound, produce either superoxide anion or hydrogen peroxide. These compounds appear to be potential hazards for cells although protection of the functional integrity of the cells is assumed to be afforded within limits by the combined activities of superoxide dismutase, glutathione peroxidase, and catalase. The most disruptive forms of oxidative stress in cells appear to be produced during the metabolism of certain xenobiotic compounds such as paracetamol (5) and CCl_4 in liver tissue (6), chaotropic agents in heart subcellular particles (7), cis-stilbene metabolism by liver microsomes (8),

[1]Supported in part by NIH Grant AM08397.

ozone in lung (9), adriamycin in cardiac tissue (10), among many others that could be cited.

While the mechanisms for reducing oxidative stress due to superoxide anion and hydrogen peroxide formation normally produced in cell metabolism appear to be adequately controlled in animal tissues by the enzymes mentioned above, the introduction into animal systems of certain substances from the environment results in tissue damage which is correlated with evidence that lipid peroxidation had also occurred in those tissues. Whether or not lipid peroxidation is a cause or a consequence of events leading to tissue damage in such cases is unclear and arguments for either situation can be made (11-13). It does appear that free radical events are involved in tissue damage caused by some toxic compounds such as CCl_4. Part of the evidence for this comes from experiments which demonstrate that prior treatment of animals with antioxidant compounds reduces tissue injury caused by that compound (14). Other studies have demonstrated that trichloromethyl ($\cdot CCl_3$) radicals are formed in the primary target organ (liver) of animals treated with CCl_4. Since prior treatment of animals with antioxidant compounds also inhibits lipid peroxidation caused by metabolism of CCl_4 by liver microsomes in vitro (11), there are valid reasons for considering the hypothesis that free radical events in the liver cell, including the initiation of lipid peroxidation, are essential for the injury which results from exposure to CCl_4. There undoubtedly is free radical damage to other cellular components in addition to lipids. Radioactive labeling of either the chlorine or carbon atom of CCl_4 has provided the opportunity to demonstrate that administration of either the chorine- or carbon-labeled form of CCl_4 results in covalent binding of the compound to protein and lipid components of the endoplasmic reticulum (15). This labeling is not affected by prior treatment of the animals with antioxidants although the animals are protected against cell injury. This suggests that free radical production from CCl_4 per se and the interaction of these radicals with cell components may not constitute an insult to the cell but that subsequent reactions which may be initiated by the $\cdot CCl_3$ radicals (such as lipid peroxidation) may be essential for the toxicity of CCl_4 to be expressed.

Reactions involving free radicals clearly pose a threat to the chemical structure of unsaturated lipids in membranes since lipid peroxidation is easily initiated in these structures under conditions imposed in vitro. However, the lipid peroxidation which is sufficient to cause significant biological effects in vivo may only occur under extreme conditions involving a severe deficiency of lipid-soluble free radical scavenging compounds in the diet (which is unlikely to occur naturally) or to the action of toxic substances in the

environment. A major part of the stability of membranous organelles of cells with respect to oxidative stress (in spite
of the high unsaturated lipid content of such membranes) appears to be due to a heat-labile factor in tissues which is
not fully characterized, but is distinct from superoxide dismutase, catalase and glutathione peroxidase. The factor, which
very effectively inhibits lipid peroxidation catalyzed by both
enzymic and non-enzymic processes, has been found in the
105,000 x g supernatant fraction of the cytosolic fraction of
all rat and guinea pig tissues which have been studied thus
far (liver, lung, heart). A similar (and perhaps identical)
factor has also been found in mitochondria. This article
summarizes the current status of information concerning
this factor.

II. INITIAL OBSERVATIONS ON THE CYTOSOLIC SYSTEM WHICH
     INHIBITS LIPID PEROXIDATION

     Awareness that this system existed developed as follows.
The observations of Christophersen (16,17) and of Flohé and
Zimmermann (18) demonstrated that liver cytosol catalyzed a
glutathione-dependent reduction of fatty acid hydroperoxides
to corresponding alcohols. In addition, Christophersen's
studies demonstrated that malondialdehyde production in systems in which lipid peroxidation was occurring was inhibited
under these conditions (33). It was reasonable to assume, at
that time, that glutathione peroxidase was responsible for
these effects and that one of the functions of glutathione
peroxidase was to reduce lipid peroxides which may be formed
in biological membranes to lipid alcohols, thereby preventing
oxidative chain scission of the fatty acyl group and associated formation of malondialdehyde. Because of our interest in
the mechanisms of lipid peroxidation, we began to utilize the
glutathione-dependent system as part of a system of probes for
investigating events associated with lipid peroxidation. The
question arose: how does a cytosolic enzyme system gain
access to peroxy groups which, according to current concepts
of membrane structure, would be deep within the hydrophobic
portion of the membrane? Initial experiments were performed
which attempted to demonstrate the formation of hydroxy fatty
acids in microsomal membranes in which significant levels of
lipid peroxides had been formed. The lipid peroxides were
produced by incubating the microsomes in the presence of NADPH
and Fe^{3+} (as an $ADP-Fe^{3+}$ complex) (19). Maximum lipid peroxide formation in the membrane occurs in approximately 15
minutes (Table I). The lipid peroxides in the microsomes were
sufficiently stable to allow centrifugation and resuspension

TABLE I. Lipid Peroxide Content of Liver Microsomes as a Function of
Incubation Time in the Presence of NADPH and Fe^{3+} [a]

Incubation system	Lipid peroxide content µmoles/mg microsomal protein				
	4 min.	15 min.	25 min.	35 min.	45 min.
Minus NADPH	0.15	0.27	0.15	0.15	0.15
Plus NADPH	0.28	1.01	1.13	0.52	0.35

[a]Lipid peroxide formation at various times during NADPH oxidation by
liver microsomes in the presence of 4.0 mM ADP, 0.12 mM Fe^{3+}, 0.3 mM NADPH,
and 1 mg microsomal protein ml of reaction system, all in 0.1 M Tris-HCl
buffer, pH 7.5, which had previously been bubbled with O_2. Incubation was
carried out at 37° in a shaking waterbath. Peroxide determination on extracted microsomal lipid was determined iodometrically (19).

in an incubation medium containing dialyzed liver cytosol (as
a source of glutathione peroxidase) and glutathione (20).
After 45 minutes, total lipids were extracted from the system
and subjected to treatment with phospholipase A. The β-
position fatty acids of phospholipids released by this treatment would contain essentially all of the peroxidized fatty
acyl groups in the membrane phospholipids and, consequently,
any hydroxy fatty acids formed by incubation in the presence
of cytosolic glutathione peroxidase and glutathione. Thin-
layer chromatographic analyses of the β-position fatty acids
failed to show any trace of hydroxy fatty acid production even
though there was sufficient lipid peroxide present in the
membrane to produce a major hydroxy fatty acid component in
the chromatographed lipid had the reduction of peroxy groups
occurred (20). The failure of the cytosol + glutathione
system to form hydroxy fatty acids in membranes containing
considerable amounts of lipid peroxide was observed repeatedly
and raised the question about the requirements of the cytosolic
system to produce hydroxy fatty acids from peroxy fatty acids.
The following experiment was performed. Four peroxide substrates were assayed in a system containing cytosol, glutathione, and NADPH. If the peroxide substrate was reduced to a
hydroxyl group by the action of glutathione and cytosol,
oxidized glutathione would be formed. Oxidized glutathione
would be reduced to glutathione (GSH) again by NADPH and
glutathione reductase present in the cytosol. Therefore, the
rate of reduction of the peroxide functions to hydroxy groups
can be determined by measuring NADPH oxidation. Fig. 1 shows
that both free fatty acid peroxides and cumene hydroperoxide
are good substrates for the cytosolic system, while neither
microsomes containing lipid peroxides nor micellar suspensions

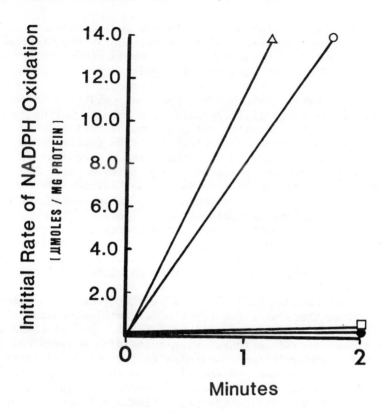

Fig. 1. Reduction of various peroxides by the glutathione dependent cytosolic factor was measured by the initial rate of NADPH oxidation in the reduction of oxidized glutathione catalyzed by excess glutathione reductase (added to the system). The assay system contained 50 mM K HPO buffer, pH 7.0, 1.0 mM EDTA, 1.0 mM NaN$_3$, 0.2 mM NADPH, 1.0 enzyme unit ml GSSG-reductase, 1.0 mM GSH, 0.1 ml dialyzed liver cytosol, and 1.0 ml peroxide substrate. Total volume of assay system was 1.0 ml. Total initial peroxide content of each substrate added was: cumene hydroperoxide, 0.75 μmoles; peroxidized microsomal lipid 0.78 μmoles; unperoxidized microsomal lipid 0.16 μmoles; and peroxidized free fatty acids, 0.84 μmoles.
△ cumene hydroperoxide   ○ peroxidized free fatty acid
□ peroxidized microsomal lipid   ● unperoxidized microsomal lipid

of microsomal lipid peroxides show any reactivity. Thus, the observation by Christophersen (16,17) and by Flohé and Zimmermann (18) that lipid peroxides are reduced to lipid alcohols appears to apply to free fatty acids but not to complex lipids such as phospholipid. At this time we then

noted that when cytosol and glutathione were present in the
system, the membrane fatty acid composition of the microsomes
remained stable under conditions that would otherwise promote
rapid peroxidation of unsaturated lipids (20). It was then established that the inclusion of the cytosolic factor (thought
at that time to be glutathione peroxidase) and glutathione in
the system prevented lipid peroxidation from occurring in
microsomal systems in which extensive peroxidation would
ordinarily occur. It became clear that the glutathione-
dependent cytosolic system was inhibiting a peroxidative
attack on the membrane polyunsaturated fatty acids. One
consequence of this phenomenon is that no malondialdehyde
would be formed under these conditions, in agreement with the
earlier observations of Christophersen (33).

   Since these investigations demonstrated that lipid
peroxidation was being inhibited by the glutathione-dependent
cytosolic factor, the question arose concerning the nature of
the cytosolic factor responsible for inhibition of lipid
peroxidation. Reports by Lawrence and Burk (21) and by
Prohaska and Ganther (22) appeared at that time which demonstrated that there were two types of glutathione-dependent
peroxidase activities in rat liver cytosol. One of these was
glutathione peroxidase (a selenium-containing enzyme) and the
other was a non-selenium-containing factor which was later
shown to be one or more of the glutathione-S-transferases(21).
Burk et al. investigated glutathione-dependent inhibition
by cytosol of malondialdehyde production by microsomes subjected to peroxidizing conditions (23). The results of their
studies on fractionation of liver cytosol indicated that the
inhibitory activity was not associated with the selenium-
containing glutathione peroxidase. Their data suggested that
the cytosolic factor might be one or more of the glutathione-
S-transferases present in cytosol. Ammonium sulfate fractionation of cytosol produced a fraction containing most of the
glutathione-S-transferase activity as well as the glutathione-
dependent inhibitor of lipid peroxidation (23). Our studies
with epoxy-activated Sepharose 6B affinity column chromatographic purification of the glutathione-S-transferases (according
to the procedure of Simons and Vander Jagt (24)), indicated
that the cytosolic factor was very likely not one of the
glutathione-S-transferases per se (25). Fig. 2 shows that
essentially all of the glutathione-S-transferase activity is
separated from rat liver cytosol by this procedure. However,
the factor which inhibits lipid peroxidation was not associated
with the glutathione-S-transferase preparation although the
latter was very active toward 1-chloro-2,4-dinitrobenzene as
substrate. All of the inhibitor activity was found in the
cytosolic fraction not retained by the column and which had
essentially no glutathione-S-transferase activity (Fig. 2).

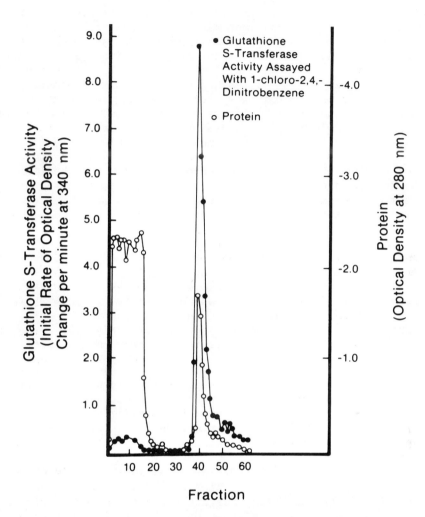

Fig. 2.   Isolation of liver cytosol glutathione-
S-transferase by affinity chromatography on epoxy-activated
Sepharose 6B.

Hence, the studies of Burk et al. (23) together with our
investigations, indicate that the cytosolic factor which
inhibits lipid peroxidation is neither glutathione peroxidase
nor a glutathione-S-transferase.
    Other investigators have observed inhibition of lipid
peroxidation by liver cytosol also, but because these studies
were done without prior dialysis of the soluble fraction, the
requirement for glutathione for the inhibition was not

observed.   Kotake et al. (26) observed that lipid peroxidation in rat liver microsomes was correlated with an inhibition
of the cytochrome P-450-dependent monooxygenase system.   Using
malondialdehyde formation as an assay for lipid peroxidation,
Kotake et al. observed that preincubation of liver microsomes
with NADPH resulted in a loss of ethyl morphine N-demethylase
activity measured immediately following the preincubation.
The decrease was roughly inversely proportional to the amount
of lipid peroxidation which had occurred during the preincubation period.   However, the inclusion of soluble fraction
from the 105,000 x g supernatant from liver in the preincubation system inhibited both lipid peroxidation and the loss of
the demethylase activity (26).   It is well-known that lipid
peroxidation in microsomes results in a loss of cytochrome P-
450 components which are required for monooxygenase activities
(27).

     The results of Kotake et al. (26) described above
confirmed earlier observations of Kamataki et al. (28), who
also demonstrated the presence of the soluble fraction in
systems containing liver microsomes and NADPH inhibited the
extent of lipid peroxidation (as measured by malondialdehyde
formation) that occurred during the demethylation of either
ethylmorphine or aminopyrine.   The latter activities were also
maintained linearly for a longer period than were demethylase
systems without the added soluble fraction.   More recently,
Kamataki et al. published a report describing an induction by
CCl_4 of an inhibitor of lipid peroxidation in the cytosolic
fraction of rat liver (29).   These investigators treated rats
with CCl_4 orally at a dose level of 0.5 ml/kg.   Higher dose
levels (2.5 ml/kg) had no influence on the cytosolic inhibitor
factor.   These workers fractionated the cytosol with different
concentrations of polyethylene glycol followed by chromatography of the most active fraction on a hydroxyapatite column.
The most active preparation which was obtained was essentially
free of glutathione peroxidase activity, in agreement with our
investigations as well as those of Burk et al., mentioned
above. Since the cytosol preparations used in the studies of
Kamataki et al. were either undialyzed or were fractionated
preparations containing dithiothreitol, the dependence of the
inhibitor factor on glutathione was not observed.   The significance of this aspect will be discussed later in this
article.

     Some very interesting studies by Talcott et al. (30)
demonstrated that the capacity of liver cytosol to inhibit
lipid peroxidation is markedly influenced by the dietary state
of the animal from which the cytosol was prepared.   These
investigators fed rats on a fat-free, high carbohydrate diet.
This diet, fed for twelve days, resulted in a liver cytosol
preparation that was quite significantly more inhibitory than

cytosol from animals fed a commercial laboratory ration. Of further interest was the observation that fasting of animals which had been fed the partially purified diet for 24 hours resulted in complete loss of the inhibitory effect of the liver cytosol on lipid peroxidation. These workers also examined the possibility that glutathione was involved in the inhibitory activity of the cytosol and concluded that the inhibitory activity of the cytosol fraction was not correlated with the glutathione concentration of the system. It appears possible, however, that the range of glutathione concentrations employed in these studies were not of a magnitude that would have indicated the requirement for glutathione. Nevertheless, the studies of Talcott et al. (30) are extremely interesting because of the implications of the dietary effects, and particularly of fasting, on the capacity of the liver cytosol to control lipid peroxidation in biological membranes. Their studies suggest that a cytosolic factor which may bind α-tocopherol (and possibly other free radical scavenging agents) may act as a carrier for these compounds into the microsomal membrane and provide protection from lipid peroxidation until the cytosolic factor is depleted of the inhibitors of lipid peroxidation. It would be of interest to know the concentration of α-tocopherol in the partially purified diet of Talcott et al. (30) as compared to that of the commercial rations employed by their laboratory.

We have noted that the dialyzed cytosolic factor is stable for about 48 hours at 4° after which its activity begins to diminish. Studies to determine how to maintain the activity of the factor for longer periods have not been done but studies on the specificity of the requirement for glutathione in this system (which are described later in this article) suggest that certain oxidizable moieties are essential for the activity of the cytosolic factor. As already mentioned, the cytosolic factor is quite labile to temperatures above 80°.

III. THE MITOCHONDRIAL HEAT-LABILE INHIBITOR OF LIPID PEROXIDATION

In the course of our studies on the inhibitor of lipid peroxidation in the cytosolic fraction of liver in which mitochondria were employed as the membranous particle subjected to peroxidation, it became clear that the liver mitochondria themselves contained a heat-labile, glutathione-dependent factor which inhibited lipid peroxidation (20). We observed that mitochondria incubated with glutathione without added cytosol were protected against peroxidative   attack catalyzed by ascorbate in the presence of ADP-Fe^{3+}. When the mitochondria were heated above 80°, however, the capacity of glutathione alone to prevent lipid peroxidation in that organelle was totally lost (20). At present there is no additional information concerning the nature of this inhibitory factor. The following aspects are to be examined: a) whether or not it is the same factor as the cytosolic inhibitor; b) if it is a soluble factor, whether or not it is in the inter or inner mitochondrial compartment (or both), and c) if it is membrane bound, whether it is bound to the outer or inner mitochondrial membrane or both. The mitochondrial factor appears to be as effective in preventing lipid peroxidation in that organelle as the cytosolic protein is in preventing peroxidative attack on microsomal lipids.

IV. INFLUENCE OF THE CONCENTRATION OF GLUTATHIONE ON THE
     INHIBITION OF LIPID PEROXIDATION BY THE CYTOSOLIC FACTOR

In order to learn more about the cytosolic system which inhibits lipid peroxidation, we investigated the influence of glutathione concentration on the effectiveness of the inhibition of enzyme-catalyzed lipid peroxidation (Table II).

TABLE II. Effect of Glutathione Concentration on the Capacity
           of the Cytosolic Factor To Inhibit Lipid Peroxidation[a]

Peroxidation system	Glutathione concentration				
	10 mM	5 mM	1 mM	0.5 mM	0.1 mM
	% inhibition of lipid peroxidation				
A. Enzyme-catalyzed lipid peroxidation	82.5	77.0	79.0	80.1	76.6
B. Non-enzymic lipid peroxidation	92.4	85.8	73.7	24.3	0.1

[a]Incubation systems contained approximately 1.0 mg microsomal protein 4.0 mM ADP, 12 μM Fe^{3+} 0.4 ml dialyzed liver cytosol (equivalent to 100 mg fresh liver tissue) and glutathione as indicated above, all in 0.15 M potassium phosphate buffer, pH 7.5. To initiate reaction in the enzymic lipid peroxidation system, a regenerating source of NADPH was added. This consisted of 5.0 mM glucose 6-phosphate, 0.3 mM NADP + 0.5 Kornberg units of glucose-6-phosphate dehydrogenase. To initiate peroxidation in the non-enzymic system, 0.66 mM ascorbate was added to an incubation system containing 1.0 mg microsomal protein, 4.0 mM ADP, 12 μM Fe^{3+}, dialyzed liver cytosol (equivalent to 100 mg fresh liver tissue), and glutathione as indicated above, all in 0.15 M phosphate buffer, pH 7.5. In the non-enzymic system, the microsomes were heated for 10 min at 100° and cooled before addition to the incubation system. Total volume of the systems was 1.0 ml. Incubations were carried out at 37° for 10 min. Percent inhibition of peroxidation was determined by dividing the amount of malondialdehyde formation in systems without glutathione by the amount formed in the complete system.

At the cytosol concentration employed (each ml of incubation
system contained the cytosol from 100 mg liver), glutathione
concentrations down to 0.5 mM were still sufficient to bring
about an 80% inhibition of enzyme-catalyzed lipid peroxidation.
It should be noted, however, that the enzymic system contained
NADPH (required for the enzyme-catalyzed peroxidation of
microsomal lipid). Hence, any glutathione (GSH) which was
oxidized to G-S-S-G would immediately be reduced back to GSH.
In a non-enzymatic lipid peroxidation system (Table II), it
was observed that the effectiveness of the inhibition by the
cytosolic system is a function of the glutathione concentration. A concentration of 0.5 mM produced only a modest inhibition (24.3%) in this system, while 10 mM resulted in 92.4%
inhibition (Table II). The range of glutathione concentrations for liver is 4-10 μmoles per gram of wet tissue, or
approximately 4-10 mM (31). G-S-S-G would not be reduced back
to GSH in this system.

V. SPECIFICITY OF THE REQUIREMENT FOR GLUTATHIONE FOR THE
    INHIBITION OF LIPID PEROXIDATION BY THE CYTOSOLIC FACTOR

This aspect of the system was investigated in an effort
to gain insights into the mechanism of the inhibition of

TABLE III. Effect of Various Cofactors on Inhibition of
            Enzyme-Catalyzed Peroxidation by the Non-Dialyzable
            Liver Cytosolic Factor[a]

Cofactor	Final concentration of added cofactor		
	10 mM	1 mM	0.5 mM
	% inhibition of lipid peroxidation		
Oxidized glutathione	84.5	84.3	81.8
β-mercaptoethanol	75.4	47.5	37.6
Dithiothreitol	89.1	85.5	86.5
Cysteine	90.2	79.0	56.5
Cystine	64.4	70.0	78.3
Methionine	5.8	5.1	11.3
$Na_2S_2O_4$	84.8	6.0	3.4
$Na_2S_2O_5$	85.6	69.8	41.9

[a]Incubation conditions were the same as in Table II for enzyme-
catalyzed lipid peroxidation except that one of the cofactors listed
above was substituted for GSH at three different concentrations. Percent inhibition of lipid peroxidation was determined as in Table II.

lipid peroxidation by the cytosolic factor. Compounds containing -SH groups were tested first. Table III shows that
β-mercaptoethanol, dithiothreitol and cysteine each could substitute for glutathione in the inhibition of enzyme-catalyzed
lipid peroxidation by the cytosolic factor, but only dithiothreitol was as effective as glutathione at the lowest concentration tested (0.5 mM). β-Mercaptoethanol, which has the
equivalent -SH content as glutathione, provided less than
half the activity of glutathione at 0.5 mM (Table III).
Oxidized glutathione (G-S-S-G), which is rapidly reduced to
GSH in the enzyme-catalyzed lipid peroxidation system, was as
effective as glutathione at all concentrations tested.

In a similar study in which a non-enzymic lipid peroxidation system was employed, β-mercaptoethanol, dithiothreitol,
and cysteine were effective at all concentrations tested, but
glutathione was relatively ineffective at the lowest concentration (0.5 mM) (Table IV). Also, oxidized glutathione,
which is not reduced to GSH in this peroxidation system, was
completely ineffective in the inhibition assay.

Because all of the thiol compounds tested were able to
substitute for glutathione in the inhibition of both enzymic
and non-enzymic lipid peroxidation, it appeared that the -SH
moiety was the essential feature required for the cytosolic
factor to exert its effect. However, the observation that
cystine (cys-S-S-cys) was equally as effective as glutathione
in the enzymic peroxidation system (Table III) and more
effective than glutathione in the non-enzymic system at 0.5 mM
(Table IV) was not consistent with that hypothesis. The fact
that oxidized glutathione was not effective in the non-enzymic
lipid peroxidation system (Table IV) (in which it was not
reduced to glutathione), suggests that cystine may be reduced
to cysteine in both the enzymic and non-enzymic lipid peroxidation systems. This has not been investigated at this time.

Methionine was not able to substitute for glutathione in
the inhibition of enzyme-catalyzed lipid peroxidation (Table
III), suggesting that the thiol group was necessary for the
inhibition. But additional studies in which either $Na_2S_2O_4$
or $Na_2S_2O_5$ was substituted for glutathione demonstrated that
both compounds, particularly $Na_2S_2O_5$, were able to substitute
for glutathione (Tables III and IV). Dithionite ($Na_2S_2O_4$) was
consistently effective only at a concentration of 10 mM.
Neither sodium selenite nor sodium selenate was effective
substitutes for glutathione.

All compounds which could substitute for glutathione in
the inhibition of peroxidation by the cytosolic factor (with
the exception of oxidized glutathione as explained above) were
effective in the inhibition of lipid peroxidation catalyzed by
both enzymic and non-enzymic processes, suggesting that an
activity common to both types of peroxidation is being

TABLE IV. Effect of Various Cofactors on Inhibition of Non-Enzymic
Lipid Peroxidation by the Cytosolic Factor[a]

Cofactor	Final concentration of added cofactor		
	10 mM	1 mM	0.5 mM
	% inhibition of lipid peroxidation		
Oxidized glutathione	5.3	3.9	2.3
β-mercaptoethanol	72.4	52.0	11.0
Dithiothreitol	95.0	94.6	85.3
Cysteine	93.6	90.8	88.9
Cystine	87.7	83.4	75.2
$Na_2S_2O_4$	91.9	6.6	12.4
$Na_2S_2O_5$	92.0	88.9	81.4

[a]Incubation conditions were the same as in Table II for the non-
enzyme lipid peroxidation except that one of the cofactors listed above
was substituted for GSH at three different concentrations. Percent
inhibition of lipid peroxidation was determined as in Table II.

influenced. Further consideration of the mechanism of the
inhibition is undertaken in the concluding section.

VI. EFFECT OF TREATING THE CYTOSOL AND MICROSOMES WITH
     IODOACETIC ACID ON THE CAPACITY OF THE CYTOSOLIC SYSTEM
     TO INHIBIT LIPID PEROXIDATION

     Dialyzed cytosol was treated with iodoacetic acid as des-
cribed in Table V and then assayed for its capacity to inhibit
either enzymic or non-enzymic lipid peroxidation. Table V
shows that when iodoacetic acid-treated cytosol was assayed
for its capacity to inhibit lipid peroxidation, decreased
effectiveness was observed, particularly at the lower concentration of glutathione (1.0 mM). In the non-enzymic lipid
peroxidation system the capacity of the iodoacetic acid-
treated cytosol to inhibit lipid peroxidation was almost
completely abolished at all concentrations of glutathione
tested.
     Treating the microsomes with iodoacetic acid before
addition to the reaction system showed that the capacity of
the microsomes to peroxidize as well as the ability of the cytosolic factor (untreated) to inhibit lipid peroxidation was
unaffected by prior treatment of the microsomes with the sulf-
hydryl reagent (Table V). The results suggest that thiol
groups in the cytosolic factor are required for the inhibition
of lipid peroxidation in the presence of glutathione. Thiol

TABLE V.   Effect of Iodoacetic Acid on Inhibition of Lipid Peroxidation
by the Non-Dialyzable, Heat-Labile Cytosolic Factor[a]

Peroxidation system	Glutathione concentration		
	10 mM	5 mM	1 mM
	% inhibition of lipid peroxidation		
A. Enzyme-catalyzed lipid peroxidation			
untreated cytosol	86.1	82.8	60.1
iodoacetate-treated cytosol	61.2	43.7	25.9
B. Non-enzymic lipid peroxidation			
untreated cytosol	88.4	86.9	82.9
iodoacetate-treated cytosol	16.5	8.7	0.8
C. Non-enzymic lipid peroxidation			
untreated microsomes	85.5	84.5	83.5
iodoacetate-treated microsomes	89.3	86.5	81.0

[a]Liver cytosol was treated with β-mercaptoethanol, 0.25 mM final concentration, for 15 min at 4°. This was then followed by incubation with
iodoacetic acid, 5.0 mM final concentration, for 2 hr at 4°. The cytosol
was dialyzed overnight against 0.15 M phosphate buffer, pH 7.5. The resulting preparation was used as the source of the non-dialyzable, heat-labile
factor in both enzyme-catalyzed and non-enzyme catalyzed lipid peroxidation
systems as describedin Table II. Heat-treated microsomes (see Table II)
were treated with β-mercaptoethanol and iodoacetic acid as described above,
and then washed twice in 0.15 M phosphate buffer, pH 7.5. The treated
microsomes were suspended in the 0.15 M phosphate buffer, pH 7.5, and added
to the non-enzyme catalyzed incubation system as described in Table II.
The percent inhibition of lipid peroxidation was also determined as
described in Table II.

groups in the microsome either are not involved or are easily
regenerated when the cytosol and glutathione are added to the
system.

At present, there is no explanation for the observation
that iodoacetic acid treatment of cytosol has a more profound
effect on the inhibition of lipid peroxidation in the non-
enzymic peroxidative system as compared to the enzyme-catalyzed
one. Possibly the former exerts a greater oxidative stress on
the microsomal membrane than does the latter, so that an impairment of the inhibitory properties of the cytosolic factor
would render it significantly less effective in the non-
enzymic system. It is also possible that the two types of
peroxidative systems involve sufficiently different mechanisms
that the effect of iodoacetic acid might exert the differential
influence observed. A third possibility is that the fresh
microsomes used in the enzyme-catalyzed peroxidation system
contain some of the heat-labile factor which inhibits lipid
peroxidation in the presence of glutathione. A number of
experiments indicate this may be the case because addition of
glutathione alone results in some inhibition of peroxidation.
In the non-enzymic systems, which employed heat-inactivated
microsomes, no such effect of glutathione alone is observed.

VII. INHIBITION BY THE CYTOSOLIC SYSTEM OF LIPID PEROXIDATION DURING THE METABOLISM OF CARBON TETRACHLORIDE (CCl_4)

In an attempt to elucidate the mechanism of action of the cytosolic factor in the inhibition of lipid peroxidation, the influence of the cytosolic system on the metabolism of CCl_4 by liver microsomes was undertaken. The rationale for this study was to determine if the cytosolic factor could inhibit lipid peroxidation associated with the metabolism of this halo-carbon. In addition, since an electron spin resonance tech-nique has been developed to detect the $^•CCl_3$ radical (32), we wished to determine the influence of the cytosolic system on the formation of the latter since this radical is a likely candidate for initiating at least part of the lipid peroxida-tion associated with the metabolism of CCl_4. Microsomal sys-tems for metabolizing CCl_4 were constituted as described in Table VI. Formation of malondialdehyde was assayed both in the presence and absence of the cytosolic system. The addi-tion of CCl_4 to a reaction mixture containing NADPH and mi-crosome stimulated the formation of malondialdehyde approxi-mately four-fold (Table VI). Addition of cytosol alone or glutathione alone to the CCl_4-metabolizing system had little effect on the formation of malondialdehyde. However, the ad-dition of both cytosol and glutathione to the system resulted in a marked inhibition of malondialdehyde formation. It was established earlier in this article that the inhibition of peroxidation was due to prevention of an oxidative attack on

TABLE VI. Malondialdehyde Formation During the Metabolism of Carbon Tetrachloride by Rat Liver Microsomes[a]

System	Malondialdehyde formation
	mM/mg Protein
Microsomes + NADPH	3.78
Microsomes + CCl_4 + NADPH	15.47
Microsomes + CCl_4 + Cytosol + NADPH	13.99
Microsomes + CCl_4 + GSH + NADPH	13.21
Microsomes + CCl_4 + Cytosol + GSH + NADPH	1.33

[a]Where indicated, the incubation system contained 1.5 mg microsomal protein/ml reaction system, an NADPH generating system (see Table II), 12 mg cytosolic protein/ml reaction system, 7.5 mM glutathione, 10 µl CCl_4/ml reaction system, in 0.05 M phosphate buffer, pH 7.4. The incu-bations were carried out at 37° for 30 min.

the polyunsaturated fatty acids of the microsomal membrane. In view of this nearly complete suppression of peroxidation, we investigated the effect of the cytosolic system on the formation of the $^\cdot CCl_3$ radical in the incubation system. Fig. 3 shows electron spin resonance spectra obtained from the same systems employed to assay malondialdehyde formation in Table VI excepting that the spin-trapping agent, phenyl-t-butyl-nitrone (PBN) was added to trap the $^\cdot CCl_3$ radicals in a stable radical adduct. The top spectrum (A) shows the ESR signal generated by the microsomal system in which CCl_4 is being metabolized. The major components of the signal is derived from $^\cdot CCl_3$ radical adducts (32). Spectrum (B) shows that the addition of cytosol had no effect on signal generation. Similarly, spectrum (C) shows that the addition of glutathione to the system has only a small but reproducible effect on the signal. Spectrum (D), however, shows that the addition of both cytosol and glutathione to the system nearly completely suppresses the signal. A problem arises in interpreting this result.

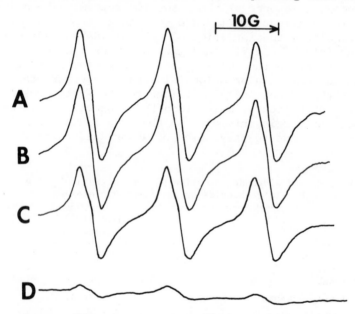

Fig. 3. Electron spin resonance spectra observed during the metabolism of CCl_4 by rat liver microsomes in the presence of a spin-trapping agent (μM phenyl-t-butyl nitrone (PBN)). Concentrations of components in the various systems and conditions of incubations are as in Table VI. (A) Microsomes + NADPH + PBN + CCl_4. (B) Microsomes + NADPH + PBN + CCl_4 + cytosol. (C) Microsomes + NADPH + PBN + CCl_4 + glutathione. (D) Microsomes + NADPH + PBN + CCl_4 + cytosol + glutathione.

There are several possibilities. One is that the cytosolic system may inhibit the metabolism of CCl_4 to $^\cdot CCl_3$ radicals. Another possibility is that the cytosolic system may catalyze the reduction of $^\cdot CCl_3$ radicals to chloroform ($CHCl_3$). Also, the PBN radical adduct of $^\cdot CCl_3$ may be modified by the cytosolic system to quench the radical. These possibilities can be sorted out by appropriate experiments. At this time, however, it is not known which of these alternatives, if any, is the case.

VIII. CONCLUDING REMARKS

From the various experiments which have been described, it appears that the inhibition of lipid peroxidation by the cytosolic system may involve some type of radical scavenging properties. Admittedly, this possibility will be difficult to distinguish from an inhibition of radical formation, given the present lack of knowledge as to what factor (or factors) is (are) involved in the phenomenon. The effect of iodoacetic acid on the cytosolic factor points to a role for thiol groups in the inhibition of peroxidation. The common feature of the cofactor requirement is that all of the compounds contain a sulfur atom. But this may only be an apparent correlation since the studies concentrated on sulfur-containing substances. A wider range of compounds may reveal if sulfur per se is important, or if it is some other property, such as the oxidation-reduction potential of the cofactor.

It is certain that the inhibition of lipid peroxidation by the cytosolic factor is a real effect since the polyunsaturated fatty acid components of the membrane remain unaltered under conditions which otherwise would cause their rapid peroxidation.

It seems virtually certain that the inhibitory factor in cytosol is neither glutathione peroxidase nor one of the glutathione transferases. Hence, the task now remains to purify the factor (or factors) responsible for this effect for further investigation of the mechanism of the effect.

Because the capacity of this cytosolic system to inhibit both non-enzymic as well as enzyme-catalyzed lipid peroxidation appears to be considerable, and because it has been found to be present in all tissues tested so far, it may represent one of the most potent protection mechanisms against oxidative stress in animal tissues.

ACKNOWLEDGMENTS

The authors wish to acknowledge the interest and advice
of Dr. K. Roger Hornbrook in these investigations, and the
assistance in experimental studies of Dr. Kuo-Lan Fong,
Edward Lai, and Larry Olson. We also wish to thank Wanda
Honeycutt for her help in the preparation of this manuscript.

REFERENCES

1. McCay, P.B.   J. Biol. Chem. 241, 2333 (1966).
2. Fong, K.-L., McCay, P.B., Poyer, J.L., Keele, B.B., and
    Misra, H.   J. Biol. Chem. 248, 7792 (1973).
3. Pederson, T.C., Buege, J.A., and Aust, S.D.   J. Biol.
    Chem. 248, 7134, (1973).
4. Pfeifer, P.M., and McCay, P.B.   J. Biol. Chem. 247, 6763,
    (1971).
5. Wendel, A., Feurenstein, S., and Konz, K.-H.   Biochem.
    Pharm. 28, 2051 (1979).
6. Comporti, M., Saccocci, C., and Dianzani, M.W.
    Enzymologia 29, 185 (1965).
7. Hatefi, Y., and Hanstein, W.G.   Arch. Bioch. Biophys.
    138, 73 (1970).
8. Watabe, T., and Akamatsu, K.   Biochem. Pharm. 23, 1979
    (1974).
9. Goldstein, B.D., Lodi, C., Collison, C., and
    Balchum, O.J.   Arch. Envir. Health 18, 631 (1969).
10. Myers, C.E., McGuire, W.P., Liss, R.H., Ifrim, I.,
    Grotzinger, K., and Young, R.C.   Science 197, 165 (1977).
11. Slater, T.F., in "Free Radical Mechanisms of Tissue
    Injury" (T.F. Slater, ed.), p. 91.   Pion Publ.,
    London (1972).
12. deToranzo, E.G.D., Diaz Gomez, M.I., and Castro, J.A.
    Res. Commun. Chem. Path. Pharm. 19, 347 (1978).
13. Weddle, C.C., Hornbrook, K.R., and McCay, P.B.   J. Biol.
    Chem. 251, 4973 (1976).
14. Sawyer, B.C., and Slater, T.F.   Biochem. Soc. Trans. 5,
    1029 (1977).
15. Reynolds, E.S.   J. Pharm. Exp. Therap. 155, 117 (1967).
16. Christophersen, B.O.   Biochim. Biophys. Acta 176,
    463 (1969).
17. Christophersen, B.O.   Biochim. Biophys. Acta 164, 35
    (1968).
18. Flohé, L., and Zimmermann, R.   Biochim. Biophys. Acta
    223, 210 (1970).
19. Tam, B.K., and McCay, P.B.   J. Biol. Chem. 245, 2295
    (1970).

20. McCay, P.B., Gibson, D.D., Fong, K.-L., and Hornbrook, K.R. Biochim. Biophys. Acta 431, 459 (1976).
21. Lawrence, R.A., and Burk, R.F. Biochem. Biophys. Res. Commun. 71, 952 (1976).
22. Prohaska, J.R., and Ganther, H.E. J. Neurochem. 27, 1379 (1976).
23. Burk, R.F., Trumble, M.J., and Lawrence, R.A. Biochim. Biophys. Acta 618, 35 (1980).
24. Simons, P.C., and Vander Jagt, D.L. Anal. Biochem. 82, 334 (1977).
25. Gibson, D.D., McCay, P.B., and Hornbrook, Biochim. Biophys. Acta, 620, 572 (1980).
26. Kotake, A.N., Deloria, L.B., Abbott, V.S., and Mannering, G.J. Biochem. Biophys. Res. Commun. 63, 209 (1975).
27. Levin, W., Lu, A.Y.H., Jacobson, M., Kuntzman, R., Poyer, J.L., and McCay, P.B. Arch. Biochem. 158, 842 (1973).
28. Kamataki, T., Ozawa, N., Kitada, M., and Kitagawa, H. Biochem. Pharm. 28, 2485 (1974).
29. Kamataki, T., Sugita, O., Kitada, M., Naminohira, S., and Kitagawa, H. Res. Commun. Chem. Path. Pharm. 17, 265 (1977).
30. Talcott, R.E., Denk, H., Eckerstorfer, R., and Schenkman, J.B. Chem.-Biol. Interact. 12, 355 (1976).
31. Jocelyn, P.C. Biochem. J. 85, 480 (1962).
32. Poyer, J.L., McCay, P.B., Lai, E.K., Janzen, E.G., and Davis, E.R. Biochem. Biophys. Res. Commun. 94, 1154 (1980).
33. Christophersen, B.O. Biochem. J. 106, 515 (1968).

PRODUCTION OF ETHANE AND PENTANE DURING LIPID PEROXIDATION: BIPHASIC EFFECT OF OXYGEN

Gerald Cohen

Mount Sinai School of Medicine
of the City University of New York (CUNY)
Fifth Avenue and 100 Street
New York, New York

INTRODUCTION

Lipid peroxidation (or oxidative rancification) is an event that takes place readily in vitro in isolated tissues or homogenates. Detection of the process in vivo is a problem that has plagued biologists. Lipid peroxidation is most frequently detected with the 2-thiobarbituric acid method, which measures malondialdehyde; the malondialdehyde arises either spontaneously from lipid peroxides in tissues, or during a heat-induced breakdown of lipid peroxides during the assay procedure. However, both malondialdehyde and lipid peroxides can be lost from tissues through rapid metabolism in vivo. Moreover, in vitro, the spontaneous formation of lipid peroxides in excised tissues has raised serious questions about the biologic significance of these chemical measurements. In many instances, it is unclear whether the observed malondialdehyde is present before tissues are removed for analyses, or whether loss of anti-oxidants during experimental manipulation makes tissues more prone to artifactual lipid peroxidation during subsequent handling.

Some years ago, we introduced a novel method for detecting the process of lipid peroxidation in vivo. It consists of measuring by gas chromatography the presence of short-chain hydrocarbon gases produced during the lipid peroxidation process. The method is non-invasive and the danger of artifactual formation of lipid peroxides during tissue handling is eliminated.

In the late 1960's, efforts by plant physiologists to

LIPID PEROXIDES IN BIOLOGY AND MEDICINE

pinpoint the metabolic source of ethylene, a plant hormone, led to one suggestion, among others, that the breakdown of lipid peroxides in plants might represent a viable pathway. A chance discussion with M. Lieberman, a plant physiologist, led us to test together an idea that lipid peroxides might give rise to ethylene in animal tissues. If this were to prove true, then a method for detecting lipid peroxidation in vivo would be in hand. We found ethane and other saturated hydrocarbon gases, but very little ethylene (1). In studies in vivo, mice exhaled ethane during lipid peroxidation induced in liver by carbon tetrachloride (2). In retrospect, it appears that the formation of pentane and other volatile products during the rancification of foodstuffs had been known for a quarter century or more; however, no attempt had been made to translate these observations into a method for studying lipid peroxidation in living mammalian systems. It might be noted in passing that the source of ethylene in plants has recently been elucidated; ethylene derives, not from lipid peroxides, but from S-adenosylmethionine (3).

In this article, the initial studies which proved the value of measuring exhaled ethane are reviewed and placed into perspective with more recent observations by several other groups of investigators. In addition, a new finding that the yields of ethane and pentane during lipid peroxidation are elevated under conditions of low oxygen tension is described.

STUDIES IN VITRO

Although the first experiments demonstrating ethane formation during lipid peroxidation were actually performed in vivo, it would appear more "logical" in this presentation to begin with in vitro studies. Figure 1 compares gases arising spontaneously from a representative animal tissue and a representative plant tissue. Experimental results are qualitatively the same when rat liver, brain, kidney, etc. are substituted for calf liver, and when tomato, apple or other plants or fruits are substituted for onion.

More detailed studies with animal tissues compared the accumulation of lipid peroxides with the accumulation of ethane in a closed system (2). Figure 2 illustrates a correlation between ethane and lipid peroxides in tissue homogenates. In brain homogenates (Panel A), production of both ethane and malondialdehyde proceeded in a relatively linear course from zero time. In liver homogenates (Panel B), a lag period was evident for both but, thereafter, production of ethane paralleled production of malondialdehyde. The concordance between ethane production and lipid peroxidation

indicates that these two processes are linked together.   In
Panel C, α-tocopherol (vitamin E) was added to a concentration
of 1.0 μg/ml either at zero time or after 120 minutes of incubation.   When α-tocopherol was added at zero time, lipid peroxidation was blocked and, as shown in Panel C, no ethane was
evolved.   When the addition of α-tocopherol was delayed for
two hours, at which time active production of both lipid peroxides and ethane were under way, there was little effect on
ethane production.   This result indicates that α-tocopherol
does not act at same step subsequent to lipid peroxidation to
block formation of ethane.   Ethane appears to be a natural byproduct of the lipid peroxidation process.

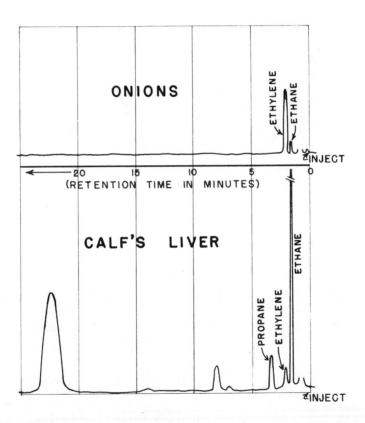

Figure 1.   Hydrocarbon gases arising spontaneously from representative animal and plant tissues.   The onion was diced and
incubated without a fluid medium at room temperature.   The
liver was diced and incubated in isotonic saline-phosphate,
pH 7.4, at 37°C.   Samples were incubated in sealed Erlenmeyer
flasks.

In other studies, ethane production was studied with intact slices of mouse liver. When carbon tetrachloride was
added, ethane production was stimulated (Fig. 3). Carbon
tetrachloride is known to provoke lipid peroxidation in liver
(4,5) via its metabolism to ·CCl₃ and ·Cl radicals by the
cytochrome P-450 system in the smooth endoplasmic reticulum.
Therefore, stimulation of ethane production to accompany the
lipid peroxidation process is expected. When mice were injected with α-tocopherol for several days, the slices were
resistant to spontaneous ethane formation and stimulation by
carbon tetrachloride was absent (Fig. 3). Treatment of the
mice _in_ _vivo_ permitted elevation of tissue levels of vitamin
E. In contrast, when α-tocopherol was added directly to the

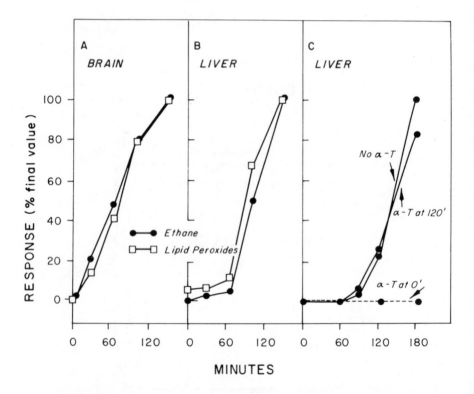

Figure 2. Correlation between ethane production and accumulation of lipid peroxides (2-thiobarbituric acid method) in
homogenates of brain and liver. The maximum lipid peroxide
values (nmol malondialdehyde per ml homogenate) were 23.8 and
14.6 in panels A and B, respectively. The maximum ethane
values (nmol per ml gas phase) were 0.14, 0.15 and 0.30 in
panels A, B and C, respectively. (From Ref. 2; copyright 1974
by the American Association for the Advancement of Science).

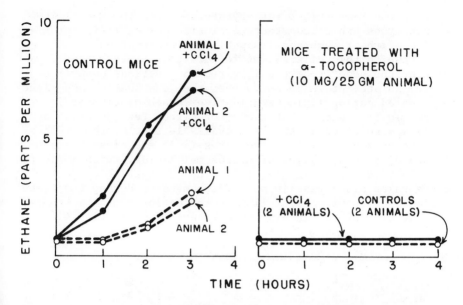

Figure 3. Ethane production from diced mouse liver (0.5 g)
incubated in isotonic saline-phosphate buffer, pH 7.4, at
37°C. Carbon tetrachloride (20 µl) was added as indicated.
Some mice were pretreated with α-tocopherol for two days.

slices rather than pretreating the mice in vivo, little or no
protective action was seen (not shown). This latter result
probably reflects the poor penetration of α-tocopherol into
the liver slices during the in vitro experiments.

STUDIES IN VIVO

It would be logical to seek evidence for hydrocarbon gas
formation during lipid peroxidation in simpler in vitro
studies before proceeding to in vivo experiments. Our first
attempts to detect hydrocarbon gas formation were conducted in
vitro with vitamin E-deficient erythrocytes from rats and from
human subjects. It was known that E-deficient erythrocytes
underwent lipid peroxidation readily when they were stressed
with hydrogen peroxide. The test system employed azide (to
inhibit catalase) and the addition of a small quantity of
hydrogen peroxide to initiate the lipid peroxidation of the
erythrocyte membrane (6). Human erythrocytes deficient in α-
tocopherol were obtained from subjects exhibiting acantho-
cytosis secondary to a-β-lipoproteinemia (7). However,

neither the E-deficient rat erythrocytes nor the E-deficient
acanthocytes exhibited ethane evolution during lipid peroxidation of their membranes (unpublished observation).

   Despite this distressing failure with erythrocytes (and
without the benefit of the in vitro observations described
above) an attempt was made to test the idea that ethane might
be exhaled during lipid peroxidative attack on tissues in
vivo. To this end, the toxicity of carbon tetrachloride on
liver was selected as the best available test system. Prior
studies by Recknagel and associates (4,5) had clearly established diene conjugation of microsomal lipids, which indicated
peroxidative attack in vivo (8). The conjugation of double
bonds takes place readily as a rearrangement during the free-
radical, chain process of lipid peroxidation.

   Results of the in vivo studies are shown in Fig. 4.

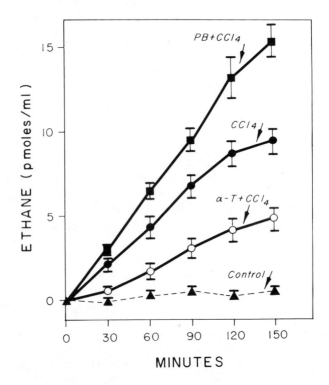

Figure 4. Ethane exhalation by mice after intraperitoneal
injection of carbon tetrachloride. Some mice had been pretreated with phenobarbital (PB, 80 mg/kg, three days) or with
α-tocopherol (α-T, 25 mg/kg, three times over two days).
(From Ref. 2; copyright by the American Association for the
Advancement of Science).

Treatment of mice with carbon tetrachloride resulted in ethane exhalation. No ethane was exhaled by control mice. In these studies, the mice were housed in a 2.4-liter desiccator and access to the atmosphere inside the chamber was provided by means of a 3-way stopcock (2).

Prior studies by others (9) had shown that α-tocopherol protects animals against the hepatotoxicity of carbon tetrachloride. In our experiments, pretreatment of mice with α-tocopherol resulted in the expected decline in the production of ethane in vivo (Fig. 4). Other investigators had shown that phenobarbital accentuates the damage produced by carbon tetrachloride (10); it is known that phenobarbital induces the liver enzymes that are responsible for the production of toxic free radicals from carbon tetrachloride. In our experiments, phenobarbital pretreatment resulted in the expected accentuation in the amounts of exhaled ethane (Fig. 4). These results indicate that exhalation of ethane can be correlated with lipid peroxidation in liver in vivo.

These studies prove the value of measuring exhaled ethane to evaluate lipid peroxidation in vivo. The detection and measurement of exhaled hydrocarbons provide the means to monitor on-going lipid peroxidation in remote organs and tissues in vivo. The trauma of biopsy is eliminated, and the danger of artifactual formation of lipid peroxides during tissue handling is avoided.

WORK DONE BY OTHERS

The past several years has seen impressive development and application of the technology of monitoring hydrocarbon gas evolution. The method has been validated and applied to studies of other cell toxins and other experimental conditions, such as vitamin E deficiency. Studies have included measurement of exhaled pentane, as well as ethane.

Special note should be made of the work of Lindstrom and Anders (11). These investigators used the carbon tetrachloride toxicity model to evaluate the correspondence between the exhalation of ethane and the presence of conjugated dienes in liver. Groups of rats were pretreated with various agents selected to either augment or protect against lipid peroxidation and, subsequently, the animals were challenged with carbon tetrachloride. The linear relationship observed on plotting exhaled ethane against conjugated dienes validates the measurement of exhaled ethane as an index of lipid peroxidation in vivo.

Tappel and his associates (12-14) have been vigorous in applying the new technology to a variety of experimental

situations. An important contribution has been the use of a
pre-column to bind exhaled gases over a timed interval (e.g.,
5 min); in this method, only gases emanating from the rat head
are collected, while the remainder of the body is excluded by
means of a specially constructed restraining harness. This
method appears important in studies of vitamin E deficiency to
avoid potential contributions from external, rancifying fats,
especially when diets are supplemented with oils that contain
unsaturated fatty acids which may adhere to the animal fur.
These investigators have tended to stress the value of measuring exhaled pentane in addition to or in place of ethane.

Hoekstra and associates (15-17) monitored ethane exhalation during combined vitamin E and selenium deficiency. Fasting induced a number of pathological changes (hematuria, lung
hemorrhage, liver necrosis), accompanied by ethane exhalation,
which increased exponentially prior to death. In studies of
carbon tetrachloride hepatotoxicity, the feeding of vitamin E,
selenium or methionine was protective in concordance with a
simultaneous decline in ethane exhalation. In other experiments, injection of 50 mg iron/kg (as iron-dextran) provoked
ethane exhalation.

It is apparent that diet can play an important role in
providing the precursors of ethane and pentane. Thus, feeding
cod liver oil, which is rich in linolenic acid, greatly enhances the yield of ethane during lipid peroxidation induced
in liver by injection of carbon tetrachloride (15). Feeding
corn oil, which is rich in linoleic acid, accentuates the
exhalation of pentane in vitamin E-deficient rat (12).

Other studies by Kappus and associates (18,19) have utilized the detection and measurement of exhaled ethane to evaluate the toxicity of halogenated hydrocarbons, as well as
ethanol, on the liver. Donovan and Menzel (20) have stressed
a role for ferrous ions in provoking pentane formation from a
model lipid hydroperoxide in vitro. Wendel et al. (21) found
an inverse correlation between liver glutathione levels and
the hepatotoxicity of paracetamol in starved mice; an important role for glutathione is the removal of hydrogen peroxide
and organic hydroperoxides through the action of glutathione
peroxidase. The latter authors suggest that ethane exhalation
may provide a convenient means to screen for hepatotoxic and
hepatoprotective agents.

The combined results of these and other studies have
been: (i) the independent confirmation of the value of measuring exhaled hydrocarbon gases during hepatotoxicity induced by
injected halocarbons, (ii) validation of the link to lipid
peroxidation by direct correlation between ethane and conjugated dienes and (iii) extension of the early results to more
sophisticated and more detailed studies of tissue lipid peroxidation in vivo under a variety of experimental conditions.

BIPHASIC EFFECT OF OXYGEN ON THE YIELD OF HYDROCARBON GASES

A strong correlation exists between conjugated dienes and exhaled ethane (11), as well as between yields of ethane and conditions that either enhance or prevent lipid peroxidation in vivo. Therefore, it might be expected that elevated oxygen concentrations, which spur lipid peroxidation, would simultaneously elevate ethane and pentane production. However, the opposite is true: Oxygen concentrations greater than that in air suppress, while nitrogen accentuates, hydrocarbon gas production. How can nitrogen accentuate hydrocarbon gas production when oxygen is required for lipid peroxide formation? The answer can be found in the following experiments.

The experimental results in Table I confirm what is expected for formation of lipid peroxides in vitro: In the absence of oxygen (nitrogen atmosphere), no lipid peroxides are formed, while in pure oxygen, an increased yield of lipid peroxides is seen compared to air. In the absence of oxygen, the initiation of the free radical chain reaction between unsaturated lipids does not occur and the production of organic peroxides is not possible.

Table II shows results of an experiment in which lipid peroxidation was first initiated in air and, then, aliquots of homogenate were removed and incubated separately under atmospheres of air, nitrogen or oxygen. In this experiment, formation of lipid peroxides and hydrocarbon gases were under way at the time the homogenate was split into aliquots. As shown

TABLE I. Lipid Peroxidation in Mouse Tissue Homogenates[a]

	Lipid peroxides (μM)		
	Nitrogen	Air	Oxygen
Brain	0.8	16.9	29.8
Liver	0.8	12.2	58.3

[a] Mouse brain or liver was homogenized in 24 volumes of isotonic saline. Aliquots of homogenate were sealed into flasks; the atmosphere in the flask was removed by vacuum and the flasks were flushed with either air, oxygen or nitrogen. The process was repeated three times. The flasks were then incubated at 37°C for 3 hours. Lipid peroxides were measured with the 2-thiobarbituric acid method.

in Table II, nitrogen accentuates the yield of hydrocarbon gas
product, while oxygen suppresses the yield. An exception may
be present in the very weak effect of oxygen on pentane production in liver. In other experiments (Table III), the unsaturated fatty acids, linolenic acid and linoleic acid, were
added to liver homogenates. It had been shown previously that
linolenic acid serves as a relatively selective precursor for
ethane, while linoleic acid serves as precursor for pentane
(13,15,20,22). The results in Table III show that nitrogen
enhances formation of ethane from linolenic acid and formation
of pentane from linoleic acid, while oxygen suppresses the
yield of ethane, but not pentane.

The apparent diversion between conditions leading to formation of lipid peroxides and those leading to formation of
ethane and pentane can be understood. In the absence of oxygen, no lipid peroxides or hydrocarbon gases can be formed.
However, once the free-radical chain process is under way, the
concentration of oxygen regulates the yield of hydrocarbon gas
from free radical precursors.

Other authors have presented reaction schemes for the

TABLE II. Spontaneous Production of Ethane and Pentane in
Mouse Tissue Homogenates[a]

Atmosphere	Brain	Liver
	Ethane (nmol/g/hr)	
Nitrogen	3.7 ± 0.4	0.8 ± 0.1
Air	1.4 ± 0.3	0.5 ± 0.1
Oxygen	0.7 ± 0.2	0.1 ± 0.0
	Pentane (nmol/g/hr)	
Nitrogen	4.6 ± 0.3	1.7 ± 0.3
Air	2.7 ± 0.4	1.0 ± 0.2
Oxygen	2.1 ± 0.2	0.9 ± 0.3

[a] Mouse brain or liver was homogenized in 9 volumes of
   isotonic saline (buffered with phosphate to pH 7.4).
   Tissue homogenates were first incubated in air for
   3 hours at 37°C. Subsequently, aliquots were removed
   and added to flasks which were flushed with either air,
   oxygen or nitrogen. The flasks were sealed with pliable
   rubber caps and they were incubated for an additional
   hour at 37°C. Samples of gas phase were removed by
   syringe for gas chromatographic assay of ethane and
   pentane.

production of ethane and pentane during lipid peroxidation
(12,15,20). It is assumed that simple alkyl radicals ($\cdot$RH,
ethyl radical or pentyl radical) are the immediate precursors
of ethane or pentane. It is also suggested that organic
hydroperoxides (lipid peroxides) serve to generate required
alkoxy radicals ($\cdot$OR) that fragment to ethyl or pentyl radicals. However, the intermediacy of a formal organic hydroperoxide need not be necessary and requires more study because
fragmentation of larger radicals to smaller radical species
during the lipid peroxidation process is a distinct possibility. Nonetheless, if an ethyl or pentyl radical were the immediate precursor of ethane or pentane, a role for oxygen could
be projected as shown in equations 1 and 2 below:

$$O_2 + \cdot RH \longrightarrow \cdot O_2RH \tag{1}$$

$$O_2 + \cdot RH \longrightarrow \cdot O_2H + R \tag{2}$$

In equation 1, oxygen adds to $\cdot$RH to form an alkyl peroxy
radical. This appears to be the likely explanation for suppression of ethane and pentane by oxygen. The organic hydroperoxy radical (e.g., ethylperoxy) would not yield ethane,
but would give rise instead to other products (e.g., ethyl

TABLE III. Ethane and Pentane Production from Linolenic Acid
and Linoleic Acid Added to Mouse Liver Homogenates[a]

Atmosphere	Linolenic Acid	Linoleic Acid
	Ethane (nmol/g/hr)	
Nitrogen	30.4 ± 7.9	
Air	12.4 ± 1.6	Less than
Oxygen	5.0 ± 0.8	1.2
	Pentane (nmol/g/hr)	
Nitrogen		11.4 ± 1.4
Air	Less than	3.9 ± 1.0
Oxygen	1.8	4.0 ± 1.1

a Aliquots of mouse liver homogenates (1:10 in isotonic
 saline-phosphate, pH 7.4) with added linolenic acid or
 linoleic acid (15 mM) were preincubated at 37°C for two
 hours. Subsequently, the tubes were flushed with air, oxygen or nitrogen, and resealed. Incubation was continued at
 37°C for an additional hour. Samples for gas phase were
 removed by syringe for gas chromatographic analysis of
 ethane and pentane.

hydroperoxide). In equation 2, oxygen abstracts a hydrogen
atom from ·RH to yield an oxidized, stable product (R: ethylene, pentene); the second product is the hydroperoxy radical
(which, at neutral pH, would ionize to the superoxide radical
anion). This reaction is unlikely as an explanation because
suppression of ethane production by elevated oxygen tension
in the experiments described above did not result in a corresponding increase in yield of ethylene.

A likely interpretation of the current data is that the
three classes of products, namely, hydrocarbon gases, lipid
peroxides and conjugated dienes, derive from a common precursor (or precursors). Thus, the formation of ethane and
pentane, and the formation of conjugated dienes, would be
expected to correlate with one another. However, the relative
concentration of oxygen in the microenvironment of the lipid
peroxidation process controls the overall yield of hydrocarbon
gas. This may or may not result in a distortion of the quantitative relationship between hydrocarbon gas production and
diene conjugation; this remains to be seen. It is of strong
interest that a recent report (23) has described a decrease in
the yield of exhaled ethane when rats are treated with carbon
tetrachloride in a pure oxygen environment. This observation
tends to indicate that the in vitro observations described
above can have a parallel with events that take place in vivo.
Conditions of relative tissue anoxia, either naturally occurring or experimentally controlled, may facilitate the detection of free radical tissue pathology (lipid peroxidation) in
vivo.

REFERENCES

1. Cohen, G., Riely, C. A., and Lieberman, M. Fed. Proc. 27,
    648 (Abstr. #2436)(1968).
2. Riely, C. A., Cohen, G., and Lieberman, M. Science 183,
    208 (1974).
3. Lieberman, M. Ann. Rev. Plant Physiol. 30, 533 (1979).
4. Recknagel, R. O., and Ghoshal, A. K. Lab. Invest. 15,
    132 (1966).
5. Hashimoto, S., Glende, Jr., E. A., and Recknagel, R. New
    Engl. J. Med. 279, 1082 (1968).
6. Cohen, G. Progr. Clin. Biol. Res. 1, 685 (1975).
7. Dodge, J. T., Cohen, G., Kayden, H. J., and Phillips, G.
    B. J. Clin. Invest. 46, 357 (1967).
8. Comporti, M., Benedetti, A., and Casini, A. Biochem.
    Pharmacol. 23, 421 (1974).
9. Gallagher, C. H. Nature (London) 192, 881 (1961).
10. Garner, R. C., and McLean, A. E. M. Biochem. Pharmacol.

18, 645 (1969).
11. Lindstrom, T. D., and Anders, M. W. Biochem. Pharmacol. 27, 563 (1978).
12. Dillard, C. J., Dumelin, E. E., and Tappel, A. L. Lipids 12, 109 (1977).
13. Dumelin, E. E., and Tappel, A. L. Lipids 12, 775 (1977).
14. Sagai, M., and Tappel, A. L. Toxicol. Appl. Pharmacol. 49, 283 (1979).
15. Hafeman, D. G., and Hoekstra, W. G. J. Nutr. 107, 656 (1977).
16. Hafeman, D. G., and Hoekstra, W. G. J. Nutr. 107, 666 (1977).
17. Dougherty, J. J., and Hoekstra, W. G. Fed. Proc. 36, 1151 (Abstr. #4652)(1977).
18. Koster, U., Albrecht, D., and Kappus, H. Toxicol. Appl. Pharmacol. 41, 639 (1977).
19. Koster-Albrecht, D., Koster, U., and Kappus, H. Toxicol. Lett. 3, 363 (1979).
20. Donovan, D. H., and Menzel, D. B. Experientia 34, 775 (1978).
21. Wendel, A., Feuerstein, S., and Konz, K.-H. Biochem. Pharmacol. 28, 2051 (1979).
22. Cohen, G. in "Oxygen Free Radicals and Tissue Damage", (D. FitzSimons, ed.), Ciba Foundation Symp. No. 65, (New Series), p. 177. Excerpta Medica, Amsterdam (1979).
23. Kieczka, H., and Kappus, H. Toxicol. Lett. 5, 191 (1980).

MEASUREMENT OF <u>IN VIVO</u> LIPID PEROXIDATION
VIA EXHALED PENTANE AND PROTECTION BY VITAMIN E[1]

Al Tappel

Department of Food Science and Technology
University of California
Davis, California

I. VOLATILE HYDROCARBONS FROM LIPID PEROXIDATION: MECHANISM
    OF FORMATION AND METHODS OF MEASUREMENT

Firstly, it is important to consider the broad-scale
scientific needs for techniques to measure <u>in vivo</u> lipid peroxidation. The individual and concerted influences of nutri-
ents, oxidants, and antioxidants on lipid peroxidation <u>in
vivo</u> and <u>in vitro</u> have been studied for many years. A recent
review on the importance of these interacting compounds in
lipid peroxidation (1) and reviews on enzymatically catalyzed
reactions in the endoplasmic reticulum (2,3) implicate
involvement of lipid peroxidation in some normal and pathological reactions <u>in vivo</u>. Researchers recognize that these
reactions in vivo are very complex. Results obtained with
peroxidizing systems <u>in vitro</u> cannot accurately describe all
reactions that occur during lipid peroxidation <u>in vivo</u>. The
interactions in lipid peroxidation of the various components
need continued investigation since lipid peroxidation is a
basic deteriorative reaction that is involved in many disease
processes and chemical toxicities.

[1]*Supported by grant AM 09933 from the National Institute
of Arthritis, Metabolism and Digestive Diseases and grant
ES 00628-05A1 from the National Institute of Environmental
Health Sciences and the Environmental Protection Agency.*

Lipid peroxidation in biological tissues is usually investigated by measurement of the major peroxidation products,
lipid hydroperoxides and conjugated dienes, and of the minor
products, malonaldehyde, hexanal, fluorescent carbonyl-amine
products, and volatile hydrocarbons. Conjugated dienes are
formed in direct proportion to the formation of lipid hydroperoxides. Lipid hydroperoxides may be metabolized by
selenium-glutathione peroxidase, and in other reactions
alkoxy radicals may undergo β-scission to produce small
amounts of volatile hydrocarbons. The amounts of hydroperoxides formed in vivo are low, but the great sensitivity of
gas chromatography makes it possible to measure the trace
amounts of volatile hydrocarbons formed when hydroperoxides
decompose. The decomposition of ω3- and ω6-unsaturated fatty
acid hydroperoxides leads to the formation of ethane and
pentane, respectively (4,5). Pentane was estimated in an
in vivo system to be formed at a level of about 0.002 mol per
mol of hydroperoxide formed after rats were injected with a
potent halogenated hydrocarbon (6). The following reactions
indicate the mechanism for pentane formation from an ω6-
unsaturated fatty acid:

$$CH_3(CH_2)_4\underset{\overset{|}{OOH}}{C}-R + Fe^{2+} \longrightarrow CH_3(CH_2)_4\underset{\overset{|}{O\cdot}}{C}-R + Fe^{3+} + OH^- \quad [1]$$

$$CH_3(CH_2)_4\underset{\overset{|}{O\cdot}}{C}-R \xrightarrow[\text{β-SCISSION}]{} CH_3(CH_2)_3\overset{\bullet}{C}H_2 + \underset{\overset{\|}{O}}{C}R \quad [2]$$

$$CH_3(CH_2)_3\overset{\bullet}{C}H_2 \xrightarrow[\text{H ABSTRACTION}]{} CH_3(CH_2)_3CH_3 \quad [3]$$

As shown in [1], hydroperoxides formed during lipid peroxidation can be decomposed by lower oxidation state metals.
Since the in vivo concentrations of iron, hematin, and copper
are high, it is assumed that they catalyze β-scission of
alkoxy radicals. As shown in equation [3], the five-carbon
alkoxy free radical abstracts a hydrogen to form pentane.

A gas chromatographic method was developed in this
laboratory to measure volatile hydrocarbon products of in
vivo lipid peroxidation (7,8). An individual rat is placed
in a chamber with the head separated from the body by rubber
and Teflon gaskets. Hydrocarbon-free air is supplied to the
rat at a known flow rate. Expired air is quantitatively
collected via a manifold onto a gas-sample loop partially
filled with alumina and immersed in a semi-hard liquid nitro-
gen-ethanol bath. After collection, the sample is injected
onto the column via a 6-way gas-sample valve. The picomoles
of volatile hydrocarbons are calculated from the recorded

peak areas, which are compared with the peak areas obtained
with standards. The following computation is applied: (pmol
hydrocarbon)(flowrate of air in ml/min)/(g rat body weight X
10^{-2})(sample volume in ml) = pmol hydrocarbon/100 g body
weight/min.

Other research groups also use volatile hydrocarbons as a
measure of in vivo lipid peroxidation. These groups use
different methods for collection of samples to be measured by
gas chromatography. In one system, unrestrained animals are
placed in whole-body chambers. This system requires that
respired gases accumulate to concentrations sufficient for
measurement in a relatively small sample of chamber head
space. This type of system is limited in that early effects
of toxicants after their injection into animals cannot be
measured. Hafeman and Hoekstra (9) described a life-support
apparatus for doing studies of volatile hydrocarbons from
animals, and Köster et al. (10) described a system that
supplies ethane-free oxygen to animals held in desiccators.
Our experience has shown that an individual animal chamber
that exposes only the animal's head to hydrocarbon-free air
is the best type of chamber to use to follow the early time-
course of toxicant-initiated lipid peroxidation in vivo.

II.   EVIDENCE THAT VOLATILE HYDROCARBONS ARE PRODUCTS OF
      LIPID PEROXIDATION

A major line of evidence that the volatile hydrocarbons
ethane and pentane arise following peroxidation of unsaturated fats was provided by in vitro studies (4,5). Metal-
catalyzed decomposition of preformed ω3-unsaturated fatty
acid hydroperoxides in vivo yielded ethane as the major
volatile hydrocarbon and decomposition of ω6-unsaturated
fatty acid hydroperoxides yielded pentane as the major volatile hydrocarbon.

Another line of evidence that volatile hydrocarbons arise
during lipid peroxidation is shown by results of in vivo
studies on the effects of dietary antioxidants and lipid
sources (7,11,12). Pentane and/or ethane production by rats
is inversely proportional to the amount of vitamin E or
selenium in the diet and also is related to the amount and
type of dietary polyunsaturated fat fed to the animals.
Hafeman and Hoekstra (9,12) studied the effects of vitamin E
and selenium deficiency in the rat. Pathological signs in
the rats and records of their death were the gross deficiency
syndromes that correlated with ethane production as a measure
of in vivo lipid peroxidation (12).

In addition to the antioxidant protection provided to animals by vitamin E, protection is provided by an enzyme system that involves selenium-glutathione peroxidase. This enzyme reduces lipid hydroperoxides and hydrogen peroxide to hydroxy fatty acids and water, respectively. The metabolism of hydroperoxides in vivo was reviewed recently by Chance et al. (13). The influence of selenium on glutathione peroxidase has been investigated primarily by dietary studies. As reviewed (14), the active site of selenium-glutathione peroxidase is selenocysteine, which is located in the polypeptide backbone of the enzyme.

We have fed weanling rats a basal Torula yeast diet with varying amounts of vitamin E and selenium, and with either 10% stripped corn oil, stripped lard, or coconut oil (11). After 7 weeks, the rats fed a vitamin E- and selenium-deficient diet that contained corn oil produced twice as much pentane as did rats fed a vitamin E- and selenium-deficient diet that contained coconut oil or lard. Rats fed the doubly deficient diet that contained corn oil produced six times more pentane than did rats fed 40 I.U. of vitamin E/kg of the same basal diet. The level of plasma vitamin E was six times higher in the vitamin E-supplemented rats than in the nonsupplemented rats. Pentane production by rats fed 0.1 ppm selenium as sodium selenite was one-half that by rats fed neither vitamin E nor selenium.

Figure 1 summarizes some of the data on ethane and pentane production by rats in the two studies described above (11,12). There is general agreement between the results of these two studies. Both studies agree with the concept that in vivo lipid peroxidation is inhibited by vitamin E and selenium-glutathione peroxidase, that it is directly proportional to the polyunsaturated fatty acid content of the diet, and that it is inversely proportional to the amount of the dietary chain-breaking antioxidant, vitamin E. Hydroperoxides produced in vivo should be reduced by selenium-glutathione peroxidase, and the amount of hydroperoxides reduced should be proportional to the amount of the enzyme present.

A third line of evidence that volatile hydrocarbons arise during lipid peroxidation was shown by a study in which the amount of pentane produced by vitamin E-deficient rats was reversibly decreased when the nonbiological antioxidant p-diphenylenediamine was fed to the rats (15). In this same study, the level of pentane in the breath was inversely proportional to the log of the dietary vitamin E concentration.

A fourth line of evidence that volatile hydrocarbons are products of lipid peroxidation was shown by a study in vivo in which the amount of pentane exhaled by rats was directly proportional to the amount of conjugated dienes measured in the livers 30 min after the animals were injected with

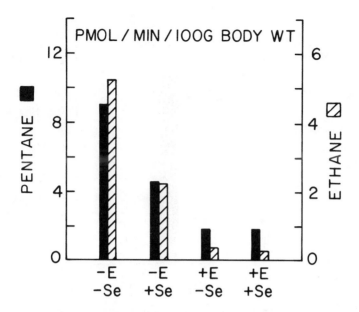

*FIGURE 1.   Comparisons of the results of Dillard et al.
(11) for the measurement of pentane and Hafeman and Hoekstra
(12) for the measurement of ethane.   The data show the
effects of dietary vitamin E (E) and selenium (Se) on the
production of volatile hydrocarbons.*

halogenated hydrocarbons (6). When rats were injected with
bromotrichloromethane, carbon tetrachloride, or chloroform,
the time-response relationships showed that the maximum
production of pentane occurred by 15-30 min following the
injections. When conjugated dienes were measured following
administration of carbon tetrachloride and bromotrichloromethane, 85 and 94%, respectively, of the total conjugated
dienes measured in the rat tissues was found in the liver.
The amount of pentane produced by the rats during the 30-min
period following injection of these two compounds corresponded to about 0.2% of the liver lipid peroxides measured as
conjugated dienes at 30 min. Liver is thus the most likely
principal site of pentane production in vivo.
      Köster et al. (10) did similar studies with rats injected
with carbon tetrachloride and bromotrichloromethane, except
that they measured ethane as an index of lipid peroxidation.
Figure 2 shows that the results of the experiments done by
Sagai and Tappel (6) and by Köster et al. (10) were similar,
even though sample collection methods and the time courses
were different. In both studies, a small dose of bromo-

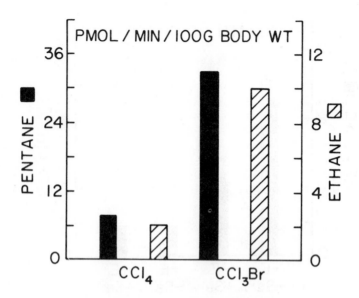

FIGURE 2.   Comparisons of the results of Sagai and Tappel (6) for the measurement of pentane and Köster et al. (10) for the measurement of ethane.   The data show the relative effects of CCl_4 and CCl_3Br.

trichloromethane caused rats to exhale one-half as much volatile hydrocarbons as did a ten-times larger dose of carbon tetrachloride. The data shown in Figure 2 are normalized to show the results for doses equivalent to 10 μg of halogenated hydrocarbon.

III. APPLICATIONS OF THE METHOD FOR ANALYSIS OF PENTANE

As indicated above, manipulation of dietary antioxidant levels and administration of strong chemical oxidants to animals are useful means to study lipid peroxidation in vivo. Recknagel et al. (16) and Bus and Gibson (17) recently reviewed the role of lipid peroxidation in the toxicity of halogenated hydrocarbons and a number of other xenobiotics. Table I summarizes the results of a number of studies that used pentane as an index of lipid peroxidation in vivo. The extent of peroxidation varied with the strength of the oxidizing agent used. The conditions used to initiate peroxidation included the extremes of injecting a very potent oxidizing agent, methyl ethyl ketone peroxide, and exposing

TABLE I. Lipid Peroxidation as a Function of Dietary Antioxidants or Treatment with Oxidants

Treatment	Antioxidant status of 10% corn oil diet	Pentane (pmol/100 g body weight/min)	Time
Methyl ethyl ketone peroxide	+ Vitamin E	50	20 Min
Methyl ethyl ketone peroxide	− Vitamin E	600	20 Min
Carbon tetrachloride	− Vitamin E	30	15 Min
Excess iron	− Vitamin E	70	30 Days
Vitamin E− and selenium-deficient	− Vitamin E, − Se	10	7 Weeks
Vitamin E-deficient	− Vitamin E	5	7 Weeks
1 ppm ozone exposure	− Vitamin E	12	1 Hour

rats to a transient and mildly oxidizing level of ozone.
Some of the studies summarized in the table are described
below. The results of studies on the effects of nutritional
deficiencies of antioxidants and on the effects of injection
of carbon tetrachloride are described above.

In this laboratory, R. Litov measured pentane production
by rats injected with methyl ethyl ketone peroxide at a dose
of 0.8 of the LD_{50}. The measurements were made 20 min
following the i.p. injection of 50 mg of methyl ethyl ketone
peroxide in mineral oil/kg of body weight. The level of
exhaled pentane averaged 50 and 600 pmol/100 g body weight/
min, respectively, for rats fed vitamin E-supplemented and
vitamin E-deficient diets for 6-9 weeks.

Iron was reported to increase lipid peroxidation in
vitamin E- and selenium-deficient rats, as measured by ethane
evolution (18). In another study, a large amount of iron,
4.6 g/kg of body weight, injected as iron dextran over a 28-
day time period elevated pentane in exhaled breath of rats
fed a standard laboratory diet to a level four-times higher
than that of noninjected rats (19). In this laboratory, J.
Downey is studying the effect on pentane production of injecting iron dextran into vitamin E-deficient rats. Injections of 2 g of iron/kg of body weight at the beginning of
a 30-day time period increased pentane production five-fold
as compared with that of the nontreated vitamin E-deficient
control rats or as compared with iron-injected rats fed 40 mg
of vitamin E/kg of diet.

The mechanism(s) by which ozone mediates lipid peroxidation is not known with certainty. It is known that ozone
reacts directly with carbon-carbon bonds in unsaturated fatty
acids (20) and that free radicals can be detected by electron
paramagnetic resonance following the reaction of ozone with
linoleic acid. According to Pryor (21), ozone acts as an
initiator of free radical autoxidation. In a study of the
effect of ozone on lipid peroxidation in rats, pentane and
ethane production was measured during exposure of rats to 1
ppm ozone for 1 hour (8). For 8 weeks the rats had been fed
a vitamin E-deficient diet to which had been added either 0,
11, or 40 I.U. of dl-alpha tocopherol acetate/kg. The effect
of dietary vitamin E on the production of pentane that was
directly related to exposure of the rats to ozone was shown
by the increment of pentane produced over the 1-hour exposure
time. In terms of pmol pentane/100 g body weight/hour, these
increments were 89, 12, and 0, respectively, for the rats fed
0, 11, and 40 I.U. of vitamin E/kg of diet. Analysis of the
data by the paired $\underline{t}$ test showed that pentane production was
significantly increased in only the rats fed the vitamin E-

deficient diet. The actual level of pentane production by vitamin E-deficient rats exposed to ozone, as shown in Table I, was 12 pmol/100 g body weight/min.

Another study (22), not summarized in Table I, was done to determine whether an effect of ozone on lipid peroxidation in humans could be detected by measurement of exhaled pentane. Human subjects exercised for 1 hour on a bicycle ergometer while breathing 0.3 ppm ozone. Analyses of expired breath were done at various time intervals as the subjects exercised while being exposed to ozone or hydrocarbon-scrubbed air. These analyses showed that there was no increase in pentane production as a result of exposure to 0.3 ppm ozone. Although ozone had no effect on the levels of expired pentane, exercise did increase pentane production. There was a significant decrease in the amount of pentane expired during exercise following daily supplementation of the subjects with 1200 I.U. dl-alpha tocopherol for 2 weeks. From this study it was concluded that lipid peroxidation occurs during vigorous exercise, that lipid peroxidation is modifiable by ingestion of vitamin E, and that dietary levels of vitamin E protect humans against lipid peroxidation that might otherwise be induced by exposure to low levels of ozone.

This review of the measurement of lipid peroxidation in vivo points out that there are many different conditions or compounds that serve as lipid peroxidation initiators and that vitamin E and selenium-glutathione peroxidase protect against lipid peroxidation. The newest technique to measure lipid peroxidation in vivo is the analysis of breath for volatile hydrocarbons that form when unsaturated lipid hydroperoxides decompose. The major volatile hydrocarbons measured in current research are ethane and pentane. Results reported by several laboratories throughout the world that use this technique have been similar when similar experimental conditions have been used. The technique should find wide application in future studies of lipid peroxidation.

REFERENCES

1. Tappel, A. L., *in* "Free Radicals in Biology" (W. A. Pryor, ed.), Vol. IV, p. 1. Academic Press, New York (1980).
2. McCay, P. B., and Poyer, J. L., *in* "Enzymes of Biological Membranes" (A. Martonosi, ed.), p. 239. Plenum, New York (1976).

3. Plaa, G. L., and Witschi, H., *Ann. Rev. Pharmacol. Toxicol. 16*, 125 (1976).

4. Donovan, D. H., and Menzel, D. B. *Experientia 34*, 775 (1978).

5. Dumelin, E. E., and Tappel, A. L. *Lipids 12*, 894 (1977).

6. Sagai, M., and Tappel, A. L. *Toxicol. Appl. Pharmacol. 49*, 283 (1979).

7. Dillard, C. J., Dumelin, E. E., and Tappel, A. L. *Lipids 12*, 109 (1977).

8. Dumelin, E. E., Dillard, C. J., and Tappel, A. L. *Arch. Environ. Health 33*, 129 (1978).

9. Hafeman, D. G., and Hoekstra, W. G. *J. Nutr. 107*, 656 (1977).

10. Köster, U., Albrecht, D., and Kappus, H. *Toxicol. Appl. Pharmacol. 41*, 639 (1977).

11. Dillard, C. J., Litov, R. E., and Tappel, A. L. *Lipids 13*, 396 (1978).

12. Hafeman, D. G., and Hoekstra, W. G. *J. Nutr. 107*, 666 (1977).

13. Chance, B., Sies, H., and Boveris, A. *Physiol. Rev. 59*, 527 (1979).

14. Tappel, A. L., *in* "Biochemical and Clinical Aspects of Oxygen" (W. S. Caughey, ed.), p. 679. Academic Press, New York (1979).

15. Downey, J. E., Irving, D. H., and Tappel, A. L. *Lipids 13*, 403 (1978).

16. Recknagel, R. O., Glende, Jr., E. A., and Hruszkewycz, A. M., *in* "Free Radicals in Biology" (W. A. Pryor, ed.), Vol. III, p. 97. Academic Press, New York (1977).

17. Bus, J. S., and Gibson, J. E. *Rev. Biochem. Toxicol. 1*, 125 (1979).

18. Dougherty, J. J., and Hoekstra, W. G. *Fed. Proc. 36*, 1151 (1977).

19. Dillard, C. J., and Tappel, A. L. *Lipids 14*, 989 (1979).

20. Menzel, D. B., *in* "Free Radicals in Biology" (W. A. Pryor, ed.), Vol. II, p. 181. Academic Press, New York (1976).

21. Pryor, W. A., *in* "Free Radicals in Biology" (W. A. Pryor, ed.), Vol. I, p. 1. Academic Press, New York (1976).

22. Dillard, C. J., Litov, R. E., Savin, W. M., Dumelin, E. E., and Tappel, A. L. *J. Appl. Physiol.: Respirat. Environ. Exercise Physiol. 45*, 927 (1978).

ASSAY FOR SERUM LIPID PEROXIDE LEVEL
AND ITS CLINICAL SIGNIFICANCE

Kunio Yagi

Institute of Biochemistry
Faculty of Medicine
University of Nagoya
Nagoya

I. INTRODUCTION

In 1952, Glavind et al. (1) reported that lipid peroxides
occurred in the plaques of human atheroma and that the degree
of atheroma correlated with the extent of lipid peroxidation
in the plaques. Since the plaques always contact with the
blood, some correlation between them could be suspected. That
was the main reason why we intended to measure the level of
lipid peroxides in the blood. Because of the lack of reports
on the determination of blood lipid peroxides at that time, we
intended to devise a reliable method. We thought that the a-
mount of lipid peroxides might be small and that the procedure
must be simple to avoid artifact due to the peroxidation during the assay procedure, and we adopted thiobarbituric acid
(TBA) reaction. TBA reaction was known to be simple and sensitive (2,3), but also known to have low specificity. Accordingly, lipid peroxides in the blood should be isolated from
other TBA-reacting substances by a simple procedure. After a
systematic investigation, we devised a method to determine
lipid peroxides in blood serum (4). The principle of this
method is to isolate lipids by precipitating them with serum
protein using phosphotungstic acid–sulfuric acid system and
to determine their amount by TBA reaction in an acetic acid
solution. By this procedure, TBA-reacting substances other
than lipid peroxides are easily eliminated and the reaction
product assayed by the absorption at 532 nm. This method was
rather widely used, especially in Japan. However, it required at least 1 ml of the blood to obtain reliable data. This

prevents its application to a small amount of the blood and, therefore, to infants or small animals. Because of increasing interests in deleterious effects of lipid peroxides in our body, the necessity of devising a micromethod for the determination of lipid peroxide level of the serum was increased. Since the author found that the product of TBA reaction is fluorescent, this property was utilized and a simple fluorometric method for the determination of lipid peroxide level of blood plasma or serum was devised (5). Using this method, numerous clinical reports appeared especially in Japan (see Review by Goto (6) for some of them). This paper deals with the detailed description of the method and some data indicating the significance of serum lipid peroxide level in clinical fields.

II. ASSAY FOR SERUM LIPID PEROXIDE LEVEL BY TBA REACTION

1. Basic data for assay procedure

As was demonstrated by Bernheim et al. (2), TBA reacts with lipid peroxides, and the product of the reaction was found to have the structure shown in Fig. 1 (7). Malondialdehyde gives the same product upon reaction with TBA.

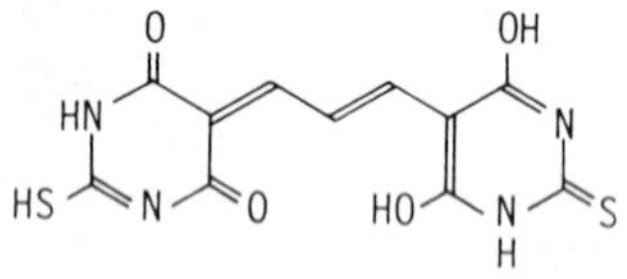

Figure 1. Structure of the product from the reaction of lipid peroxides with TBA.

a. Absorption and fluorescence spectra of the reaction product

The absorption spectrum of this product is shown in Fig. 2. This red pigment was found to fluoresce (5). The excitation and emission spectra are shown in Fig. 3. When tetraethoxypropane or tetramethoxypropane, which converts quantitatively to malondialdehyde during the reaction procedure, was reacted with TBA, the fluorescence intensity of the reaction product ran parallel with its concentration. These results showed that TBA reaction can be followed by fluorometry.

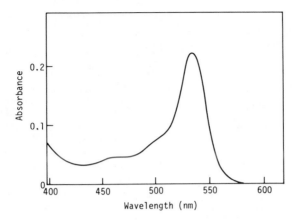

Figure 2. Absorption spectrum of the product from the reaction of linoleic acid hydroperoxide with TBA.

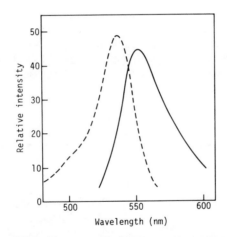

Figure 3. Excitation and emission spectra of the product from the reaction of malondialdehyde with TBA. Dotted line: excitation; solid line: emission.

b. Reaction pH and temperature

Since it is known that the reaction of lipid hydroperoxides with TBA required heating at nearly 100°C, the effect of pH on the reaction was observed by heating the reaction mixture at a different pH at 95°C for 60 min. As shown in Fig. 4, the maximum formation of the product was found at around pH 3.5 for the reaction of hydroperoxide of linoleic, linolenic or arachidonic acid with TBA (8).

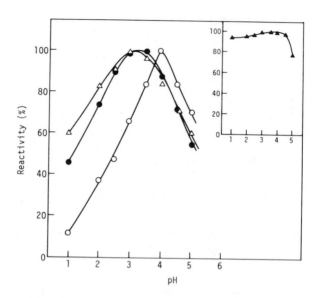

Figure 4. pH profile of the reaction of lipid peroxides with
TBA. Open circle: linoleic acid hydroperoxide; closed circle:
linolenic acid hydroperoxide; open triangle: arachidonic acid
hydroperoxide; closed triangle: tetramethoxypropane.

In an aqueous medium, the reaction rate became higher
upon elevating temperature. Accordingly, 95°C was adopted for
the reaction. Time course of the reaction of lipid hydroperoxides with TBA was obtained as shown in Fig. 5. It seems
that 60 min is required for the maximum degree of reaction.

c. Elimination of substances that react with TBA other than
lipid peroxides

Water-soluble substances, which react with TBA to yield
the same product as that from lipid peroxides, must be removed
by a simple procedure prior to the reaction. After many examinations, we reached a conclusion that the separation can
be attained by precipitating lipid hydroperoxides along with
serum protein with phosphotungstic acid-sulfuric acid system.
Another substance that reacts with TBA is sialic acid.
It was found that sialic acid cannot react with TBA in acetic
acid solution, though it reacts strongly with TBA in trichloroacetic acid solution, as shown in Fig. 6. This indicates that trichloroacetic acid adopted by some researchers
(9,10) is not suitable for samples which contain sialic acid.
Bilirubin was found to react with TBA. Figure 7 shows
the absorption spectrum of the product formed from the

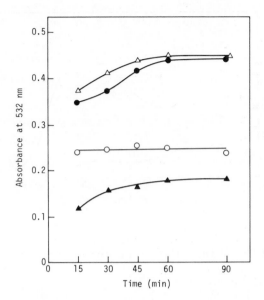

Figure 5.   Time course of the reaction of lipid hydroperoxides
with TBA.   Open circle: linoleic acid hydroperoxide; closed
circle: linolenic acid hydroperoxide; open triangle: arachidonic acid hydroperoxide; closed triangle: tetramethoxypropane.

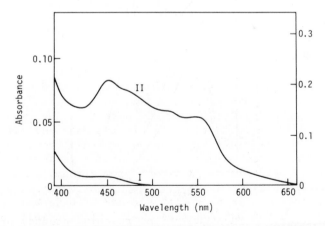

Figure 6.   Absorption spectrum of the product from the reaction of sialic acid with TBA.   I: reaction in acetic acid
solution (left scale); II: reaction in trichloroacetic acid
solution (right scale).   The concentration of sialic acid:
0.3 μg/reaction mixture.

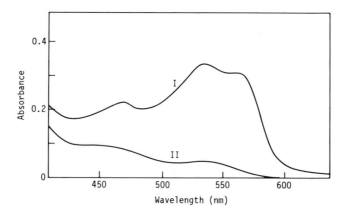

Figure 7. Absorption spectrum of the reaction product of
bilirubin with TBA.   I: bilirubin bound to albumin; II: albumin without bilirubin.   The reaction mixture contained 2 µg
bilirubin and 11 mg albumin.   Albumin contained a minute
amount of lipid peroxides.

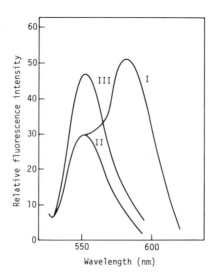

Figure 8. Emission spectrum of the reaction product of
bilirubin with TBA.   I: bilirubin bound to albumin; II: albumin without bilirubin; III: tetramethoxypropane.   Albumin
contained a minute amount of lipid peroxides.

TABLE I. Effect of Platelet Aggregation on Assay of Lipid Peroxide Level in Blood

	Lipid peroxide (nmol)	
	Total	Precipitates[a]
Serum[b]	0.061	0.042
Platelet[c]	0.053	–
Platelet + NEM	0.146	–
Serum + platelet	0.120	0.044
Serum + platelet + NEM	0.190	0.046

[a] Obtained with phosphotungstic acid – sulfuric acid.
[b] Serum 20 µl.
[c] Platelet suspension $(3.98 \times 10^6/ml)$ 80 µl.

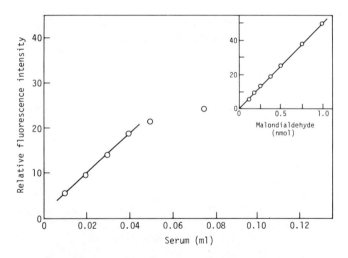

Figure 9. Correlation between the amount of serum or malondi-aldehyde and fluorescence intensity. Relative fluorescence intensity of the product from the TBA reaction was measured at 553 nm with excitation at 515 nm.

reaction of bilirubin with TBA. It was found, however, that the fluorescence of this product was different from the fluo-rescence of the reaction product of lipid peroxides with TBA, as shown in Fig. 8. Accordingly, bilirubin does not disturb the analysis of lipid peroxides by TBA reaction, if the fluorescence intensity of the product is determined at 553 nm.

It was anticipated that platelet aggregation, if it occurred during the drawing of the blood, would liberate the
TBA-reacting substances. To check this point, the effect of
the addition of platelets to the serum was measured. As shown
in Table I, the effect of their aggregation caused by N-ethyl-
maleimide (NEM) was found to be eliminated by the treatment
with phosphotungstic acid-sulfuric acid system.

d. Linearity between lipid peroxide level and fluorescence
intensity

On the basis of the above examinations, the linearity
between the amount of malondialdehyde or lipid peroxides and
the fluorescence intensity of the reaction product was checked. As shown in Fig. 9, linearity was found in their definite
concentration range.

2. Standard procedure

Taking into account the data mentioned above, the following procedure is recommended as standard method for the micro-
determination of lipid peroxides in blood plasma or serum.

1. Using a pipet for determination of blood cells, 0.05 ml
of the blood is taken (e.g. from the ear lobe).
2. The blood is put into 1.0 ml of physiological saline in
a centrifuge tube, and shaken gently.
3. After centrifugation at 3,000 rpm for 10 min, 0.5 ml of
the supernatant is transferred to another centrifuge tube. In
the case of the serum, 20 µl of the specimen are taken.
4. To this solution, 4.0 ml of N/12 H_2SO_4 are added and
mixed gently.
5. Then, 0.5 ml of 10% phosphotungstic acid is added and
mixed. After allowing to stand at room temperature for 5 min,
the mixture is centrifuged at 3,000 rpm for 10 min.
6. The supernatant is discarded, and the sediment is mixed
with 2.0 ml of N/12 H_2SO_4 and 0.3 ml of 10% phosphotungstic
acid. The mixture is centrifuged at 3,000 rpm for 10 min.
7. The sediment is suspended in 4.0 ml of distilled water,
and 1.0 ml of TBA reagent is added. TBA reagent is a mixture
of equal volumes of 0.67% TBA aqueous solution and glacial
acetic acid. The reaction mixture is heated for 60 min at
95°C in an oil bath.
8. After cooling with tap water, 5.0 ml of n-butanol are
added and the mixture is shaken vigorously.
9. After centrifugation at 3,000 rpm for 15 min, the n-
butanol layer is taken for fluorometric measurement at 553 nm

with 515 nm excitation.

10. Taking the fluorescence intensity of the standard solution, which is obtained by reacting 0.5 nmol of tetramethoxypropane with TBA by steps 7-9, as $\underline{F}$ and that of the sample as
$\underline{f}$, the lipid peroxide level (Lp) can be expressed in terms of
malondialdehyde:

$$\text{Plasma Lp} = 0.5 \times \frac{\underline{f}}{\underline{F}} \times \frac{1.05}{0.05} \times \frac{1.0}{0.5} = \frac{\underline{f}}{\underline{F}} \times 21 \ (\text{nmol/ml of blood})$$

$$\text{Serum Lp} = 0.5 \times \frac{\underline{f}}{\underline{F}} \times \frac{1.0}{0.02} = \frac{\underline{f}}{\underline{F}} \times 25 \ (\text{nmol/ml of serum})$$

III. CLINICAL SIGNIFICANCE

1. Lipid peroxide levels of normal subjects

Eighty-two healthy subjects were selected on the basis of
general physical examination, and they showed no abnormal laboratory findings. Serum lipid peroxide levels of these subjects are shown in Table II (11). As can be seen from the
table, the level at age 11-20 is significantly higher than
that at age -10, and that at age 31-40 is significantly higher
than that at age 21-30. It tends to increase with age and
tends to decrease at age more than 71.

2. Diabetes

Plasma lipid peroxide levels of the patients suffering
from diabetes mellitus are shown in Table III (12). As can be
seen from the table, lipid peroxide levels of the patients
with stable diabetes are significantly higher than those of
normal subjects. The patients with angiopathy were compared
with those without angiopathy. The diabetic patients with
angiopathy were those who had retinopathy, nephropathy, or
atherosclerosis such as coronary insufficiency and cerebrovascular accidents or combinations of them. It is obvious
that the levels of plasma lipid peroxides of the patients
with angiopathy are higher than those of the patients without
angiopathy. This fact suggests that the elevation of lipid
peroxide levels in the blood is a cause of angiopathy.

3. Burn injury

From the results of many investigators on "toxic factor"
after burn injury, we suspected that burn toxin might be

TABLE II. Serum Lipid Peroxide Levels of Normal Subjects

Age (years)	Lipid peroxide level			
	Male		Female	
−10	1.86 ± 0.60	(10)	2.04 ± 0.48	(7)
11–20	2.64 ± 0.60	(10)*	2.64 ± 0.54	(9)
21–30	3.14 ± 0.56	(10)	2.98 ± 0.50	(9)
31–40	3.76 ± 0.52	(11)**	3.06 ± 0.50	(9)***
41–50	3.94 ± 0.60	(11)	3.16 ± 0.54	(10)***
51–60	3.92 ± 0.92	(8)	3.30 ± 0.74	(10)
61–70	3.94 ± 0.70	(10)	3.46 ± 0.72	(10)
71–	3.76 ± 0.76	(12)	3.30 ± 0.78	(10)
Mean	3.42 ± 0.94	(82)	3.10 ± 0.62	(75)***

Lipid peroxide level is expressed in terms of malondialdehyde (nmol/ml of serum). Mean ± SD. Number is in parenthesis.
  *$p < 0.05$ when compared with the group age −10.
 **$p < 0.05$ when compared with the group age 21–30.
***$p < 0.05$ when compared with the corresponding group of
   males.

TABLE III. Plasma Lipid Peroxide Levels of Diabetic Patients and Normal Subjects.

	Number	Age	Lipid peroxide level
Normal subjects	331	40.8 ± 0.7	3.74 ± 0.13
Diabetic patients	110	49.9 ± 1.2	5.30 ± 0.71*
Without angiopathy	57	47.3 ± 1.7	3.82 ± 0.53
With angiopathy	47	52.6 ± 1.9	7.15 ± 0.71**

Lipid peroxide level is expressed in terms of malondialdehyde (nmol/ml of plasma). Mean ± SE.
  *$p < 0.001$ when compared with normal subjects.
 **$p < 0.001$ when compared with diabetic patients without angio-
pathy.

similar to lipid peroxides, and measured the levels of lipid
peroxides in animal tissues and serum of postburn period (13).
By pouring boiling water on the back of a rat, a burn injury
was made on the skin. The change in lipid peroxide level in
the skin measured by TBA method (14) after burn injury is
shown in Fig. 10. One hour after the injury, lipid peroxide
level in the burned skin had already increased significantly.
Three hours after the injury, it reached the level six times
higher than that of the control. Such a high level continued
for 1 day, then decreased and reached the level of the control
after 7 days. As also shown in Fig. 10, lipid peroxide level
in the serum was parallel with that of the burned skin. This
suggests that lipid peroxides formed in the burned skin are
released into the bloodstream, causing the increment of serum
lipid peroxide level. The lipid peroxide level in the spleen
significantly increased 7 days after burn injury (Fig. 10).
This increment after the abnormal increase of serum lipid peroxide level suggests that serum lipid peroxides induced the
elevation of lipid peroxide level in the organ. In the liver
and kidney the increase was statistically not significant.
    These results suggest that the increased lipid peroxides
in the serum would have some deleterious effects on intact
tissues and organs. To check this possibility, the enzymes

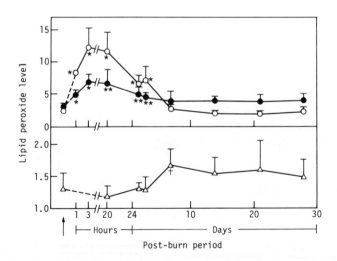

Figure 10. Changes in lipid peroxide levels in the skin,
serum and spleen. Lipid peroxide levels are expressed in
terms of malondialdehyde (nmol/ml for the serum and nmol/mg
protein for the skin and spleen). Mean ± SD (n = 6). Arrow
shows the control. *p < 0.001; **p < 0.005; †p < 0.02. Open
circle: skin; closed circle: serum; open triangle: spleen.

released from the organs into the serum were measured. Of
enzyme activities measured, those of glutamate oxaloacetate
transaminase, alkaline phosphatase, and lactate dehydrogenase
significantly increased after burn injury, but that of glutamate pyruvate transaminase did not increase. From these
results, it is considered that the formation of lipid peroxides in the burned skin followed by the release of lipid peroxides into bloodstream results in the damage to the membranes
of various organs, such as the spleen, liver, and kidney, permitting the leakage of the enzymes into the bloodstream. Accordingly, it must be emphasized that the increase in lipid
peroxide level in the blood is deleterious to intact organs or
tissues.

4. Retinopathy

     It is known that the retina is one of the tissues which
are susceptible to oxygen. Noell (15) reported that the exposure of adult rabbits to a high concentration of oxygen at
ambient pressure for a few days induced retinal degeneration
which is similar to that induced by X-ray irradiation.
Noell's work was based on the earlier results of Gerschman et
al. (16), who reported that irradiation and oxygen poisoning

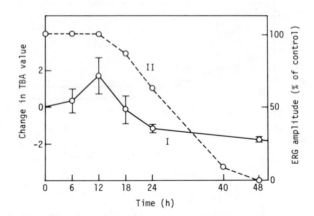

Figure 11. Changes in lipid peroxide level and ERG of the
retina of rabbit upon exposure to a high concentration of oxygen. Lipid peroxide levels measured by TBA method are represented by the difference between those with and without oxygen
exposure. Bars represent SD calculated from the difference
(n = 4). The amplitudes of ERG are expressed by percent to
those of the controls. Solid line: lipid peroxide level;
dotted line: ERG amplitude.

produce their lethal effects through a common mechanism, possibly through the formation of oxidizing free radicals. As pointed out by Haugaard (17), peroxidation may play a role in the toxic effects of high concentrations of oxygen. On the basis of these reports, we decided to see if lipid peroxides are produced in the retina of rabbit during oxygen-mediated degeneration of the retina (18).

The adult albino rabbits weighing 2–3 kg were exposed to 90–95% oxygen at ambient pressure for different periods, and control animals were exposed to air under the same conditions. As shown in Fig. 11, the exposure to oxygen for 12 h resulted in the increase in lipid peroxides in the retina as compared with the control (p < 0.02). Upon continuing the exposure to oxygen, the content of lipid peroxides began to decrease and reached a level lower than that of the control (at 24 h, p < 0.01; at 48 h, p < 0.001). After the exposure to oxygen for 12 h, the electro-retinogram (ERG) amplitude began to decrease and became non-recordable after the exposure for 48 h. It should be noted that ERG began to decrease in accordance with the increase in lipid peroxides in the retina.

To observe the histological change in the retinal tissue, a light microscopic observation was made. A pronounced degeneration in the visual cell layers was found in the case of oxygen exposure for 40 h.

These results suggest the possibility that the formation of lipid peroxides in the retina is induced by a high

TABLE IV. Changes in Lipid Peroxide Levels in Blood Plasma and Retina of Chick Embryo of the 14th-Day upon Exposure to a High Concentration of Oxygen for Different Intervals.

Exposure (h)	Lipid peroxide level	
	Plasma	Retina
3	100	100
6	230*	109
12	233*	138**
24	118	126
48	138	93

The lipid peroxide levels were measured with TBA method and are expressed as percent of the control.
*p < 0.02, **p < 0.05.

concentration of oxygen and the lipid peroxides formed dena-
ture the associated proteins, resulting in an inability of
retinal function as observed by ERG and the change in structure as observed by light microscope.

In relation to this retinopathy, the etiology of retinopathy of prematurity was examined, since this was accelerated
by the administration of a high concentration of oxygen to immature infants (19). To check this problem, a model experiment using chick embryo was made (20).

When chick embryos at various stages were exposed to a
high concentration of oxygen under 2 atm pressure for 6 h, the
lipid peroxide levels in the plasma and retina increased at
each stage. In the liver, the increase in lipid peroxides was
not significant except for a slight increase at the 9th-day
embryo. To check the effect under conditions similar to those
of immature infants, the changes in lipid peroxide levels in
the plasma and retina were followed with the 14th-day chick
embryo under milder conditions; 95% oxygen gas flow at ambient
pressure. As shown in Table IV, the lipid peroxide level in
the plasma did not change after exposure for 3 h, but it was
elevated after exposure for 6 h. The elevation was also found
at 12 h of exposure, but it was not observed at 24 h and 48 h
of exposure. It should be noted that the elevation in the

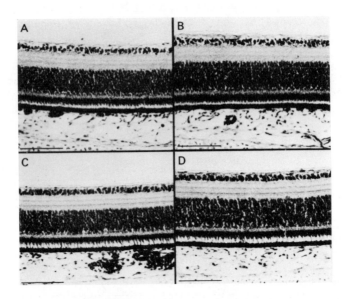

Figure 12. Microphotograph of the retina of chick embryo
(14th-day) exposed to a high concentration of oxygen. A: control for B; B: exposed to oxygen for 6 h; C: control for D;
D: exposed to oxygen for 48 h. Bars show 100 μm.

retina occurred after the elevation in the blood. No significant elevation in the liver was observed at any time of the
exposure.

Histological examination of the retinal tissue of chick
embryo exposed to oxygen under the milder conditions mentioned
above was made by light microscopy. Figures 12B and 12D show
the microphotographs of the retinas of chick embryo exposed to
a high concentration of oxygen for 6 h and 48 h, respectively.
By comparing Fig. 12D with the control (Fig. 12C), it is clear
that the blood vessels became narrow. A loose arrangement and
broadening of the bipolar cell layer are also observed. These
changes can also be seen in the retina of chick embryo after
6 h of exposure (compare Fig. 12B with 12A), though the
changes are less pronounced. These features are less remak-
able as compared with those obtained in the rabbit (18), but
could be regarded as the initial stage of the degeneration of
the visual cell layer.

Assuming that a high concentration of oxygen increases
the amount of activated oxygen in the lung, the primary site
of lipid peroxidation in animals exposed to a high concentration of oxygen would be the lung. If this is valid, the present study implies that lipid peroxides formed in the lung are
transferred to the retina via bloodstream and cause the degeneration of the retina. Considered in this way, it is reasonable to recommend to physicians that the concentration of oxygen should be controlled by monitoring lipid peroxide level in
the blood. For this purpose, the micro method for the assay
of lipid peroxides in the blood (5) is useful.

It might be added that the degeneration of the retina in
ocular siderosis is also due to the increased lipid peroxides
in the retina provoked by iron ions liberated from the iron
piece broken into the eyeball (21).

5. Lesion of aortic intima caused by intravenous injection of
linoleic acid hydroperoxide

As described in the previous section, the increased lipid
peroxides in the blood would be deleterious to the blood vessels and other intact organs and tissues. To verify this
view, a model experiment was carried out with rabbits (22).
Linoleic acid hydroperoxide was administered by a single
intravenous injection at a dose of 25 mg (peroxide value:
2,000 meq/kg)/kg body weight through the ear vein to an adult
male rabbit over a 5-min interval, and the blood was drawn
from the opposite ear vein at timed intervals after injection.
Figure 13, curve I, shows the pattern of the change in the
level of TBA-reacting substances in the serum after injection
of linoleic acid hydroperoxide. The level increased

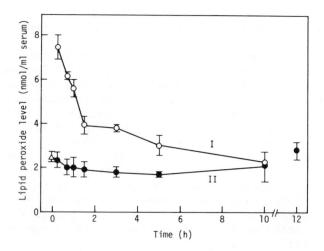

Figure 13.   Change in the level of TBA-reacting substances in
the serum after injection of linoleic acid hydroperoxide or
linoleic acid.   The level of TBA-reacting substances in the
serum is expressed in terms of malondialdehyde (nmol/ml·serum).
Mean ± SE (n = 4).   I: after injection of linoleic acid hydroperoxide; II: after injection of linoleic acid.   Open triangle
shows the level before injection.

TABLE V.   Lipid Peroxide Levels in Various Tissues 24 h after
Injection of Linoleic Acid Hydroperoxide.

	Lipid peroxide level	
	Linoleic acid hydroperoxide	Linoleic acid
Aorta	3.54 ± 0.27*	1.62 ± 0.30
Vein	7.99 ± 0.87	7.20 ± 0.59
Retina	14.0 ± 1.76	14.3 ± 1.64
Liver	6.78 ± 1.60	8.49 ± 1.22
Lung	7.65 ± 0.23	8.70 ± 1.08
Spleen	15.8 ± 1.83	16.3 ± 2.29

Lipid peroxide levels are expressed in terms of malondialdehyde (nmol/100 mg wet weight).   Mean ± SE (n = 4).
*p < 0.01.

immediately after injection and then decreased gradually to
reach a steady state at 1.5 h after injection. This level was
maintained for about 10 h. The half-lifetime of the injected
hydroperoxide in the serum was calculated to be approximately
50 min. In contrast, the injection of linoleic acid (25 mg/kg
body weight) did not affect the level in the serum (Fig. 13,
curve II).

After treatment of the serum with phosphotungstic acid-
sulfuric acid, the precipitates consisting of serum proteins
and lipids were also assayed. The pattern was similar to that
estimated for the untreated serum in principle, but the level
at 3 h after injection almost reached the initial level. The
difference in these patterns would indicate the existence of
free hydroperoxide for about 10 h.

The level of lipid peroxides complexed with protein in

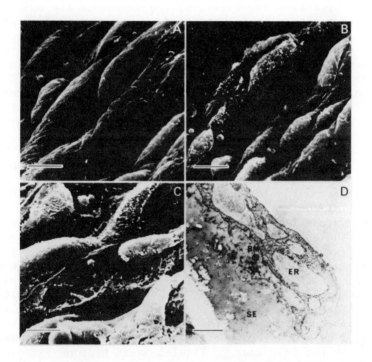

Figure 14. Electron micrographs of aortic endothelium after
administration of linoleic acid hydroperoxide or linoleic
acid. A: before injection; B: 24 h after injection of linoleic acid; C and D: 24 h after injection of linoleic acid
hydroperoxide. A, B, and C: scanning electron micrograph;
bars: 5 μm. D: transmission electron micrograph; ER: endoplasmic reticulum; BM: endothelial basement membrane; SE:
subendothelial space; bar: 1 μm.

the serum 24 h after injection was higher than that before
injection, indicating that the secondary damage to the other
tissues occurred as observed in the case of thermal injury
(13). Among the tissues examined with TBA method, the lipid
peroxide level increased only in the aorta 24 h after injection (Table V). This suggests that the hydroperoxide was
transferred to the aorta through the blood.

In Fig. 14, the profiles of the endothelial surface of
the thoracic aorta in scanning electron micrograph are shown.
In the rabbit without injection, the endothelial surface was
normal and smooth; rhomboidal endothelial cells were arranged
along the longitudinal axis oriented in the direction of blood
flow and nuclear bulges were seen at the center of cells (Fig.
14A). Upon injection of linoleic acid, endothelial arrangement became partially irregular, and some holes were seen at
the sites of endothelial nuclei (Fig. 14B). After linoleic
acid hydroperoxide was injected, the damage to the endothelial
cells was markedly pronounced (Fig. 14C). The surface was
ulcerated deeply and stripped diffusely. Many holes were seen
and some cells were enucleated. In some areas, subendothelial
fibrous tissue was exposed. To this fibrous tissue, platelets
appeared to be adhering and exhibited pseudopodia.

In a transmission electron micrograph of an endothelial
cell of the aorta of a rabbit injected with linoleic acid
hydroperoxide (Fig. 14D), the cell was found to be deformed by
an enormous expansion of rough surfaced endoplasmic reticulum
which contains some membranous or amorphous materials. Endothelial basement membrane was disintegrated, and the subendothelial space was filled with plasma fluid.

Similar morphological changes were also observed in the
endothelial surface of the pulmonary artery, though the
changes were slight as compared with those in the thoracic
aorta. These morphological changes were not observed in other
tissues examined.

Taking into account the features of the early stage of
atherogenesis in rabbits described hitherto, the views obtained by scanning electron microscopy of the aorta of rabbits
after injection of linoleic acid hydroperoxide could be regarded as the features of the initial change in atherogenesis.
It should be emphasized that such drastic changes were only
induced by the injection of linoleic acid hydroperoxide, but
not by linoleic acid.

Since the injection of linoleic acid hydroperoxide increased the lipid peroxide concentration in the serum and
caused morphological changes only in the aorta, it appears
that the hydroperoxide moiety of linoleic acid hydroperoxide
directly caused the lesions in the aorta of the rabbit. Although Cutler and Schneider (23) already mentioned this possibility, the present data seem to provide clearer evidence,

since these changes were caused by only a single injection of the hydroperoxide. It is plausible that the damage to the intima of the aorta provoked with lipid peroxides is the initial event in pathogenesis of human atherosclerosis.

REFERENCES

1. Glavind, J., Hartmann, S., Clemmesen, J., Jessen, K. E., and Dam, H. (1952) Acta Pathol. 30, 1.
2. Bernheim, F., Bernheim, M. L. C., and Wilbur, K. M. (1948) J. Biol. Chem. 174, 257.
3. Wilbur, K. M., Bernheim, F., and Shapiro, O. W. (1949) Arch. Biochem. 24, 305.
4. Yagi, K., Nishigaki, I., and Ōhama, H. (1968) Vitamins 39, 105.
5. Yagi, K. (1976) Biochem. Med. 15, 212.
6. Goto, Y. (1981) in this volume, 295.
7. Sinnhuber, R. O., Yu, T. C., and Yu, T. C. (1958) Food Res. 23, 626.
8. Ohkawa, H., Ohishi, N., and Yagi, K. (1978) J. Lipid Res. 19, 1053.
9. Kohn, H. I., and Liversedge, M. (1944) J. Pharmacol. Exp. Therap. 82, 292.
10. Slater, T. F. (1968) Biochem. J. 106, 155.
11. Suematsu, T., Kamada, T., Abe, H., Kikuchi, S., and Yagi, K. (1977) Clin. Chim. Acta 79, 267.
12. Sato, Y., Hotta, N., Sakamoto, N., Matsuoka, S., Ohishi, N., and Yagi, K. (1979) Biochem. Med. 21, 104.
13. Nishigaki, I., Hagihara, M., Hiramatsu, M., Izawa, Y., and Yagi, K. (1980) Biochem. Med. 24, 185.
14. Ohkawa, H., Ohishi, N., and Yagi, K. (1979) Anal. Biochem. 95, 351.
15. Noell, W. K. (1958) Arch. Ophthal. 60, 702.
16. Gerschman, R., Gilbert, D. L., Nye, S. W., Dwyer, P., and Fenn, W. O. (1954) Science 119, 623.
17. Haugaard, N. (1968) Physiol. Rev. 48, 411.
18. Hiramitsu, T., Hasegawa, Y., Hirata, K., Nishigaki, I., and Yagi, K. (1976) Experientia 32, 622.
19. Patz, A., Hoeck, L. E., and Cruz, E. D. L. (1952) Am. J. Ophthal. 35, 1248.
20. Yagi, K., Matsuoka, S., Ohkawa, H., Ohishi, N., Takeuchi, Y. K., and Sakai, H. (1977) Clin. Chim. Acta 80, 355.
21. Hiramitsu, T., Majima, Y., Hasegawa, Y., Hirata, K., and Yagi, K. (1976) Experientia 32, 1324.
22. Yagi, K., Ohkawa, H., Ohishi, N., Yamashita, M., and Nakashima, T. (1981) J. Appl. Biochem. 3, 58.

23. Cutler, M. G., and Schneider, R. (1974) Atherosclerosis
     20, 383.

THE ROLE OF LIPID PEROXIDATION ON THE DEVELOPMENT
OF PHOTOSENSITIVE SYNDROME BY PHEOPHORBIDE a

Shuichi Kimura
Toshihiko Isobe
Hiroaki Sai
Yuji Takahashi

Laboratory of Nutrition
Faculty of Agriculture
Tohoku University
Sendai

Pheophorbide a, which is one of the decomposition products from chlorophyll, is produced by elimination of phytol
and magnesium (Fig. 1). These reactions are performed respectively by chlorophyllase hydrolysis and by acid treatment.
The decomposition products from chlorophyll are known to cause
photosensitivity in both sheep and cattle (1,2). In 1955,
Clare (3,4) reported that the photosensitivity observed in
rats fed on dried corn was due to pheophorbide a and pyro-
pheophorbide a. Among some fishermen in the Tohoku district
of Japan, there is an old saying that in the beginning of
Spring, the viscera of abalones become toxic to cats. This
toxicity was demonstrated by collapse of the cats ears.
Takenaka and his co-workers (5) reported that a severe dermatitis, especially edema and erythema, occurred in men who ate
abalone viscera. Hashimoto et al. (6-8) have shown that this
poisoning was a characteristic photosensitivity reaction.
They proved that the syndrome was caused by pyropheophorbides
which came from chlorophyll in abalones. It has been known
that chlorophyllases, which convert chlorophylls to pheophorbides, are widely distributed in plant tissues. Additionally,
there are reports that the pheophorbides are gradually produced in the process of preserving salted vegetables. Yamada
et al. (9) observed the photosensitive syndrome in rats fed
on salted Nosawana and salted Takana under irradiation.
Recently, photosensitive dermatitis has been found to
occur in people who had ingested large quantities of chlorella

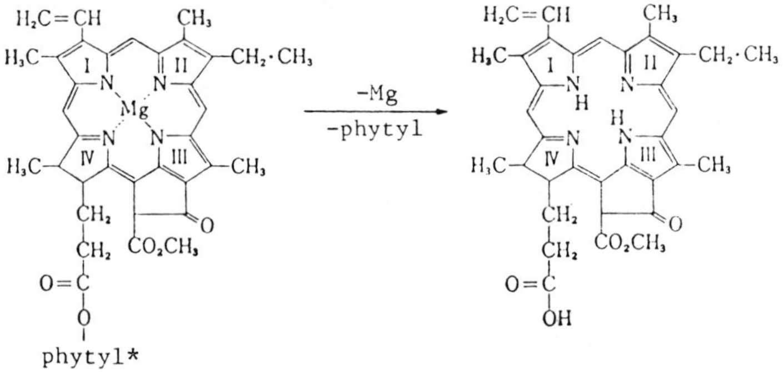

chlorophyll *a* pheophorbide *a*

*phytyl

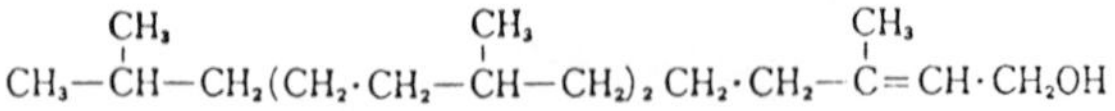

Figure 1. Chemical formulas of chlorophyll and pheophorbide.

tablets produced by a certain company. This syndrome is ascribed to the large amounts of pheophorbide contained in some lots of these tablets (10,11). However, little is known about the mechanism of the development of this photosensitivity. Therefore, we have studied this from biochemical and physiological view points. Recently, we have shown that in the presence of pheophorbide *a*, visible light irradiation leads to the production of an active oxygen. This, in turn, results in cell damage by oxidation of lipids, especially in the cell membranes (12-15). Further, we discovered that some kinds of vitamins display a preventive action against the photosensitivity caused by pheophorbide *a*. In this paper, a brief account of our work concerning the mechanism of development of photosensitivity caused by pheophorbide *a* is given.

I. STUDY OF CONDITIONS PRODUCING PHOTOSENSITIVITY IN RATS

We first examined the conditions which led to the production of acute and chronic photosensitivities in rats.

1. Development of photosensitive dermatitis in rats

Wistar strain rats were fed on a diet containing definite amounts of pheophorbide *a* from 8:00 p.m. to 8:00 a.m. They were irradiated under 25,000 lux from 8:00 a.m. to 6:00 p.m.

every day, using the irradiation apparatus shown in Fig. 2.
After 6 days under these conditions, dermatitis at the edge of
the ear lobe was seen in rats fed on the diet containing
pheophorbide *a* at a level of 100 mg per 100 g diet. A pro-
longed irradiation caused a fall of the ears completely. Dermatitis was also observed on the back of the body. Microscopic observation of the skin of the ear lobe showed a typical
photosensitized necrosis at the surface (Figs. 3,4).

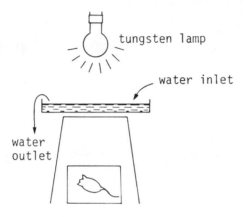

Figure 2. Apparatus to test photosensitivity.

Figure 3. Macroscopic observation of the photosensitized rat.
Dermatitis at the edge of ear lobe was seen in rat fed on the
diet containing pheophorbide *a* at the level of 100 mg per 100
g diet.

Figure 4. Microscopic observation on the ear lobe skin of a
rat photosensitized by pheophorbide a. Typical photosensitized necrosis occurred on the surface.

TABLE I. Appearance of hypersensitivity in different
species (3 males and 3 females)

Species	Survival time (hr)
dd-Mouse	2-3
C57BL/6j Mouse	2-3
Mastomys	3-4
Hamster	3-4
Rat	5-6

Light: 10,000 lux; temperature: $25°C \pm 1$; amount of
pheophorbide a; 1 mg/30 g body weight.

2. A shock-like death occurred in acute photosensitized rats

It was shown that acute photosensitivity occurred in rats
which had received an intraperitoneal injection of pheophorbide a under irradiation. The time required for the shock-
like death depended on the quantity of pheophorbide a when
the irradiation illuminance was fixed. Notably, a shock-like
death occurred after 5 or 6 hours in rats injected with 5 mg
or more pheophorbide a under the condition of 10,000 lux
irradiation. As expected, no change was observed in rats with
neither pheophorbide a nor visible light irradiation. Some

other species of animals also showed a shock-like death when
injected with the same amount of pheophorbide *a* per body
weight as in the experimental rats. The time required for
acute photosensitivity induced shock-like death (namely the
survival time), under the experimental conditions is shown in
Table I. If we allowed more than 6 hours in the dark after
the injection of pheophorbide *a*, shock-like death was not
induced. Therefore it was suggested that pheophorbide *a* was
metabolized quickly.

II. MECHANISM OF THE DEVELOPMENT OF SHOCK-LIKE DEATH BY ACUTE PHOTOSENSITIVITY

1. Investigation of the immediate cause of shock-like death

To clarify the mechanism of shock-like death of photosensitivity in rats, studies were made from a physiological
view point. The changes of rectal temperature and respiratory
ratio were examined during the process of acute photosensitivity. Both values immediately decreased in similar manner
after irradiation (Fig. 5). Next, the electroencephalographic
and electrocardiographic changes were examined. The electro-
encephalogram was found to be normal, but the electrocardio-
gram showed an abnormal pattern. This pattern was similar to
that of hyperkalemia, which sometimes results in shock-like
death. This was checked by testing the serum electrolytes:
A high potassium level in the serum was confirmed. Since
potassium is distributed inside cells in a high concentration,
the acute leakage of potassium into the serum strongly suggested that cell damage must have occurred in some tissues or
organs. Therefore, these hypersensitive rats were examined
histologically, and great changes occurred in the spleen,
where phagocytosis of red blood cells was observed. The blood
of photosensitized rats showed a severe hemolysis. We assumed
that photodynamic action with pheophorbide *a* produced hemolysis first, and secondly potassium was eluted out from erythrocyte to serum. Thus hyperkalemia developed and the rats'
death ensued.

2. The mechanism of pheophorbide-photosensitized hemolysis

As mentioned above, hemolysis was recognized as an important factor for the occurrence of photosensitivity. We performed in vitro experiments using blood cells, to elucidate
the mechanism of this hemolysis. We observed hemolysis only
in the case where red blood cells were irradiated in the
presence of pheophorbide *a* (the same conditions as in our in

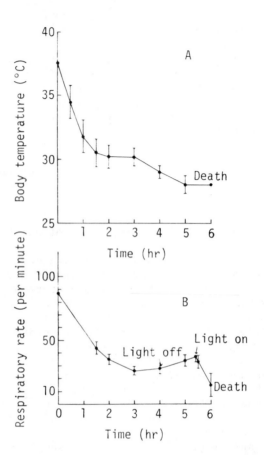

Figure 5. The changes of rectal temperature and respiratory ratio during the process of acute photosensitivity.
A: Changes of body temperature in the hypersensitive rats;
B: Changes of respiratory rate (experiment 2).
Symbols indicate the geometrical means and S.D. of three rats.

vivo experiment). That is to say, the photodynamic action of pheophorbide a was also shown in an in vitro system.

It was unclear why this hemolysis occurred in the process of photodynamic action. To answer this question we examined in vitro the changes occurring in the red blood cell-membrane of rats. We found that thiobarbituric acid (TBA) values recognized as an index of the amount of lipid peroxide were much higher in the membrane irradiated in the presence of pheophorbide a (Table II). Further examination of the fatty acid composition of phospholipid in the membrane showed dramatically decreased level of arachidonic acid in the membrane

TABLE II.   Changes in thiobarbituric acid (TBA) values of red blood cell membrane irradiated in vitro

Addition	TBA values
RBC membrane	0.537
RBC membrane + irradiation	0.565
RBC membrane + pheophorbide *a*	0.573
RBC membrane + pheophorbide *a* + irradiation	4.865

Red blood cell (RBC) membrane, 5 g; pheophorbide *a*, 5 mg; irradiation, 10,000 lux for 5 hr.

TABLE III.   Fatty acid composition of phospholipids in the red blood cell membrane photosensitized by pheophorbide *a*

Fatty acid		I	II	III	IV	V	VI
14 : 0		0.4	0.4	0.3	0.6	0.4	0.4
16 : 0		40.4	38.5	39.9	45.2	34.8	33.4
	1	2.6	3.1	2.6	7.0	2.6	2.8
18 : 0		12.5	13.7	14.4	13.2	17.5	17.9
	1	11.8	11.7	11.7	15.7	14.2	14.1
	2	9.2	8.8	8.3	9.4	9.4	9.1
20 : 4		20.1	20.2	19.2	8.2	19.7	20.9
22 : 4		1.5	1.7	1.3	0.7	0.7	0.7
	5	1.3	0.7	1.0	–	0.4	0.4
	6	1.8	1.1	1.2	–	0.4	0.4

   I: red blood cell membrane (RBC)
  II: RBC membrane
 III: RBC membrane + pheophorbide *a*
  IV: RBC membrane + pheophorbide *a*
   V: RBC membrane + pheophorbide *a* + α-tocopherol
  VI: RBC membrane + pheophorbide *a* + UQ_{10}
RBC membrane, 5 g; pheophorbide *a*, 5 mg; α-tocopherol, 4 mg; UQ_{10}, 4 mg.
Experiment system II, IV, V, and VI were irradiated with 10,000 lux for 5 hr.

irradiated in the presence of pheophorbide a (Table III).
These results suggested that some kind of active oxygen was
produced by photodynamic action of pheophorbide a, which might
produce lipid peroxide in this system. Further the effect of
α-tocopherol on the pattern of membrane fatty acid composition
was examined, because this compound is considered to be a free
radical scavenger or singlet oxygen quencher. This demonstrated that α-tocopherol had a preventive effect on the change of
fatty acid patterns caused by this photosensitive action in
the presence of pheophorbide a as shown in Table III. UQ_{10}
had the same effect as α-tocopherol in this experiment. From
these data we assumed that lipid peroxidation in the membrane
was important for the development of this photosensitivity.

III. PREVENTIVE EFFECTS OF SOME VITAMINS ON THE DEVELOPMENT
      OF PHOTOSENSITIVITY BY PHEOPHORBIDE a

As mentioned above, we showed that the photosensitivity
on the whole body of animals by pheophorbide a was developed
dermatitis or shock-like death, and that a major event occurred in this photosensitization was the lipid peroxidation in
the biomembrane. These results were supported by the hemolytic test and the measurement of the TBA value of red blood cell
membrane in vitro system. Using these systems we showed that
α-tocopherol and UQ_{10} might protect against the photodynamic
action of pheophorbide a.
We tried to test the preventive effects of other vitamins,
which are recognized as inhibitors against the lipid peroxidation.
The preventive effect of some vitamins against photosensitivity caused by pheophorbide a was examined by adding
them to fresh blood under irradiation (Table IV). α-Tocopherol, pantethine and vitamin B_2-butyrate had a preventive effect
on the hemolysis.
Recently we have developed a microbioassay method using
cultured myocardial cells obtained from the heart of chick
embryos. The development of the photodynamic action was
examined in this cultured myocardial cells.
Cultured myocardial cells beat in the petri dish. We
examined the influence of pheophorbide a and visible light
irradiation on cultured myocardial cells. However, addition
of pheophorbide a or irradiation with visible-light did not
affect the beating of these cells. Irradiation in the presence of pheophorbide a made the cells stop the beating and
death eventually ensued. We found that these cells were so
sensitive to the photodynamic action of pheophorbide a, and
that their beating stopped after irradiation for only 2 hours

TABLE IV. Effect of vitamins on pheophorbide-
photosensitized hemolysis

Addition	Relative hemolysis %
Control solvent (DMSO)	4.2
Pheophorbide *a*	100.0
Pheophorbide *a* + pantethine	13.7
Pheophorbide *a* + α-tocopherol	10.9
Pheophorbide *a* + B$_2$-butyrate	64.7

Blood from rats, 1 ml; pheophorbide *a*, 50 µg;
pantethine, 100 µg; α-tocopherol, 100 µg;
vitamin B$_2$-butyrate, 100 µg.
All experimental systems were irradiated with
10,000 lux for 2 hr.

TABLE V. Photodynamic action of pheophorbide *a* against cultured myocardial cells and protective effect of vitamins

Addition	Percent of beating cells after irradiation	TBA values µg/mg protein	Relative TBA values
Control	100	0.125	–
Pheophorbide *a* + irradiation	0	0.706	100
Pheophorbide *a* + irradiation + pantethine	50–55	0.230	35
Pheophorbide *a* + irradiation + CoA	95–100	0.201	13
Pheophorbide *a* + irradiation + α-Toc	80–85	0.253	22
Pheophorbide *a* + irradiation + B$_2$-butyrate	25–30	0.620	85

Pheophorbide *a*, 1.03×10^{-8} M; α-tocopherol, 100 µg/ml;
pantethine, 100 µg/ml; Coenzyme A, 100 µg/ml; vitamin B$_2$-
butyrate, 100 µg/ml; irradiation, 10,000 lux for 2 hr.

(10,000 lux) with pheophorbide a $(1.03 \times 10^{-8}$ M). TBA values
of the cells irradiated in the presence of pheophorbide a were
also demonstrable (Table V).

Because it is possible to test the photodynamic action in
a micro amounts, we thought that this bioassay system was very
useful for screening of protectors against photodynamic action. Using this method, preventive effects of α-tocopherol,
pantethine and vitamin B_2-butyrate were examined. α-Tocopherol and pantethine protected cultured myocardial cells from
photodynamic action. This protective effect was also estimated using TBA values. These results suggested the relation
between photodynamic action and lipid peroxidation. Vitamin
B_2-butyrate had a weak effect as a protector. The results
obtained from cultured myocardial cells corresponded to those
from hemolysis. Since pantethine is a precursor of Coenzyme
A, the effect might have occurred after synthesis of Coenzyme
A by the cells. Thus we examined the preventive action of
Coenzyme A and recognized it as a very effective agent. We
guessed that α-tocopherol might play a role as a free radical
scavenger or singlet oxygen quencher, and that Coenzyme A
might act as a sweeper of lipid peroxide through the lipid
metabolism. But this assumption should be examined in the
future.

IV. DETERMINATION OF SPECIES OF ACTIVE OXYGEN WHICH INFLU-
ENCED ON PHOTODYNAMIC ACTION BY PHEOPHORBIDE a

The finding previously noted strongly suggested that
active oxygens participated in the development of photosensitivity with pheophorbide a. We attemped to determine the
species of active oxygen. Some specific deactivators of
active oxygen were added in the photosensitized hemolysis
system. Singlet oxygen quencher, hydroxy radical scavenger
and catalase had a preventive effect on the photosensitized
hemolysis in the presence of pheophorbide a. On the contrary,
superoxide dismutase had no effect on this system (Fig. 6).
In order to check the effect of the singlet oxygen quencher,
the effect of β-carotene, which is well known as a singlet
oxygen quencher, was investigated. In this experiment
lecithin-cholesterol liposomes were used and the preventive
effect of this compound on photosensitized lipid peroxidation
was determined. We found that β-carotene had a marked effect.

In conclusion, experimental results indicated that in the
presence of pheophorbide a, irradiation with visible light
leads to the production of active oxygen, which might be
singlet oxygen, hydroxy radical or H_2O_2. This active oxygen
causes damage to the cell through lipid peroxidation.

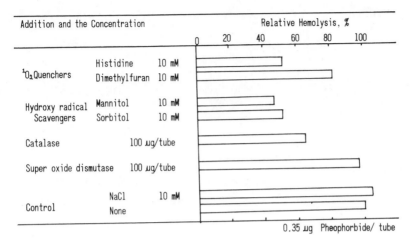

Addition and the Concentration			Relative Hemolysis, %

Figure 6.   Effect of $^1O_2$ quenchers, hydroxy radical scavengers, catalase and SOD on pheophorbide-photosensitized hemolysis.

REFERENCES

1. Fischer, H., Bartholomäus, E., and Röse, H. (1913) Z. Physiol. Chem. 84, 262.
2. Rimington, C., and Quin, J. I. (1933) Nature 29, 178.
3. Clare, N. T. (1955) Adv. Vet. Sci. 2, 182.
4. Clare, N. T. (1953) J. New Zealand Inst. Chem. 17, 57.
5. Takenaka, S., Sawada, G., and Yoshioka, M. (1899) Tokyo Iji Shinshi, No. 1114, 1359.
6. Hashimoto, Y., Naito, K., and Tsutsumi, J. (1960) Nippon Suisan Gakkaishi (Bull. Jpn. Soc. Sci. Fish.) 26, 1216.
7. Hashimoto, Y., and Tsutsumi, J. (1961) Nippon Suisan Gakkaishi (Bull. Jpn. Soc. Sci. Fish.) 27, 859.
8. Tsutsumi, J., and Hashimoto, Y. (1964) Agric. Biol. Chem. 28, 467.
9. Yamada, K., and Nakamura, N. (1972) Eiyo To Shokuryo (J. Jpn. Soc. Food. Nutr.) 25, 466.
10. Amano, R., Ike, K., and Uchiyama, M. (1978) Shokuhin Eisei Kenkyu 28, 739.
11. Tamura, Y., Maki, T., Shimamura, Y., Nishigaki, S., and Naoi, K. (1979) Shokuhin Eiseigaku Zasshi 20, 173.
12. Isobe, A., and Kimura, S. (1976) Eiyo To Shokuryo (J. Jpn. Soc. Food. Nutr.) 29, 225.
13. Isobe, A., and Kimura, S. (1976) Eiyo To Shokuryo (J. Jpn. Soc. Food. Nutr.) 29, 221.
14. Isobe, A., Sasaki, R., and Kimura, S. (1977) Eiyo To Shokuryo (J. Jpn. Soc. Food. Nutr.) 30, 99.

15. Kimura, S., Isobe, A., and Sai, T. (1979) Vitamins 53, 543.

Note added in proof: The species of the active oxygen participated primarily in this reaction was found to be singlet oxygen (Kimura, S., and Takahashi, Y. (1981) Photomedicine Photobiology 3, 73).

ADRIAMYCIN-INDUCED LIPID PEROXIDATION
AND ITS PROTECTION

Ryohei Ogura

Department of Medical Biochemistry
Kurume University School of Medicine
Kurume

I. INTRODUCTION

Adriamycin (ADM) is one of the most widely used antitumor agents in the field of cancer chemotherapy. It is
highly active against a variety of solid tumors and malignant
hematological processes. However, chronic cardiotoxicity has
limited the clinical use of ADM as an antineoplastic agent
(1,2). A morphologically distinct form of cardiotoxicity is
characterized by sarcoplasmic vascular degeneration, myocyto-
lysis, atrophy of myocytes, interstitial edema and fibrosis
(2,3). The incidence of this side effect increases significantly when the cumulative doses exceed 550 mg/m^2 body surface
area (4). The pathogenesis of ADM-induced cardiotoxicity is
not known, but several mechanisms have been proposed. Myocar-
dium is the most energy-consuming tissue, and the energy is
generated and supplied through mitochondrial respiration.
Therefore, the purpose of the present study is to find out the
functional and morphological disorders in cardiac mitochondria
affected with ADM.
Recently, Myers et al. (5) have indicated that tissue
damage by ADM may involve at least two mechanisms: one which
involves binding to DNA, and the other which involves lipid
peroxidation. In the present paper, the protective effect of
antioxidant for the prevention of developmental changes in
contractile behavior of cardiac mitochondria is to be also
discussed.

II. MATERIALS

Adriamycin (doxorubicin hydrochloride) was purchased from Kyowa Hakko Kogyo Company Ltd. (Tokyo). ADM was dissolved in distilled water before injection. Ubiquinone (CoQ-10) was supplied by Eisai Pharmaceutical Company Ltd. (Tokyo). Solutions of CoQ-10 were prepared in polyoxyethylene (60)-hydrogenated castor oil derivative (Nikkol HCO-60). Vitamin B_2 butyrate (riboflavin-2',3',4',5'-tetrabutyrate, B_2 butyrate) was supplied by Tokyo Tanabe Company Ltd. (Tokyo). B_2 butyrate was dissolved in 10% ethanol and diluted with olive oil to the final concentration of 0.8% B_2 butyrate.

III. EXPERIMENTAL PROCEDURES

A short-term animal experiment was conducted with four groups of ten rats with a mean weight of 250 g body weight. The first group was treated with ADM only, the second group with both of ADM and antioxidant, and the third group with antioxidant only. The last group was the control animals. ADM was injected daily intraperitoneally with a dose of 4 mg per kg body weight. Antioxidant was administered daily with an intramuscular injection of CoQ-10 (25 mg/kg body weight) or with an intraperitoneal injection of B_2 butyrate (20 mg/kg).

The animals were killed by decapitation on the eighth day following treatment with ADM and/or antioxidant. Mitochondria were prepared from heart homogenate according to the method of differential centrifugation after Nagarse digestion (6).

For transmission electron microscopy mitochondrial pellets were prepared by the method of Shimada et al. (7). After staining with uranyl acetate and lead citrate, thin sections were examined in a Hitachi H-500 transmission electron microscope. In addition to the examination by the transmission electron microscopy, some of the specimens were freeze-fractured prior to the critical point drying (8). The fractured surfaces of mitochondria were sputter-coated with gold, and examined under a Hitachi field emission scanning electron microscope.

Respiratory responses of isolated mitochondria were determined polarographically with a Clark oxygen electrode, as described previously (7). The rate of oxygen consumption was expressed as nanoatoms of oxygen per min per mg of mitochondrial protein (mean ± SD), using succinate as a substrate. ADP-stimulated respiration and ADP-limited respiration were designated as State 3 and State 4 respiration, respectively. The respiratory control ratio (RCR) and ADP/O ratio were also calculated.

Lipid peroxide contents of mitochondria were determined using a TBA reaction that had been established as being opti-

mal for isolated mitochondria (9).   The content was expressed
as nanomoles of malonaldehyde per mg of mitochondrial protein.
    For the determination of CoQ homologues in mitochondria,
CoQ homologues were extracted from isolated mitochondria with
n-hexane after saponification, and analyzed on LiChrosorb RP-
18 column (particle size 5 μm, E. Merck, Darmstadt) using
ethanol: water (97 : 3 v/v) as mobile phase by the method of
Toyama et al. (10).   2,3,6-Trimethyl-5-nonaprenyl-1,4-benzo-
quinone (TQ-9) in dioxane was used as an internal standard.
    Protein contents of mitochondrial suspension were deter-
mined using a biuret reaction (11).

IV.   RESULTS

1.   Lipid peroxide contents in cardiac mitochondria

    The lipid peroxide contents of mitochondria from the
control animals and those from ADM-treated animals were
0.95 ± 0.08 and 1.92 ± 0.14 nanomoles per mg protein, respectively, indicating that administration of ADM caused the
elevation of lipid peroxide content in mitochondria.

2.   Electron microscopy of cardiac mitochondria

    As shown in Fig. 1, the transmission electron microgram
of mitochondria prepared from heart homogenate, demonstrated
the mitochondria to be round in shape with a smooth outer
surface.   Fine, well-developed cristae were clearly apparent,

Figure 1.   Transmission electron micrograph of cardiac mitochondria from the control rat.

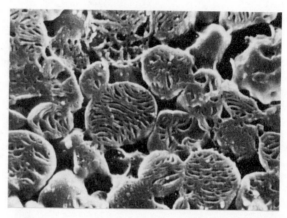

Figure 2. Fractured surface image (scanning electron micrograph) of cardiac mitochondria from the control rat.

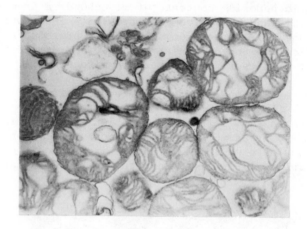

Figure 3. Transmission electron micrograph of cardiac mitochondria from rat treated daily with adriamycin (4 mg/kg)
intraperitoneally for one week.

oriented parallel. Fig. 2 shows the interior of intact mitochondria using a method of freeze-fracture, scanning electron
microscope. A double membrane and well-developed cristae in
the isolated mitochondria can be seen. From these electron
microscopic examinations, the mitochondria isolated by our
procedure are identified morphologically to be satisfactory.
    Fig. 3 shows the transmission electron microscopic analy-
sis of mitochondria from ADM-treated animal, and Fig. 4 shows
the scanning electron microscopic image of fractured surface

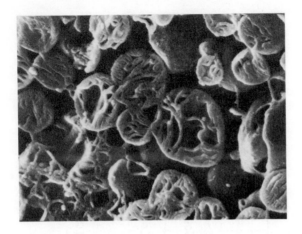

Figure 4. Fractured surface image (scanning electron micrograph) of cardiac mitochondria from rat treated daily with
adriamycin (4 mg/kg) intraperitoneally for one week.

of mitochondria affected with ADM. Mitochondria were larger
in size and irregular in shape than those from the control
animals. The mitochondrial cristae were oriented irregularly,
and the intracristal spaces were substantially enlarged.
Changes in the mitochondria affected with ADM are characterized by swelling, separating of cristae and condensation of
the outer membrane.

3. Respiratory responses of cardiac mitochondria

The rates of oxygen consumption by mitochondria from the
control animals and ADM-treated animals are shown in Table I.
Respiratory control ratio (RCR) and ADP/O ratio were found to
approximate the theoretical value for succinate as substrate.
It is considered that the Nagarse treatment in our isolation
process did not affect the respiratory responses and that the
isolated mitochondria were intact biochemically. As shown in
the ADM-treated animals, the oxygen consumption of State 4
increased and that of State 3 decreased as compared with those
of the control animals. And, both RCR and ADP/O ratio decreased in the ADM-treated animals. It is considered that the
oxidative phosphorylation is not coupled with the electron
transport system in mitochondria affected with adriamycin.

4. Distribution of CoQ homologues in cardiac mitochondria

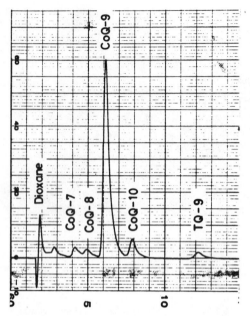

Figure 5. Chromatographic profile of CoQ homologues from cardiac mitochondria on LiChrosorb RP-18 column (4.6 × 150 mm).
Mobile phase : ethanol water (97 : 3, v/v). Flow rate: 0.5 ml/
min. The effluent was monitored at 275 nm. Internal standard: 2,3,6-trimethyl-5-nonaprenyl-1,4-benzoquinone (TQ-9).

   The chromatographic separation of the CoQ homologues from
rat cardiac mitochondria is shown in Fig. 5. Quantitative
determination of the constituents was made on the basis of
the recovery of TQ-9 internal standard. The CoQ contents of
mitochondria from the control animals and ADM-treated animals
are shown in Table II. The predominant component of CoQ homologues in rat mitochondria is CoQ-9, which constitutes 90.5%
of the total. From the results shown in Table II, it is
apparent that administration of ADM resulted in a significant
decrease in mitochondrial CoQ-10.

5. Effect of antioxidant to ADM-treated animals

   a. Lipid peroxide content

   The lipid peroxide content in mitochondria from rats
treated with both ADM and antioxidant was not high as that in
mitochondria from animals treated with ADM alone. The alteration of lipid peroxide in mitochondria affected with ADM and/

TABLE I. Respiratory Responses of Cardiac Mitochondria from Rats Treated with Adriamycin[a]

	Control	Adriamycin-treated group
Oxygen consumption[b]		
State 3 respiration	102.37 ± 1.01 (3)	84.50 ± 0.72 (6)
State 4 respiration	25.10 ± 1.31 (3)	27.05 ± 0.30 (6)
RCR[c]	4.01 ± 0.10 (3)	3.19 ± 0.04 (8)
ADP/O[d]	2.26 ± 0.03 (3)	2.13 ± 0.05 (8)

[a] Respiratory responses of cardiac mitochondria from the control and adriamycin (4 mg/kg body wt intraperitoneally daily for 1 week)-treated rats. Mean $\pm$ SD. Values in parentheses indicate the number of animals used for the determination.

[b] The values given for the oxygen consumption are expressed as nanoatoms of oxygen utilized per min per mg mitochondrial protein, using succinate as substrate.

[c] Respiratory control ratio (RCR) is the quotient of the State 3 respiration value and the succeeding State 4 respiration value, and indicates the tightness of the coupling oxidative phosphorylation.

[d] ADP/O ratio is the molar ratio of the added ADP to the consumed oxygen in the State 3, using succinate as a control.

TABLE II. Content of Individual CoQ Homologues in Cardiac Mitochondria from Rats Treated with Adriamycin[a]

	Control	Adriamycin-treated group
CoQ_7	0.007 ± 0.003	0.007 ± 0.002
CoQ_8	0.21 ± 0.03	0.25 ± 0.02
CoQ_9	5.58 ± 0.71	5.63 ± 0.76
CoQ_{10}	0.37 ± 0.05	0.25 ± 0.02

[a] Values are expressed as μg/mg mitochondrial protein. Mean $\pm$ SD (n = 6).

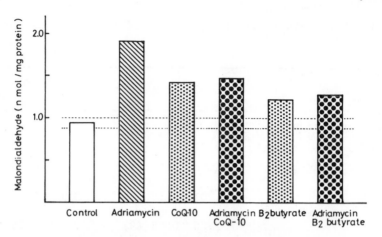

Figure 6. Lipid peroxide content of cardiac mitochondria from rats treated with adriamycin and/or antioxidant.

or antioxidant is illustrated in Fig. 6. From this evidence, it was revealed that the antioxidant could reduce the lipid peroxidation induced by ADM.

b. Transmission electron microscopy

Transmission electron micrographs of mitochondria from ADM·CoQ-10 treated animals and ADM·B$_2$ butyrate treated animals are shown in Figs. 7 and 8, respectively. Mitochondria were generally round in shape, and in many of them the mitochondrial cristae were arranged normally. Thus, this analysis confirmed that the changes in morphology resulting from administration of ADM were reduced by antioxidant administration.

c. Respiratory responses

In the ADM-antioxidant treated group of animals, the respiratory responses were found to return toward normal range. Figs. 9 and 10 show the alteration of respiratory responses of each group being expressed in terms of the percentage deviation from the control level. The respiratory disorders of cardiac mitochondria affected with ADM were improved evidently by co-administration of antioxidant.

d. CoQ homologues

The alteration of CoQ homologues in mitochondria from rats treated with ADM and/or antioxidant is shown in Figs. 11 and 12, being expressed in terms of the percentage deviation

from the control level. Exogenous CoQ-10 is incorporated into
cardiac mitochondria, and CoQ-10 content elevated remarkably.
The CoQ-10 content in mitochondria from rats treated with ADM
and CoQ-10 returned to normal levels. However, following administration of exogenous B₂ butyrate with ADM, CoQ-10 content
in cardiac mitochondria had not returned to the normal level,
unlike that found with CoQ-10 administration. B₂ butyrate has
not a direct effect for reducing CoQ-10 content of mitochondria caused by ADM treatment.

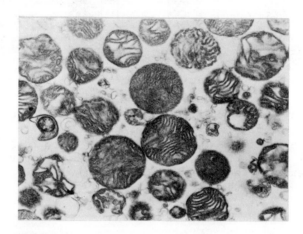

Figure 7. Transmission electron micrograph of cardiac mitochondria from rat treated daily with adriamycin (4 mg/kg)
intraperitoneally and CoQ-10 (25 mg/kg) intramuscularly for
one week.

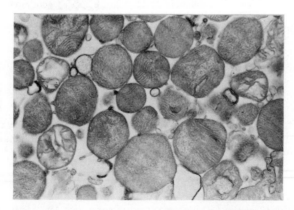

Figure 8. Transmission electron micrograph of cardiac mitochondria from rat daily treated with adriamycin (4 mg/kg)
intraperitoneally, B₂ butyrate (20 mg/kg) intraperitoneally
CoQ-10 (25 mg/kg) intramuscularly for one week.

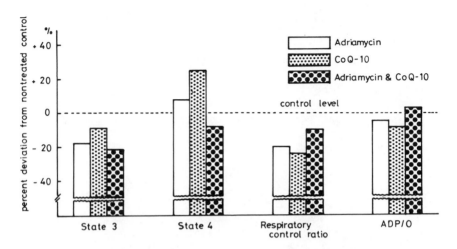

Figure 9. Alterations of respiratory responses of cardiac
mitochondria of rats following treatment daily with adriamycin
(4 mg/kg) intraperitoneally and/or CoQ-10 (25 mg/kg) intramuscularly for one week. The figure presents the individual
value in terms of the percentage deviation from the control
level as shown in Table I.

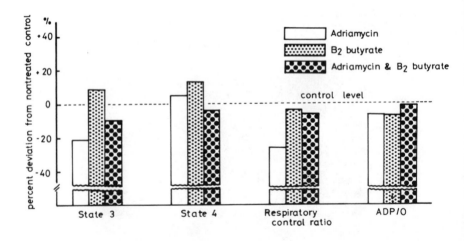

Figure 10. Alterations of respiratory responses of cardiac
mitochondria of rats following treatment daily with adriamycin
(4 mg/kg) intraperitoneally and/or vitamin B_2 butyrate (20 mg/
kg) intraperitoneally for one week. The figure presents the
individual value in terms of the percentage deviation from the
control level as shown in Table I.

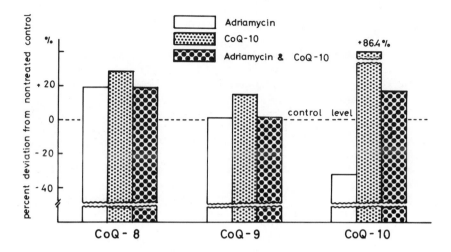

Figure 11. Alterations of CoQ homologue contents in cardiac mitochondria of rats following treatment daily with adriamycin (4 mg/kg) intraperitoneally and/or CoQ-10 (25 mg/kg) intramuscularly for one week. The figure presents the individual value in terms of the percentage deviation from the control level as shown in Table II.

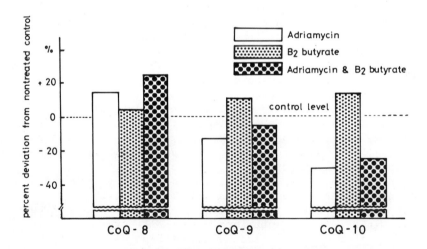

Figure 12. Alterations of CoQ homologue contents in cardiac mitochondria of rats following treatment daily with adriamycin (4 mg/kg) intraperitoneally and/or vitamin B$_2$ butyrate (20 mg/kg) intramuscularly for one week. The figure presents the individual value in terms of the percentage deviation from the control level as shown in Table II.

V. DISCUSSION

The results described in this paper firmly indicated that the cardiac mitochondria of rats are quite sensitive to adriamycin (ADM), developing accumulation of lipid peroxide, respiratory disorders and deficiency of CoQ-10 pool in the membrane. At the beginning of study, the integrity of isolated cardiac mitochondria prepared by our procedure has been confirmed morphologically to be suitable for studying the function of mitochondria affected with ADM. By means of ultrastructural study, the administration of ADM was found to cause a swelling of mitochondria with disruption of membranes and cristae. These morphological changes were reduced by coadministration of CoQ-10 or vitamin B_2 butyrate.

ADM is known to bind to mammalian DNA, causing fragmentation of DNA and inhibition of DNA synthesis (12). This DNA binding has been proposed as a mechanism for the antitumor effect of ADM. Myers et al. (5) have provided that ADM may involve lipid peroxidation as an additional mechanism of tissue damage, and that lipid peroxide appears to play a major role in the development of cardiotoxicity. Both this lipid peroxidation and cardiac toxicity of ADM were found to be reduced by prior treatment of animals with tocopherol (5). ADM-induced cardiotoxicity was considered to be associated with peroxidation of cardiac lipid in animals. Yamanaka et al. (13) also demonstrated that co-administration of CoQ-10 or tocopherol strongly inhibited the ADM-induced elevation of lipid peroxide level in serum. They suggested that the increase of serum lipid peroxide levels might be related to the lipid peroxide formed in cardiac membrane. In our laboratory, it has been confirmed that the administration of ADM increases the lipid peroxidation in cardiac mitochondria (14). The antioxidants, such as CoQ-10 (14), vitamin B_2 butyrate (15), vitamin B_2 (16), and pantethine, a derivative of pantothenic acid (17), are found to be effective in inhibiting the ADM-induced lipid peroxidation.

In our laboratory, Chinami et al. (18) have studied the influence of antioxidant on the responsiveness in vivo of Ehrlich ascites tumor to ADM. The co-administration of CoQ-10 or vitamin B_2 did not interfere with the incorporation of tritiated thymidine into ascites tumor DNA, although these antioxidants decreased lipid peroxides formed by ADM in ascites tumor cell. Myers et al. (5) have also reported that the antitumor responsiveness of P388 ascites tumor was not antagonized by tocopherol. These results indicate that treatment with antioxidant does not impair the responsiveness of tumor cell to ADM.

The exact mechanism of cardiotoxicity induced by ADM has

not been clarified. Kishi and Folkers (19) showed that ADM directly inhibits CoQ-10 enzymes, such as succinoxidase and NADH-oxidase which are present in cardiac tissue. They proposed that ADM-induced cardiotoxicity would be largely due to inhibition of CoQ-10 enzymes. In our laboratory, the CoQ homologue level of cardiac mitochondria was determined by high-pressure liquid chromatography (14,20). The CoQ-10 content was found to decrease markedly following ADM administration as described in this paper. The administration of exogenous CoQ-10 restored the CoQ-10 level to the normal, thus lowering the lipid peroxide level and relieving the respiratory disorders in cardiac mitochondria. The morphological and biochemical disorders of cardiac mitochondria were found to be closely related to the lipid peroxidation of mitochondria.

In the present study, vitamin B_2 butyrate was found to prevent the enhanced lipid peroxidation and restore the respiratory disorders of cardiac mitochondria affected with ADM. However, the deficiency of CoQ-10 pool was not restored. From this experiment, CoQ-10, which is one of the components of respiratory chain and occurs at a high concentration in mitochondria, was considered to play a role as a natural antioxidant. Takeshige et al. (21) suggested that the reduced form of CoQ-10 functions as a potent antioxidant against membrane lipid peroxidation. The peroxidation may be controlled according to the reduction level of CoQ-10. In our previous paper (22), ultraviolet light-induced mitochondrial lipid peroxidation was inhibited by exogenous antioxidants, namely vitamin B_2 butyrate, vitamin B_2 and glutathione. The natural protection mechanism against mitochondrial oxidation must be present. It is suggested that CoQ-10 protects the cellular membrane system against lipid peroxidation. The property of CoQ-10 is to be a potent antioxidant as well as one of components of the electron transport system in mitochondrial membrane.

ADM-induced alterations in the membrane merit further in depth study. In particular, protein in biomembrane exists in close molecular proximity to unsaturated fatty acid, and therefore, membrane protein will be readily damaged by lipid peroxidation. In our laboratory, the surface potential of C-side of inner membrane of mitochondria affected with ADM was determined using an anisotropic inhibitor. Katsuki et al. (23) have found that the rate of negative charge formation of C-side membrane induced by energization is inhibited by treatment with ADM. However, the co-treatment with CoQ-10 was effective in reversal of negative charge formation. Schioppacassi et al. (24) have reported that incubation of ADM in vitro with erythrocytes results in increased permeability of membrane. From this evidence, ADM is suggested to react with protein or glycoprotein in the membrane of cardiac myocytes. It is interesting to note that CoQ-10 is effective in the maintenance of structural and functional integrity of mitochondrial membrane.

ADM is known to convert to its semiquinone radical by either mitochondrial or microsomal metabolism (25). Chinami et al. (26) have observed that in the presence of cytochrome c the ESR signal of ADM is at g = 2.0048 which is assigned to that of semiquinone radical. To confirm a generation of superoxide anion, the reduction of cytochrome c coupled to ADM was determined. The rate of reduction was found to increase gradually, and was inhibited substantially by adding superoxide dismutase. Therefore, it is considered that ADM can be converted to its semiquinone radical and semiquinone can give rise to free radical formation, thus causing peroxidation of polyunsaturated fatty acids in the mitochondrial membrane.

In summary, administration of ADM induced the elevation of lipid peroxide, deficiency of CoQ-10, and uncoupled phosphorylation of cardiac mitochondria, accompanied by morphological changes in mitochondria. Since all of these influences should be responsible for the progressive impairment of cardiac function, the use of a combination of antioxidant with ADM was found to be useful in preventing or minimizing the mitochondrial disorders induced by ADM. Accordingly, the simultaneous administration of antioxidant would increase the clinical use of ADM in the field of cancer chemotherapy.

REFERENCES

1. Bristow, M.R., Billingham, M.E., Mason, J.W., and Daniel, J.R. (1978) Cancer Treat Rep. 62, 873.
2. Billingham, M.E., Mason, J.W., Bristow, M.R., and Daniel, J.R. (1978) Cancer Treat Rep. 62, 865.
3. Ferrans, V.J. (1978) Cancer Treat Rep. 62, 955.
4. Lefrak, E.A., Pitha, J., Rosenheim, S., and Gottlieb, J. A. (1973) Cancer 32, 302.
5. Myers, C.E., McGuire, W.P., Liss, R.H., Ifrim, I., Grotzinger, K., and Young, R.C. (1977) Science 197, 165.
6. Ogura, R., Toyama, H., Nagata, O., Ono, T., Shimada, T., and Murakami, M. (1979) Kurume Med. J. 26, 51.
7. Ogura, R., Sayanagi, H., Noda, K., Hara, H., Shimada, T., and Murakami, M. (1978) Kurume Med. J. 25, 65.
8. Shimada, T., and Murakami, M. (1975) J. Electron Microscop. 24, 199.
9. Nagata, O., Sayanagi, H., and Ogura, R. (1978) J. Kurume Med. Assoc. 41, 292.
10. Toyama, H., Hara, H., Ono, T., and Ogura, R. (1979) J.

Kurume Med. Assoc. 41, 164.
11. Gornal, A.G., Bardawill, C.S., and David, M.M. (1949) J. Biol. Chem. 177, 751.
12. Lee, Y.C., and Byfield, J.E. (1976) J. Natl. Cancer Inst. 57, 221.
13. Yamanaka, N., Kato, T., Nishida, K., Fujikawa, T., Fukushima, M., and Ota, K. (1980) in "Biomedical and Clinical Aspects of Coenzyme Q" (Y. Yamamura, K. Folkers and Y. Ito eds.), vol. 2, p. 213. Elsevier/North-Holland Biomedical Press, New York.
14. Toyama, H. (1979) J. Kurume Med. Assoc. 42, 603.
15. Ogura, R., Katsuki, T., Daoud, A.H., and Griffin, A.C. submitted to Cancer Letter.
16. Ogura, R., Toyama, H., and Katsuki, T. (1979) Vitamin 53, 569.
17. Toyama, H., Nagata, O., and Ogura, R. (1979) Pantethine Symposium (II), p. 163. Daiichi Pharmaceutical Co. Ltd., Tokyo.
18. Chinami, M., Okamoto, K., and Ogura, R. (1980) in Proceeding of Jap. Cancer Assoc. (Tokyo) No. 918.
19. Kishi, T., Kishi, H., and Folkers, K. (1977) in "Biomedical and Clinical Aspects of Coenzyme Q (K. Folkers and Y. Yamamura eds.), vol. 1, p. 47. Elsevier/North-Holland Biomedical Press, New York.
20. Ogura, R., Toyama, H., Shimada, T., and Murakami, M. (1979) J. Applied Biochem. 1, 325.
21. Takeshige, K., Takayanagi, R., and Minakami, S. (1980) in "Biomedical and Clinical Aspects of Coenzyme Q" (Y. Yamamura, K. Folkers and Y. Ito eds.), vol. 2, p. 15. Elsevier/North-Holland Biomedical Press, New York.
22. Sayanagi, H. (1977) J. Kurume Med. Assoc. 40, 1703.
23. Katsuki, T., Ihara, S., and Ogura, R. being prepared for publication.
24. Schioppassi, G., Kanter, P.M., and Schwartz, H.S. (1977) in Prodeedings of the Congress on Chemotherapy. Zürich.
25. Thayer, W.S. (1977) Chem.-Biol. Interact. 19, 265.
26. Chinami, M., Okamoto, K., Kato, T., and Ogura, R. being prepared for publication.

ARACHIDONATE PEROXIDATION AND FUNCTIONS OF HUMAN PLATELETS

Minoru Okuma
Hiroshi Takayama
Haruto Uchino

The First Division
Department of Internal Medicine
Faculty of Medicine
Kyoto University
Kyoto

Platelets play a pivotal role in hemostasis and thrombus formation, and aggregation is one of the most important functions of the platelets (1,2). Lipid peroxides in human platelets were found by the use of thiobarbituric acid (TBA) reaction when washed platelets were aggregated either by thrombin or by anti-platelet antibody, although the biochemical basis for the production of the TBA-reactive substance in the platelets was not clear in 1971 (3). Transformation of arachidonic acid (AA) in human platelets has recently been elucidated (Fig. 1) and its significance for the platelet function has been demonstrated (reviewed in ref. 2 and 4). In human platelets, AA is hydrolyzed from phospholipids by aggregation stimuli, and peroxidation of AA thus liberated is catalyzed by the platelet cyclo-oxygenase (PCO) and lipoxygenase (PLO) enzymes. PCO produces prostaglandin (PG) endoperoxides which are then converted to thromboxane (TX) A_2 and 12-hydroxyheptadecatrienoic acid (HHT) plus malondialdehyde (MDA) as well as small amounts of primary PGs including PGD_2, and TXB_2 is formed from unstable TXA_2. More recently, it has been reported that PGG_2 is converted to 15-hydroperoxy TXA_2 and 12-hydroperoxyheptadecatrienoic acid (HPHT) plus MDA by partially purified TX synthetase from human platelets and that 15-hydroperoxy TXB_2 is formed from 15-hydroperoxy TXA_2 (5). This metabolic pathway initiated by the PCO enzyme is called the PCO pathway. PLO produces 12-hydroperoxyeicosatetraenoic acid (HPETE) which is transformed to 12-hydroxyeicosatetraenoic acid (HETE) as a stable end-product. Furthermore, AA is converted by washed

human platelets to 8,9,12-trihydroxyeicosatrienoic acid (8,9,
12-THETA), 8,11,12-trihydroxyeicosatrienoic acid (8,11,12-
THETA) (6,7) and 10-hydroxy-11,12-epoxyeicosatrienoic acid
(EPHETA) (8), and a possible mechanism was suggested for the
conversion of these trihydroxy acids as well as EPHETA from
HPETE. This pathway initiated by the PLO enzyme can be called
the PLO pathway. Among the metabolites of the PCO pathway, PG
endoperoxides, TXA$_2$ and 15-hydroperoxy TXA$_2$ induce platelet
functions including aggregation and release reaction (2,9),
while PGD$_2$ inhibits them (2). On the other hand, the significance of PLO products for the platelet function has yet to be
completely defined, although only HPETE has been reported to
inhibit TX synthetase (10) and cyclo-oxygenase enzymes (11) as
well as platelet aggregation (11). Therefore, it is important
to elucidate the significance of the PLO pathway by investigating the activities of these peroxidative pathways in the
platelets of patients with various disorders, especially with
abnormal platelet functions, as well as in normal subjects.
    Alterations of the AA metabolism by human platelets were
demonstrated only in the PCO pathway. Hypofunctions of the

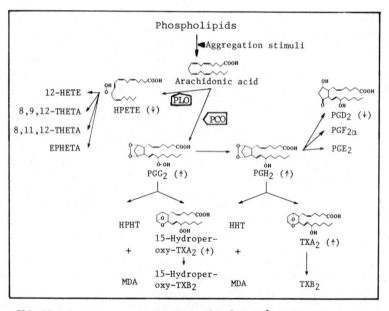

PLO, Platelet lipoxygenase ;   PCO, platelet cyclo-oxygenase
(↑), induce platelet aggregation; (↓), inhibit platelet aggregation

Figure 1. Metabolic pathways of arachidonic acid in human
platelets and the effect of products on platelet aggregation.
For abbreviations, see text.

platelet were reported in patients with PCO deficiency (12-15), while thrombosis or platelet hyperfunction in those with enhanced generation of PG endoperoxide (16) or TX (17). We have recently found PLO deficiency as well as PCO deficiency in some patients with myeloproliferative disorders (MPD) including chronic myeloid leukemia (CML), myelofibrosis (MF), polycythemia vera (PV) and essential thrombocythemia (ET), and studies on their platelet function suggested that the PLO activity could modulate platelet aggregation possibly through its effect on the production of TXA2 (18). Although the AA metabolism by the platelet has usually been studied by the use of a radioisotope technique, we have developed a simple method for estimation of these peroxidative (PLO and PCO) pathways in human platelets by the use of TBA reaction (19). This non-radioisotope technique should be useful for clinical studies on arachidonate peroxidation by human platelets. These topics form the basis for this communication.

ALTERED AA PEROXIDATION BY PLATELETS AND PLATELET AGGREGATION

Activities of PLO and PCO pathways which are initiated by AA peroxidation in human platelets were estimated by the use of ^{14}C-AA as reported previously (18). In short, after washed human platelets were routinely incubated at 37°C for 30 sec with $[1-^{14}C]AA$ (0.1 mM, f.c.), extracted lipids were separated by thin-layer chromatography at -20°C and radioactivity on the thin-layer plate was monitored by a radiochromatogram scanner or by autoradiography. In some experiments, areas or spots corresponding to PCO products (TXB2 and HHT) as well as PLO products (HPETE and HETE) were scraped off into counting vials and their radioactivities were counted by a liquid scintillation spectrometer. Percent conversion to each lipid product was calculated by relating its radioactivity to the total radioactivity recovered from the plate.

Thin-layer radiochromatograms of the products obtained for the case of the patients (CML) with a selective PLO deficiency and those for a normal subject are shown in Fig. 2. Patient's platelets formed only PCO products (i.e., TXB2 and HHT) and when aspirin-treated, i.e. cyclo-oxygenase-inhibited platelets were used, no transformation of AA was demonstrated by the patient's platelets. Only PLO products were detected by the incubation of ^{14}C-AA with aspirin-treated platelets of a normal subject.

Fig. 3 illustrates results of platelet aggregation study of this patient as compared with the result of a normal subject. Platelet aggregation was monitored using a standard nephelometric technique (20). Platelet counts in citrated

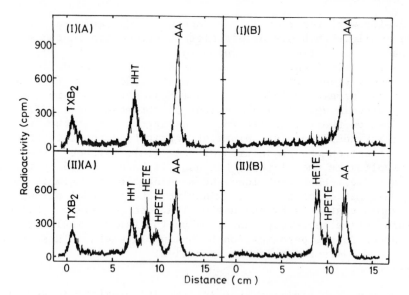

Figure 2. Thin-layer radiochromatograms. Products obtained after the incubation of ^{14}C-AA with untreated (A) or aspirin-treated (B) platelets from a CML patient with a selective PLO deficiency (I) and from a normal subject (II) are shown.

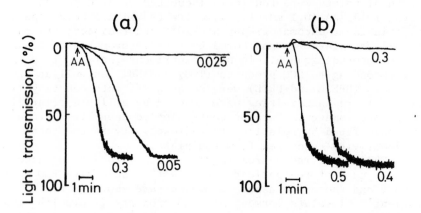

Figure 3. AA-induced platelet aggregation in a CML patient with a selective PLO deficiency (a) and in a normal subject (b). Concentrations (mM) of AA added to each platelet-rich plasma are shown below each platelet aggregation curve (18).

platelet-rich plasma were adjusted to 300,000 per microliter
and various concentrations of AA were added to determine a
threshold concentration enough to induce platelet aggregation.
Platelets of the patient were aggregated by a much lower concentration of AA (0.05 mM) than that necessary to induce normal platelet aggregation (0.4 mM). Such a high sensitivity
of the patient's platelets to AA might be explained, firstly,
by more efficient or increased availability of AA to the PCO
pathway (18,21), because no substrate was consumed by the PLO
enzyme which is deficient in this patient and, secondly, by
the defective production of HPETE which has been reported to
inhibit the cyclo-oxygenase (11) and TX synthetase enzymes
(10), thus resulting in more efficient production of TXA$_2$

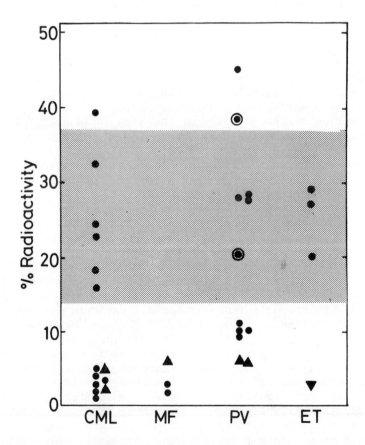

Figure 4. PLO activity and AA-induced platelet aggregation
(▲, increased; ▼, decreased; ●, normal; ◉, not studied)
in MPD patients. Normal range (dotted area): Mean ± 2 SD of
normal controls (N=19).

which is a highly potent mediator of platelet aggregation (22).
Therefore, an alteration in the AA metabolism in human
platelets secondary to selective PLO deficiency offers a
possible mechanism for hyperfunction of the platelet, which
may lead to a thrombotic tendency, one of the common features
of MPD (18).

PLO activity and AA-induced platelet aggregation in MPD
patients investigated are illustrated in Fig. 4. Although all
patients with increased platelet aggregability showed
decreased PLO activity, patients with decreased PLO activity
were not necessarily associated with increased platelet
aggregability. An ET patient with defective PCO activity as
well as decreased PLO activity showed defective AA-induced
platelet aggregation. In MPD patients, various qualitative
changes in their platelets including defective platelet lipid
peroxidation by NEM (23), changes in distribution of platelet
membrane glycoproteins (24), defective binding of thrombin to
platelets (25), platelet storage pool deficiency (26) and a
deficiency of platelet α-adrenergic receptors (27) have been
reported as the basic defects possibly contributing to hypo-
function of their platelets. Furthermore, platelet resistance
to PGD_2 was described as a possible factor contributing to a
hyperfunction of the platelet or the high incidence of thrombosis (28). These defects or abnormalities could be present
in association with the altered AA metabolism, resulting in
the various abnormal patterns of platelet aggregation in such
patients depending upon a possible various combination of
these abnormalities.

The correlation between PLO and PCO activities in all MPD
patients investigated is shown in Fig. 5. All patients with
reduced PCO activities showed reduced PLO activities, but the
patients with reduced PLO activities were not always associated either with reduced or with increased PCO activities.
Correlations between the activities of these two peroxidative
pathways were not significant for normal controls, for all
patients with PLO deficiency or for all patients with selective PLO deficiency. It was reported that in normal subjects
the released of a PLO product (HETE) was increased by the inhibition of the PCO pathway in the studies in which thrombin
(29) or AA (30) was used. Our failure to demonstrate increased amounts of PCO products even when reduced amounts of
PLO products were found might be due to a sufficient amount of
exogenous AA available to the PCO pathway irrespective of the
activity of the PLO pathway at our experimental condition in
which 0.1 mM ^{14}C-AA was added to the washed platelets in the
present investigation. If a threshold (i.e. minimal) concentration of the substrate was added to the washed platelets and
the metabolites through the two pathways were analyzed, increased amounts of the PCO products might be detectable for

patients with selective PLO deficiency. Indeed, the inhibition of HETE production by a PLO enzyme inhibitor (e.g. 15-HETE) has been reported to be accompanied by stimulation of TXB2 formation (21). Secondly, even if PCO products could be increased by a PLO deficiency in the presence of normal enzyme activities of the PCO pathway, a possibly coexistent decrease in the activity of PCO pathway in MPD patients could mask the increasing effect of a PLO deficiency on the formation of PCO products, resulting in the formation of apparently normal amounts of TXB2 and HHT. Such a "latent" deficiency in the activity in these situations might be regarded as normal (i.e. as a "selective" PLO deficiency). Another possibility involves the time course for the production of the metabolites via the two pathways. When ^{14}C-AA was incubated with normal human platelets, time courses for the production of HHT and TXB2 (PCO products) appeared to be similar and faster than those observed for the production of HPETE and HETE (PLO products). HPETE appeared only after a short incubation period,

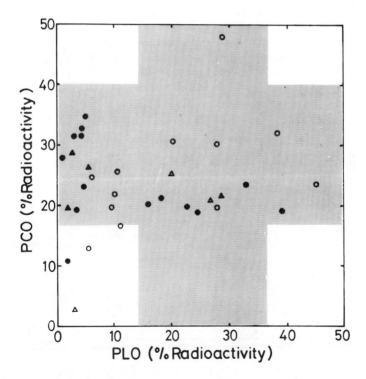

Figure 5. Correlation between PLO and PCO activities in patients with CML (●), MF (▲), PV (○) and ET (△). Normal range (dotted area): Mean ± 2 SD of normal controls (N=19) (18).

but HETE became predominant after longer periods of time (18).
Thus, the PLO pathway is activated later than the PCO pathway
and might therefore only minimally influence the latter (18).

RELATIONSHIP BETWEEN HHT (MDA) AND TXB$_2$ FORMATION BY PLATELETS

In human platelets, it was reported that formation of HHT
seemed to occur by fragmentation of the endoperoxide intermediate with expulsion of MDA (29). Both enzymatic (10,31)
and nonenzymatic (32) formations of HHT from PGH$_2$ were demonstrated. HHT as well as TXB$_2$ was enzymatically formed from
PGH$_2$ by platelet microsomes and purified TX synthetase (10,31).
Five structurally unrelated inhibitors of TX formation inhibited TXB$_2$ and HHT formation from PGH$_2$ identically, suggesting that platelet TX synthetase also catalyzes the conversion of PGH$_2$ to HHT plus MDA (33). When HHT formation was

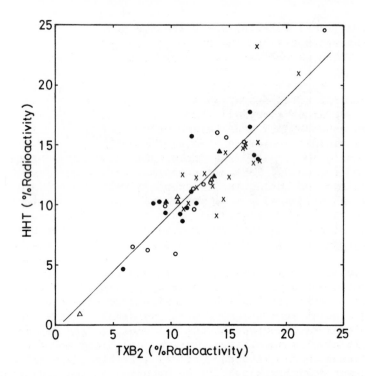

Figure 6. Correlation between HHT and TXB$_2$ formation by
platelets. ^{14}C-AA was incubated at 37°C for 30 sec with
platelets of MPD patients (●, CML; ▲, MF; ○, PV; △, ET)
and of normal subjects (×).

plotted against TXB_2 formation in all investigated normal subjects and MPD patients with various activities of the PCO
pathway, linear regression gave a slope of 0.98, an intercept
of - 0.54 and a correlation coefficient of 0.90 (Fig. 6).
Such a good correlation (p<0.001) between the formation of
these two products derived from PGH_2 is in favor of the concept that TX synthetase catalyzes the transformation of PGH_2
to both TXB_2 and HHT plus MDA. It was also reported that TXA_2
was exclusively transformed to TXB_2 and that HHT was formed
independently (33). These results have important implications
for the determination of MDA by the TBA reaction as a simple
method for estimating the PCO pathway (30,34).

A SIMPLE METHOD FOR ESTIMATION OF PLO AND PCO PATHWAYS:
USE OF THE TBA REACTION

TBA-reactive substances including MDA derive from peroxidized polyunsaturated fatty acids (35-37) and, in platelets,
peroxidation of AA is catalyzed by the lipoxygenase as well
as cyclo-oxygenase enzymes (2,29). Therefore, it was anticipated that the determination of TBA-reactive substances derived from peroxidation of AA by human platelets could be utilized as a simple method for the estimation of the PLO patyway
as well as of the PCO pathway (19), although such substances
other than MDA produced via the PCO pathway had not been
reported in human platelets. For the determination of TBA-
reactive substances, our original method (38) was slightly
modified to use a smaller amount of Bio-solv for the conveni-
ence (19). The production of TBA-reactive substance by the
incubation of AA either with aspirin-treated (i.e., cyclo-
oxygenase-inhibited) or with untreated (i.e., control)
platelets at various pH values as a function of incubation
time is shown in Fig. 7. TBA-reactive substance was not
produced within short incubation periods, especially in the
alkaline media by the platelet whose cyclo-oxygenase had been
completely inhibited by aspirin and, thereafter, gradual
increase was observed in the production over 10 to 20 min of
the incubation at pH 6.4 and 7.4. The incubation of the
aspirin-treated platelets at pH 9.0 produced a small amount
of the TBA-reactive substance. On the other hand, in experiments performed with untreated platelets, TBA-reactive substance was promptly produced, it reached almost maximal
amounts within the initial 1 - 2 min and a maximal value was
obtained at pH 7.4. Thereafter, the production at pH 6.4 was
gradually increased, whereas it leveled off at pH 7.4 and pH
9.0. Studies on the pH-dependence of ^{14}C-AA transformations
by the platelet showed that optimal pH values for the

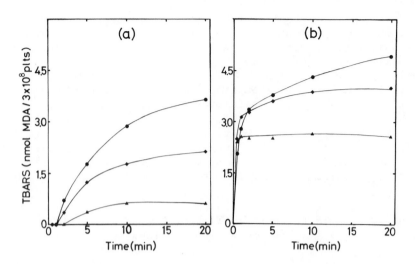

Figure 7.   TBA-reactive substances (TBARS) produced by the incubation of AA with aspirin-treated (a) or untreated (b) platelets at various pH (●—●, pH 6.4; ◆—◆ , pH 7.4; ▲—▲ , pH 9.0) as a function of incubation time (19).

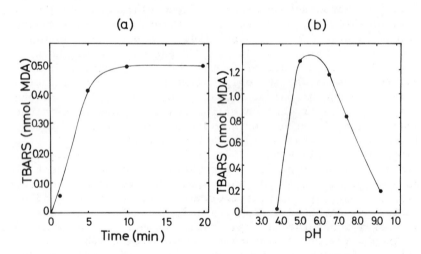

Figure 8.   Time course (a) and pH profile (b) of the production of TBA-reactive substance (TBARS) by the lipoxygenase enzyme (19).

production of HPETE and HETE (PLO products) were around 6, while those for the production of TXB_2 and HHT (PCO products) were around 7.4 (19). The incubation of the soluble fraction of broken platelets as a PLO enzyme preparation (39) with AA produced TBA-reactive substance (Fig. 8). When the incubation was performed at pH 6.5, its production increased almost linearly for 5 - 10 min and then leveled off. The use of soluble fraction which had been heated in a boiling water bath produced no TBA-reactive substance, indicating that this substance was enzymatically formed. Optimal pH for the production of this substance was around 6. An inhibitor of PLO enzyme, phenidone (40), inhibited the production of TBA-reactive substance by the incubation of AA with the PLO enzyme preparation in a concentration dependent manner (19).

Based on these experimental results, TBA-reactive substance produced by incubating AA with the cyclo-oxygenase-inhibited platelets at pH 6.5 was considered to be derived from peroxide(s) of AA produced via the PLO pathway. Although this TBA-reactive substance produced via the PLO pathway is not identified at the moment, it could be derivative(s) of lipid peroxide (e.g., HPETE) which is produced via this enzymic pathway in the platelet. The production of HPETE was faster than that of aspirin-resistant TBA-reactive substance, suggesting that some time was necessary before such derivatives were formed. TBA-reactive substances from peroxidation of γ-linolenic acid and phospholipid by soybean lipoxydase and reticulocyte lipoxygenase, respectively, have recently been reported (37,41). On the other hand, TBA-reactive substance produced by the incubation of AA with untreated platelets at pH 7.4 for a short period was considered as MDA via the PCO pathway. Therefore, such TBA-reactive substances were used as indicators for estimation of these peroxidative pathways in the platelet as follows (19). For the PLO pathway, TBA-reactive substance was determined after the incubation of 0.2 mM AA at pH 6.5 for 10 min with 10^8 platelets whose cyclo-oxygenase had completely been inhibited by the incubation with 1 mM aspirin for 5 min at 37°C. For the PCO pathway, TBA-reactive substance was determined after the incubation of 0.2 mM AA with 10^8 untreated platelets at pH 7.4 for 1 min. Activities of PLO and PCO pathways in normal subjects (N=31) as expressed in terms of nmol MDA per 10^8 platelets were 1.17 ± 0.34 (M ± SD) and 0.79 ± 0.15, respectively.

Fig. 9 illustrates effects of lipoxygenase and cyclo-oxygenase inhibitors on the activities of PLO and PCO pathways as estimated by the TBA method. Phenidone inhibited both of these enzymatic peroxidative pathways, but the inhibitory effect on the PLO pathway was much stronger than that on the PCO pathway. Furthermore, it was confirmed that aspirin inhibited the PCO pathway. When platelets were prepared from

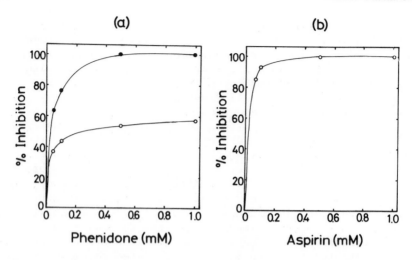

Figure 9.   Inhibitory effects of phenidone (a) and aspirin (b)
on PLO (●-●) and PCO (O-O) activities as estimated by the
TBA method (19).

PLO defective patients (18) and from aspirin-ingested normal
subjects, PLO and PCO activities, respectively, were not
detected by the TBA method.   These results confirmed the validity of the assay conditions used in the TBA method for
estimation of these peroxidative pathways in human platelets.
      Since it was reported that the TBA reaction might be used
as a convenient assay method in the evaluation of new drugs
inhibiting platelet function (30) and that the determination
of MDA in platelet-rich plasma is a simple and useful qualitative test in the study of platelet function (34), many
authors have used the TBA reaction as a simple method for
estimating the PCO pathway.   We have demonstrated that TBA-
reactive substances are produced not only by the PCO pathway,
but also by the PLO pathway when AA is incubated with washed
human platelets.   Therefore, when the TBA-reactive substance
is measured as an indicator for estimation of these peroxidative pathways in human platelets, the experimental conditions must be carefully controlled.

CONCLUDING REMARKS

      Our studies on platelet aggregation in patients with PLO
deficiency suggest that HPETE could modulate the PCO pathway
and thereby affect functions of human platelets.   The significance of HPETE has recently been suggested for the production

of TXA$_2$ and aggregation of human platelets. As the significance of the PLO pathway for human platelets has not yet been
elucidated, the estimation of the PLO pathway in various
pathologic states as well as in normal subjects must be of
great significance to clarify the biological roles of this
highly active peroxidative pathway. Since the TBA method is
simple and not time consuming as compared with the radioisotope technique, estimation of the PLO and PCO pathways by
this method should be useful for such investigations.

This research was supported in part by a grant-in-aid
(No. 557566) from the Ministry of Education of Japan.

REFERENCES

1. Weiss, H. J. (1975) New Engl. J. Med. 293, 531 and 580.
2. Marcus, A. J. (1978) J. Lipid Res. 19, 793.
3. Okuma, M., Steiner, M., and Baldini, M. (1971) J. Lab.
    Clin. Med. 77, 728.
4. Samuelsson, B. (1977) in "Biochemical Aspects of Prosta-
    glandins and Thromboxanes" (Kharash, N., and Fried, J.,
    eds.), pp. 133-155, Academic Press, New York.
5. Hammerström, S. (1980) J. Biol. Chem. 255, 518.
6. Jones, R. L., Kerry, P. J., Poyser, N. L., Walker, I. C.,
    and Wilson, N. H. (1978) Prostaglandins 16, 583.
7. Bryant, R. W., and Baily, J. M. (1979) Prostaglandins 17,
    9.
8. Walker, I. C., Jones, R. L., and Wilson, N. H. (1979)
    Prostaglandins 18, 173.
9. Hammerström, S. (1980) Prostaglandins Med. 4, 297.
10. Hammerström, S., and Falardeau, P. (1977) Proc. Natl.
    Acad. Sci. U. S. A. 74, 3691.
11. Siegel, M. I., McConnell, R. T., Abrahams, S. L., Porter,
    N. A., and Cuatrecasas, P. (1979) Biochem. Biophys. Res.
    Commun. 89, 1273.
12. Malmsten, C., Hamberg, M., Svensson, J., and Samuelsson,
    B. (1975) Proc. Natl. Acad. Sci. U. S. A. 72, 1446.
13. Lagarde, M., Byron, P. A., Vargaftig, B. B., and
    Dechavanne, M. (1978) Br. J. Haematol. 38, 251.
14. Pareti, F. I., Mannucci, P. M., D'Angelo, A., Smith, J.
    B., Santebin, L., and Galli, G. (1980) Lancet 1, 898.
15. Mestel, F., Oetliker, O., Beck, E., Felix, R., Imbach, P.,
    and Wagner, H.-P. (1980) Lancet 1, 157.
16. Lagarde, M., and Dechavanne, M. (1977) Lancet 1, 88.
17. Szczeklik, A., Gryglewski, R., Musial, J., Grodzinska, L.,
    Serwonska, M., and Mareinkiewicz, E. (1978) Thromb.
    Hemostas. 40, 66.

18. Okuma, M., and Uchino, H. (1979) Blood 54, 1258.
19. Takayama, H., Okuma, M., and Uchino, H. (1980) Thromb. Hemostas. 44, 111.
20. Born, G. V. R. (1962) Nature 194, 927.
21. Vanderhoek, J. Y., Bryant, R. W., and Bailey, J. M. (1980) J. Biol. Chem. 255, 5996.
22. Hamberg, M., Svensson, J., and Samuelsson, B. (1975) Proc. Natl. Acad. Sci. U. S. A. 72, 2994.
23. Keenan, J. P., Wharton, J., Shephard, A. J. N., and Bellingham, A. J. (1977) Br. J. Haematol. 35, 275.
24. Bolin, R. B., Okumura, T., and Jamieson, G. A. (1977) Am. J. Hematol. 3, 63.
25. Ganguly, P., Sutherland, S. B., and Bradford, H. R. (1978) Br. J. Haematol. 39, 599.
26. Gerrard, J. M., Stoddard, S. F., Shapiro, R. S., Coccia, P. F., Ramsay, N. K. C., Nesbit, M. E., Rao, G. H. R., Krivit, W., and White, J. G. (1978) Br. J. Haematol. 40, 597.
27. Kaywin, P., McDonough, M., Insel, P. A., and Shattil, S. J. (1978) New Engl. J. Med. 299, 505.
28. Cooper, B., Shafer, A. I., Puchalsky, D., and Handin, R. I. (1978) Blood 52, 618.
29. Hamberg, M., and Samuelsson, B. (1974) Proc. Natl. Acad. Sci. U. S. A. 71, 3400.
30. Hamberg, M., Svensson, J., and Samuelsson, B. (1974) Proc. Natl. Acad. Sci. U. S. A. 71, 3824.
31. Yoshimoto, T., Yamamoto, S., Okuma, M., and Hayaishi, O. (1977) J. Biol. Chem. 252, 5871.
32. Nugteren, D. H., and Hazelhof, E. (1973) Biochim. Biophys. Acta 326, 448.
33. Diczfalusy, U., Falardeau, P., and Hammerström, S. (1977) FEBS Lett. 84, 271.
34. Smith, J. B., Ingerman, C. M., and Silver, M. J. (1976) J. Lab. Clin. Med. 88, 167.
35. Patton, S., and Kurtz, G. W. (1951) J. Dairy Sci. 34, 669.
36. Dahle, L. K., Hill, E. G., and Holman, R. T. (1962) Arch. Biochem. Biophys. 98, 253.
37. Porter, N. A., Nixon, J., and Isaac, R. (1976) Biochim. Biophys. Acta 441, 605.
38. Okuma, M., Steiner, M., and Baldini, M. (1970) J. Lab. Clin. Med. 75, 283.
39. Nugteren, D. H. (1975) Biochim. Biophys. Acta 380, 299.
40. Blackwell, G. J., and Flower, R. J. (1978) Prostaglandins 16, 417.
41. Rapoport, S. M., Schewe, T., Wiesner, R., Halangk, W., Ludwig, P., Janickehöhne, M., Tannert, C., Hiebsch, C., and Klatt, D. (1979) Eur. J. Biochem. 96, 545.

LIVER AND SERUM LIPID PEROXIDE LEVELS IN PATIENTS WITH LIVER DISEASES

Toshihiko Suematsu
Hiroshi Abe

The First Department of Medicine and
the Division of Blood Transfusion
Osaka University Hospital
Osaka

The toxicity of lipoperoxide in the animal body has been amply discussed (1-3). The occurrence of a high level of lipoperoxide in animal tissues has been related to the etiology of degenerative diseases, such as atherosclerosis and retinal degeneration (4-6). Accordingly, the lipoperoxide level of the serum should be worth investigating in patients suffering from serious metabolic diseases. We studied the serum and liver lipoperoxide levels in patients suffering from liver

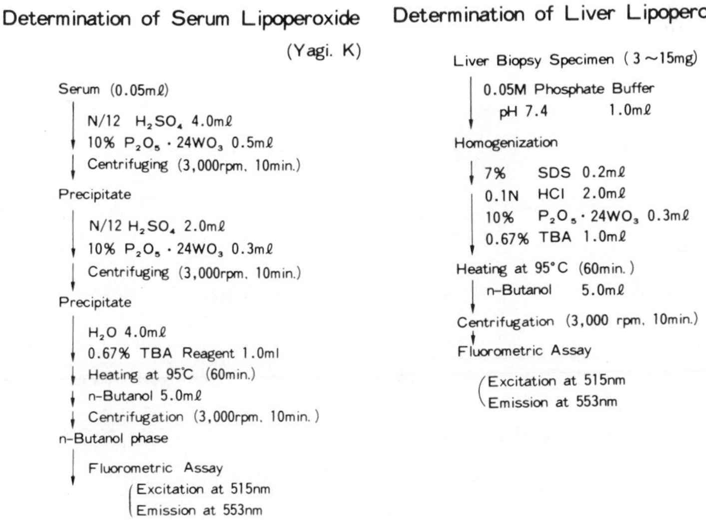

Determination of Serum Lipoperoxide

(Yagi. K)

Serum (0.05mℓ)
- N/12 H$_2$SO$_4$ 4.0mℓ
- 10% P$_2$O$_5$ · 24WO$_3$ 0.5mℓ
- Centrifuging (3,000rpm. 10min.)

Precipitate
- N/12 H$_2$SO$_4$ 2.0mℓ
- 10% P$_2$O$_5$ · 24WO$_3$ 0.3mℓ
- Centrifuging (3,000rpm. 10min.)

Precipitate
- H$_2$O 4.0mℓ
- 0.67% TBA Reagent 1.0ml
- Heating at 95℃ (60min.)
- n-Butanol 5.0mℓ
- Centrifugation (3,000rpm. 10min.)

n-Butanol phase
- Fluorometric Assay
  - Excitation at 515nm
  - Emission at 553nm

Determination of Liver Lipoperoxide

Liver Biopsy Specimen (3 ~15mg)
- 0.05M Phosphate Buffer
  pH 7.4   1.0mℓ

Homogenization
- 7%   SDS 0.2mℓ
- 0.1N   HCl 2.0mℓ
- 10%   P$_2$O$_5$ · 24WO$_3$ 0.3mℓ
- 0.67% TBA 1.0mℓ

Heating at 95°C (60min.)
- n-Butanol   5.0mℓ

Centrifugation (3,000 rpm. 10min.)

Fluorometric Assay
- Excitation at 515nm
- Emission at 553nm

Figure 1. The determination of serum and liver lipoperoxide.

LIPID PEROXIDES IN BIOLOGY AND MEDICINE

diseases using the micro assay method of Yagi (7), namely
fluorometry using thiobarbituric acid (TBA). For this purpose
0.05 ml of the serum specimen was usually used.

Fig. 1 shows the method of determination of liver lipoperoxide. The liver biopsy specimen was suspended in phosphate buffer (pH 7.4) and homogenized. Sodium dodecyl sulfate
(SDS) was then added to the homogenate to solubilize the fat
in the liver tissue. The subsequent procedure was the same as
the method of serum lipoperoxide determination.

Fig. 2 shows the serum lipoperoxide levels in healthy
subjects. Almost all values obtained were less than 4.0 nmol
in terms of malondialdehyde (MDA) per ml of serum. Therefore,
values higher than this may be considered to be abnormal.
Comparing the level in boys younger than 10 years of age with

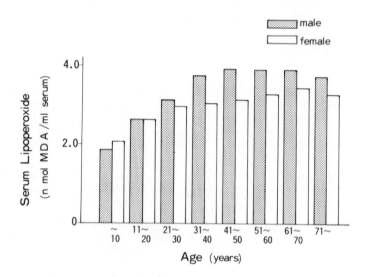

Figure 2.   Serum lipoperoxide in healthy subjects.

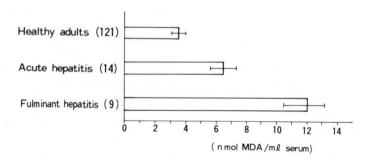

Figure 3.   Serum lipoperoxide in acute liver diseases.

that in male subjects from 11 to 20 years of age, the latter
group showed a significantly higher level. In the middle age
subjects, female subjects showed significantly lower lipoperoxide levels than male subjects. In senior subjects older
than 71 years of age, a decreasing trend in the level of serum
lipoperoxides was found.

  Fig. 3 shows the lipoperoxide level in the serum of patients suffering from acute liver disease. The levels of serum
lipoperoxide in patients with acute hepatitis and fulminant

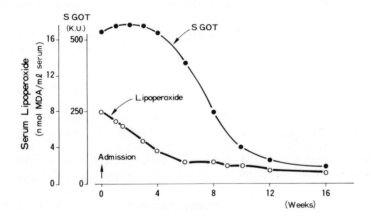

Figure 4.   Serum GOT and lipoperoxide in acute hepatitis
(S. O. 34 yr. male).

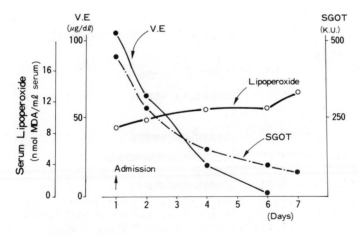

Figure 5.   Changes in serum GOT, vitamin E and lipoperoxide in
fulminant hepatitis (N. N. 45 yr, male).

hepatitis were clearly higher than those in healthy subjects.
It should be noted that extraordinary high values were obtained in fulminant hepatitis patients who were gravely ill.

Fig. 4 shows the changes in the serum glutamate oxaloacetate transaminase (GOT) and lipoperoxide levels during the
clinical course of a case of acute hepatitis following admission. Although the serum GOT level still remained markedly
high in the 1st week of admission, the serum lipoperoxide
level was observed to decrease clearly following iv administration of panthethine and other vitamins. This indicates
that the rise in the level of serum transaminase does not
always occur in parallel with the rise of the serum lipoperoxide level in acute hepatitis.

Fig. 5 shows the changes in serum GOT, vitamin E and
lipoperoxide in the clinical course of a case of fulminant
hepatitis after admission. As may be seen, the serum level of
GOT and vitamin E decreased sharply after admission, but, on
the other hand, the serum lipoperoxide level gradually rose
until death of the patient. This suggests that the inconsist-
ency between the changes in the serum transaminase and the
lipoperoxide levels in fulminant hepatitis indicates a seri-
ously ill state.

Fig. 6 shows the lipoperoxide levels in the sera and the
liver biopsy specimens of patients with chronic liver disease.
The diagnosis of chronic liver disease was made on the basis
of histological findings. In the cases of alcoholic fatty
liver, the lipoperoxide levels, both in serum and in liver
tissue, were definitely higher than the normal. The levels

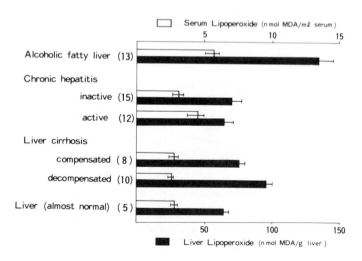

Figure 6. Serum and liver lipoperoxide in chronic liver
diseases.

of serum lipoperoxide in the cases of chronic active hepatitis
were higher than the normal level. In the cases of chronic
inactive hepatitis, however, the value of the serum could not
be concluded to be higher than the normal. The levels of
liver lipoperoxide in the cases of chronic hepatitis were
almost the same as those of healthy subjects. In liver cirrhosis, the level of serum lipoperoxide in the majority of the
patients was within the normal range, while the level of liver
lipoperoxide was significantly higher, especially in the decompensated stage of liver cirrhosis.

   Fig. 7 shows the correlation between the lipoperoxide
level in the serum and that in the liver tissue in the cases
of chronic inactive hepatitis. As can be seen, the increase
in the level of serum lipoperoxide correlated well with that
of the liver lipoperoxide. In chronic active hepatitis,

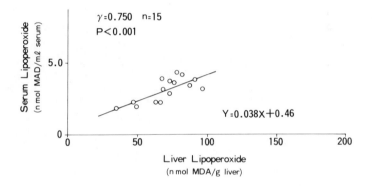

Figure 7.   Serum and liver lipoperoxide in chronic inactive
hepatitis.

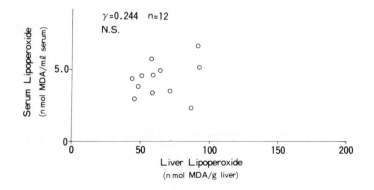

Figure 8.   Serum and liver lipoperoxide in chronic active
hepatitis.

however, the lipoperoxide level in the serum did not correlate with that in the liver tissue (Fig. 8).

The cases which were diagnosed as fatty liver are those
of particular interest for their levels of serum lipoperoxide
compared with the normal. The lipoperoxide level in the serum
in the cases of alcoholic fatty liver correlated well with
that in the liver tissue (Fig. 9). Therefore, the higher
lipoperoxide level in the serum may be considered to be a reflection of the high level in the liver tissue.

Since little change could be detected in the liver functions by the routine laboratory tests in the case of fatty
liver, the use of the serum lipoperoxide level as an index for
this disease may be clinically useful.

In the cases of compensated liver cirrhosis, the lipoperoxide level in the serum correlated significantly with that in

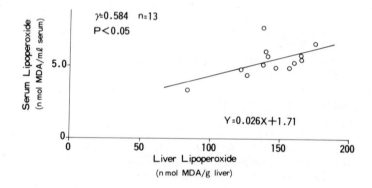

Figure 9.   Serum and liver lipoperoxide in alcoholic fatty
liver.

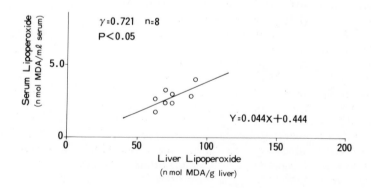

Figure 10.   Serum and liver lipoperoxide in compensated liver
cirrhosis.

the liver tissue (Fig. 10).

In decompensated liver cirrhosis, however, no significant correlation was observed between the lipoperoxide level in the serum and that in the liver tissue (Fig. 11).

In closely normal states of the liver, on the basis of histological findings, a significant correlation was observed between the lipoperoxide level in the serum and that in the liver tissue (Fig. 12).

Fig. 13 shows the relation of the lipoperoxide levels in the liver tissues with the intake of alcohol; non drinkers, light drinkers and heavy drinkers in the cases of chronic hepatitis. The lipoperoxide levels in the liver tissues of heavy drinkers were significantly higher than those of non drinkers. This indicates that the daily alcohol intake may increase the lipoperoxide in the liver.

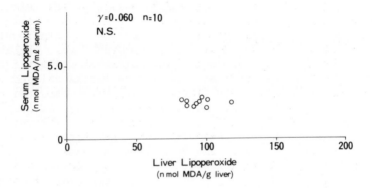

Figure 11.   Serum and liver lipoperoxide in decompensated liver cirrhosis.

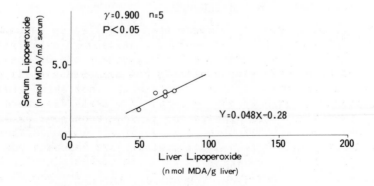

Figure 12.   Serum and liver lipoperoxide in normal liver.

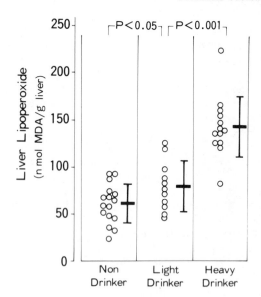

Figure 13. Alcohol intake and liver lipoperoxide levels.

    From these observations, it was concluded as follows:
1. The serum lipoperoxide levels were elevated in the cases
of acute hepatitis, fulminant hepatitis, chronic active hepatitis, and fatty liver.
2. A correlation between the lipoperoxide level in the serum
and that in the liver tissue was observed in chronic inactive
hepatitis, fatty liver and compensated liver cirrhosis.
3. Since little change could be detected in routine liver
function tests in the cases of fatty liver, the measurement of
the level of lipoperoxide in the serum may serve as an effective marker for differential diagnosis of fatty liver.
4. The level of liver lipoperoxide in heavy drinkers was
higher than that in non drinkers.
5. The occurrence of high levels of lipoperoxide in liver
tissues may be related to the etiology of liver diseases.

REFERENCES

1. Kalish, G. H., and DiLuzio, N. R. (1966) Science 152,
    1390.
2. Alpers, D. H., Solin, M., and Isselbacher, K. J. (1968)
    Mol. Pharmacol. 4, 566.
3. Tappel, A. L. (1973) Fed. Proc. 32, 1870.
4. Glavind, J., Hartmann, S., Clemmensen, J., Jensen, K. E.,

and Dam, H. (1952) Acta Pathol. Microbiol. Scand. 30, 1.
5. Hiramitsu, T., Hasegawa, Y., Hirata, K., Nishigaki, I., and Yagi, K. (1976) Experientia 32, 622.
6. Hiramitsu, T., Majima, Y., Hasegawa, Y., Hirata, K., and Yagi, K. (1976) Experientia 32, 1324.
7. Yagi, K. (1976) Biochem. Med. 15, 212.

LIPID PEROXIDES AS A CAUSE OF VASCULAR DISEASES

Yuichiro Goto

Department of Medicine
Tokai University School of Medicine
Bohseidai, Isehara

I. INTRODUCTION

It might be mentioned that the research on the relation between lipid peroxides and vascular diseases originated from the report of Glavind et al. (1), who demonstrated an intimate correlation between lipid peroxide level in the aortic wall and the degree of atherosclerosis in 1952. This finding was later confirmed by several researchers. A simple and reliable method of assay for lipid peroxides in blood serum and plasma devised by Yagi (2,3) seemed to promote the investigation by Japanese researchers in the clinical field.

In this communication, some recent data on lipid peroxides related to vascular diseases, which were obtained mainly by Japanese researchers, will be reviewed.

II. ATHEROSCLEROSIS

The active role of oxidation products of lipids in the atherosclerosis genesis suggested by Glavind et al. (1) was confirmed by Fukuzumi, and Aoyama and Iwakami; Fukuzumi demonstrated the occurrence of lipid peroxides in human atheromatous cell wall by infrared absorption technique (4) and Aoyama and Iwakami by using rabbits suffering from atherosclerosis by administration of cholesterol or lanolin (5,6). This was followed by the isolation of hydroperoxides of cholesterol linoleate from the lipids of advanced atherosclerotic plaques of human aortas by Harland et al. (7).

Since Yagi devised a simple fluorometric assay method for
lipid peroxides in blood using thiobarbituric acid reaction,
many data on lipid peroxide levels in blood of atherosclerotic
subjects have been reported.

Sato et al. (8) observed that among diabetics, the patients with various kinds of angiopathy showed higher levels
of plasma lipid peroxides in plasma. This would be, at least
partly, the cause of angiopathy in patients suffering from
diabetes.

To make this relation clear, Nishigaki et al. (9) examined the lipid peroxide levels of serum lipoprotein fractions of
diabetic patients in relation to some abnormal features of
lipid metabolism. Their results are summarized in Table I.
Among lipoprotein fractions of sera, the LDL fraction contains
a larger amount of lipid peroxides than the other fractions,
VLDL and HDL, in either normal subjects or diabetics. However,
it should be noted that the level in HDL fraction of diabetic
serum was significantly higher than that of normal serum, indicating that the increase in lipid peroxide levels observed
in the diabetic patient is due to that in the HDL fraction.
In particular, they demonstrated that the ratio of lipid peroxide levels to total lipid in LDL fraction is slightly higher
and that in the HDL fraction is markedly higher in diabetic
patients than in the normal subjects, and that in diabetic
serum, the lipid peroxide level in the HDL fraction correlated
significantly to the contents of triglyceride and phospholipid
in the same fraction, though no correlation was observed in
the HDL fraction of normal serum. These results indicate that
the increase in lipid peroxides of the sera would have some
relevance to the disorder of lipid metabolism in the patients.

It was reported by Miki et al. (10) that the blood lipid

TABLE I. Lipid Peroxide Levels of Serum Lipoprotein Fraction
of Normal and Diabetic Subjects

	n	Lipid peroxides*		
		VLDL	LDL	HDL
Normal	32	0.64 ± 0.30	1.18 ± 0.33	0.68 ± 0.16
Diabetic	31	0.68 ± 0.34	1.26 ± 0.35	1.07 ± 0.40**

*Lipid peroxide level is expressed in terms of malon-
  dialdehyde (nmol/ml serum). Mean ± SD is given.
**P < 0.001.
The data of Nishigaki et al. (9).

peroxide level of the patients suffering from atherosclerosis
was significantly higher than that of normal subjects (Fig.
1).

Dormandy et al. (11) also reported that in intermittent
claudication, atherosclerosis in the lower extremities, there
was significant correlation between some clinical findings and
the susceptibility of the red cells to autoxidation, suggesting an increase in lipid peroxide levels in the blood.

Many researchers investigated the role of lipid peroxides
in atherogenesis using experimental animals.

Hata et al. (12) demonstrated that rabbits fed on a diet
with 1% cholesterol for about 10 weeks showed a 1.5 times
higher level of serum lipid peroxides than that of rabbits
bred without cholesterol (Table II), and that the increased
lipid peroxides mostly resided on the VLDL-LDL fraction of the
serum. However, the ratio of lipid peroxide levels to total
lipid in the HDL fraction was markedly higher than that in the
VLDL-LDL fraction (Fig. 2). Furthermore, in aortic lesions of
rabbits fed on cholesterol, the lipid peroxides increased 15
times compared with the control. Yamasaki (13) also studied
the relationship between lipid peroxides and atherosclerosis
using cholesterol-fed rabbits and found that the amount of
lipid peroxides in cholesterol ester fraction of the aorta was
52.7 nmol/g wet weight of tissue, which corresponded to about
50% of total lipid peroxides in the aorta.

All these results suggest that lipid peroxides have some
role for the initiation of atherogenesis in the aorta and that
the increase in lipid peroxide level in the blood would provoke disorders in the intima of the aorta as the initial event
of atherogenesis. To make this point clear, Yagi et al. (14)

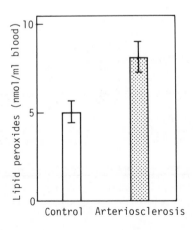

Figure 1.   Blood lipid peroxide levels of atherosclerotic
subjects.   Data of Miki et al. (10).

TABLE II.   Lipid Peroxide Levels of Rabbit Serum after
Administration of Cholesterol*

Admini-stration of cholesterol	Body weight (g)	Serum cholesterol (mg/dl)	Lipid peroxides (nmol/ml)	Surface area of lesion (%)
−	2552 ± 544 (n = 11)	51 ± 20 (n = 11)	0.91 ± 0.14 (n = 11)	0.0 ± 0.0 (n = 11)
+	3515 ± 121 (n = 8)	1097 ± 269 (n = 8)	1.47 ± 0.55 (n = 8)	14.7 ± 8.5 (n = 8)

*Male rabbits were fed on the diet containing cholesterol
(1 g/day) for 67 ± 1 days.
Data of Hata et al. (12).

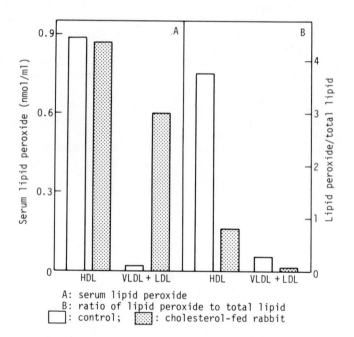

A: serum lipid peroxide
B: ratio of lipid peroxide to total lipid
☐ : control;   ▨ : cholesterol-fed rabbit

Figure 2.   Lipid peroxide levels of serum lipoprotein fractions and the ratios of lipid peroxide levels of total lipid
in each lipoprotein fraction of cholesterol-fed rabbit.   Data
of Hata et al. (12).

examined the direct effect of injected lipid peroxide on the intima of rabbit aorta by electron microscopy and by determination of intravenously administered linoleic acid hydroperoxide in the blood. They showed that drastic morphological changes in the intima and endothelial cells of the aorta were provoked by a single injection of linoleic acid hydroperoxide. The half-life of the injected hydroperoxide in the serum was about 50 min, indicating that this level of lipid peroxide in the serum was sufficient to cause the above mentioned changes in rabbit aorta.

Three possible mechanisms seem to cause atherosclerosis by elevating lipid peroxide levels in the serum.

The first possibility is that high lipid peroxide concentrations increase platelet aggregation which induces atherosclerosis. It has been known that the injection of fatty acid provokes platelet aggregation. Injection of a large amount of polyunsaturated fatty acid, especially arachidonic acid, can cause the formation of a microthrombus causing death of animal. Hydroperoxide of arachidonic acid also causes stronger aggregation of platelets than the original acid. This may be due to the changes in platelet membranes by lipid peroxides. It has been shown that such aggregation or adhesion of platelets could become a trigger of atherosclerosis through endothelial cell injury or smooth muscle cell proliferation.

The second possibility is direct disturbance of the endothelial or intimal cells by lipid peroxide. If abnormally high concentrations of lipid peroxides continue to exist in the plasma, the endothelial cells are exposed to lipid peroxides for a long time. A reaction then occurs between lipid peroxides and protein in the cells and further degeneration of membranes and change in membrane permeability could by provoked. Recent pathological evidence emphasizes the change in membrane fluidity of endothelial or intimal cells as the initial state of atherosclerosis.

The third is that high lipid peroxide concentrations influence the synthetic system of prostacyclin within the arterial tissue. Prostacyclin (PGI_2) is biosynthesized from arachidonic acid derived from plasma free fatty acid or that liberated from the β-position of phospholipid with phospholipase A_2. This reaction is catalyzed by cyclooxygenase and prostacyclin synthetase in the microsomal fraction of the arterial wall through prostaglandin endoperoxide. The activity of prostacyclin synthesis in the arterial wall is highest in the intima, followed by media and adventitia, and it is especially higher in the venous endothelium than in artery. The increase of lipid peroxides in the plasma inhibits the biosynthesis of prostacyclin in the arterial wall. Prostacyclin inhibits the aggregation of platelets and the contraction of blood vessel.

At present, the precise mechanism for atherogenesis is not known, but the above-mentioned points are though to be the future target of research to expose the relation between lipid peroxides and atherosclerosis.

III. APOPLEXY

As the main pathological changes linking to the apoplexy, it is observed that cerebrosclerosis is co-provoked by thrombus. The accumulation of some morbid changes occurs first in cerebral artery, thrombus then occurs by platelet aggregation, and further bleeding such as subarachnoid hemorrhage is provoked. At present, these changes are being pursued in relation to lipid peroxides. Expecially cerebropathy in the case of cerebral ischemia is thought to have close relation to lipid peroxides.

Tomita et al. (15) found that the level of lipid peroxides in the blood of healthy stroke-prone spontaneously hypertensive rats (SHRSP) was not significantly different from

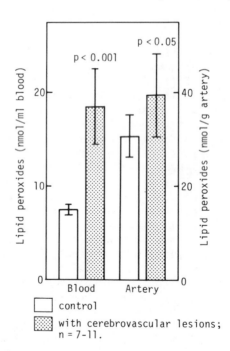

Figure 3. Lipid peroxide levels of the blood and artery of SHRSP with or without cerebrovascular lesions. Data of Tomita et al. (15).

that of normotensive Wistar Kyoto rats, while the level of
lipid peroxides of SHRSP blood with stroke was more than twice
as high as that of healthy SHRSP (Fig. 3). They also observed
that the lipid peroxide level of the blood of SHRSP with
stroke increased slightly 24 hr before stroke and that it increased abruptly simultaneously with stroke.

    Kibata et al. (16) examined the blood lipid peroxide
levels of patients suffering from cerebral apoplexy and found
that at fresh stroke, it increased twice compared with the
control of old age subjects, while at past stroke it was very
low.

    According to Kagami et al. (17), the time course of blood
lipid peroxides of the patients of cerebral apoplexy showed 2
kinds of pattern either in cerebral hemorrhage or in cerebral
infarction. In the group that died, lipid peroxide levels increased in the final stage, while in the group that survived,
it decreased gradually to the control level (Fig. 4,5).

    Yoshikawa et al. (18) examined the lipid peroxide levels
of the serum and cerebrospinal fluid of patients with cerebral
apoplexy, and found that these two levels were correlated
significantly in spite of the bleeding.

    Sasaki et al. (19) also examined the lipid peroxide
levels of cerebral spinal fluid of subjects with subarachnoid
hemorrhage. In their results, the level in patients was twice
that of the controls. In those cases without vasospasm, low
levels of lipid peroxides were demonstrated. In cases with
vasospasm, high levels were found at the initial stage of

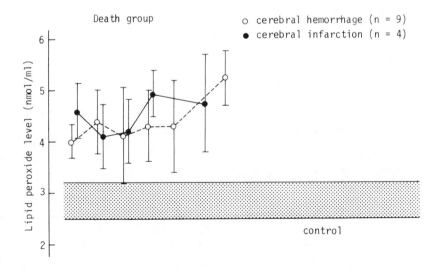

Figure 4. Changes in lipid peroxide levels in death group
after fresh stroke. Data of Kagami et al. (17).

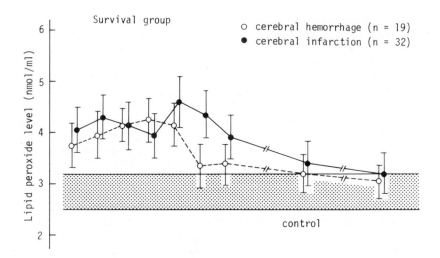

Figure 5.   Changes in lipid peroxide levels in survival group
after fresh stroke.   Data of Kagami et al. (17).

hemorrhage which decreased gradually.
     The mechanism thought to cause the cerebral ischemia was
participation of free radical formed under the conditions of
low oxygen concentration (20) and that of high lipid peroxide
levels derived by reflow of blood after ischemia (21).   In any
case, the above mentioned results suggest that the lipid peroxide levels play a notable role in the provocation of cerebral stroke.

REFERENCES

1.   Glavind, J., Hartmann, S., Clemmesen, J., Jessen, K. E.,
     and Dam, H. (1952) Acta Pathol. 30, 1.
2.   Yagi, K., Nishigaki, I., and Ōhama, H. (1968) Vitamins
     39, 105 (in Japanese).
3.   Yagi, K. (1976) Biochem. Med. 15, 212.
4.   Fukuzumi, I. (1965) Yukagaku 14, 119 (in Japanese).
5.   Aoyama, S., and Iwakami, M. (1965) Jpn Heart J. 6, 128.
6.   Iwakami, M. (1965) Nagoya, J. Med. Sci. 28, 50.
7.   Harland, W. A., Gilbert, J. D., Steil, G., and
     Brooks, C. J. W. (1971) Atherosclerosis 13, 239.
8.   Sato, Y., Hotta, N., Sakamoto, N., Matsuoka, S.,
     Ohishi, N., and Yagi, K. (1979) Biochem. Med. 21, 104.
9.   Nishigaki, I., Hagihara, M., Tsunekawa, H., Maseki, M.,
     and Yagi, K. (1981) Biochem. Med. 25, 373.

10. Miki, M., Hamada, N., Okuda, F., and Wada, K. (1981)
     Personal Communication.
11. Dormandy, J. A., Hoare, E., Khattab, A. M., Arrowsmith,
     D. E., and Dormancy, T. L. (1973) Brit. Med. J. 4, 581.
12. Hata, Y., Yamamoto, M., and Miyazaki, K. (1980)
     Domyakukoka 8, 303 (in Japanese).
13. Yamasaki, S. (1981) J. Lipid Peroxide Res. in press (in
     Japanese).
14. Yagi, K., Ohkawa, H., Ohishi, N., Yamashita, M., and
     Nakashima, T. (1981) J. Appl. Biochem. 3, 58.
15. Tomita, I., Sano, M., Serizawa, S., Ohta, K., and
     Katou, M. (1979) Stroke 10, 323.
16. Kibata, M., Shimizu, Y., Miyake, K., Shimono, M.,
     Shoji, K., Miyahara, K., Fuchimoto, T., and Nasu, Y.
     (1977) Igaku no Ayumi 101, 591 (in Japanese).
17. Kagami, M., Iwashiro, Y., Kawamoto, M., Takaoka, S.,
     Kato, T., Kasahara, N., Atarashi, J., Terashi, A.,
     Nagatsumi, A., Miyazaki, T., Katsuragi, R., and Sakai, H.
     (1978) J. Lipid Peroxide Res. 2, 115 (in Japanese).
18. Yoshikawa, T., Yamaguchi, K., Kondo, M., Minakawa, N.,
     Kamikuchi, T., Ohta, T., Yamaki, T., and Hirakawa, K.
     (1981) Proceedings of Symposium on Brain Ischemia and
     Cell Injury, pp. 76 (in Japanese).
19. Sasaki, T., Mayanagi, Y., Shimasaki, M., Kabota, M.,
     Asano, T., and Sano, K. (1980) Igaku no Ayumi 113, 970
     (in Japanese).
20. Flamm, E. S., Demopoulos, H. B., Seligman, M. L., Poser,
     R. G., and Ransohoff, J. (1978) Stroke 9, 445.
21. Kogure, K., Morooka, M., Busto, R., and Scheinberg, P.
     (1979) Neurology 29, 546.

LIPID PEROXIDE IN THE AGING PROCESS

Shunsaku Hirai
Koichi Okamoto
Mitsunori Morimatsu

Institute of Neurology and Rehabilitation
Gunma University School of Medicine
Maebashi

I. INTRODUCTION

The presence of a large amount of unsaturated lipids in cells, high susceptibility of these lipids to peroxidation and the toxicity of resultant lipid peroxide make us suggest the possible role of lipid peroxidation in various pathological states. In fact, many works on lipid peroxidation in biological systems have primarily been concerned with correlating it with various pathological conditions and aging process.

Because of the diversity of age-related phenomena, this paper mainly deals with the following two points: The first point is age-related changes in lipid peroxide contents of the brain, and the second point is vascular damage induced by lipid peroxide.

II. AGE-RELATED CHANGES IN LIPID PEROXIDE CONTENT OF THE BRAIN

The significance of lipid peroxidation in aging process is based primarily on the following facts that lipofuscin age pigments accumulate almost linearly with advancing age and these pigments presumably result from polymerization of oxidized unsaturated lipids.

This pigment has been known for many years as one of the most important age-related morphological changes since its discovery by Hannover in 1842. In 1915 Ciaccio suggested

LIPID PEROXIDES IN BIOLOGY AND MEDICINE

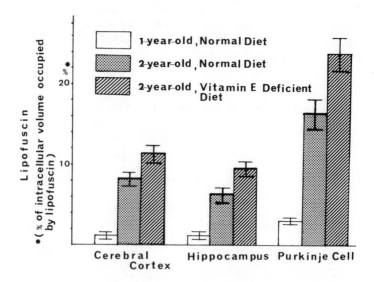

Figure 1. Relationship between age and vitamin E deficiency
on the formation of lipofuscin pigments in rat brain.

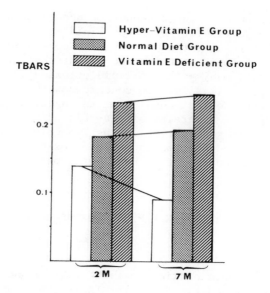

Figure 2. Effect of vitamin E on thiobarbituric acid reacting
substance (TBARS) in rat brain.

their genesis from unsaturated lipids by histochemical analysis and in 1938 Einarson and Ringsted pointed out the similarity between this pigment and ceriod pigment which appears in various tissues of vitamin E deficient animals.

From these results it has been suggested that not only ceroid, but also lipofuscin age pigments are probably derived from lipid peroxide.

As shown in Fig. 1, our experimental results also revealed that lipofuscin pigments accumulate in nerve cells of the rat brain as a function of age. These pigments increased in all areas of the brain, but in rats and guinea pigs Purkinje cells in the cerebellum showed the most prominent accumulation.

When rats were fed on vitamin E-deficient diet, these pigments accumulated much more significantly than those of age-matched controls.

Previously only histochemical approaches had been used to study the relationship between lipid peroxide and these pigments, but in 1966 we first pointed out the increase of thiobarbituric acid reacting substance (TBARS) in the brain of

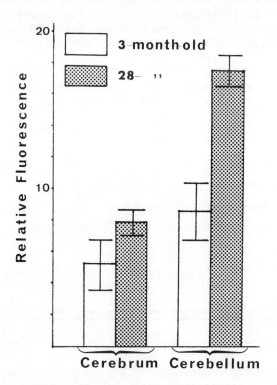

Figure 3. Increase of chloroform-methanol extractable fluorescence in guinea pig brain with aging.

rats with advancing age (1,2). When vitamin E was further
supplemented by intramuscular injection (17.5 mg/100 g body
weight/week), TBARS in the brain decreased significantly as
shown in Fig. 2. While, on the other hand, vitamin E deficiency enhanced the increase of TBARS, which correlated well
with increased accumulation of lipofuscin pigments.

   Age-related increase of TBARS has been confirmed on various tissues including liver (3), adrenal gland (4) and others
(5) thereafter.

   Estimation of TBARS had been almost the only method to
measure lipid peroxide in biological materials until a sensitive fluorometric assay method was reported by Tappel and co-
workers in 1973 (6,7). Application of this method in measurement of lipid peroxide contents of guinea pig brain revealed
that lipid solvent soluble lipid peroxide also increased with
advancing age (Fig. 3).

   This chloroform–methanol soluble fraction was then further chromatographed on LH–20 column and eluted with 1 : 9
chloroform-methanol according to the method by Csallany and
Ayaz (8). Two–milliliter fractions were collected and the
fluorescence of each of the fraction was determined with
Hitachi 204-S spectrophotofluorometer at an excitation wavelength of 365 nm and emission wavelength of 435 nm. Quinine
sulfate at a concentration of 1 μg/ml of 0.1 N H_2SO_4 was used
as a standard for fluorescence intensity.

   As illustrated in Fig. 4, only a small amount of fluorescent products was detected in 1-month-old guinea pigs. While,
it increased with advancing age and at the age of 30 months,

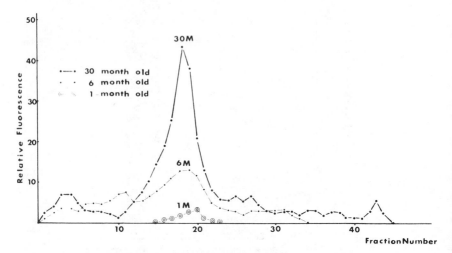

Figure 4. Elution profiles of Sephadex LH–20 column chromatography of chloroform–methanol soluble fluorescent products in
guinea pig brain.

one prominent peak could be observed among several less remarkable peaks. The excitation and emission maxima of this
fraction were 365 nm and 420 nm, respectively. Detailed
chemical nature of this fraction is not clear as yet, but, at
least, it can be said that this fraction contains the most
specific age-related lipid peroxide in the brain.

It has been proposed that fluorescent products arising
from lipid peroxidation are conjugated Schiff base fluorophores with the basic structure $-N=C-C=C-N-$, and these products can be classified into three solubility classes; water
soluble, lipid solvent soluble and insoluble (9).

The fluorescent products which characterize lipofuscin
age pigments are largely the insoluble products and they can
be produced synthetically from mixtures of protein and peroxidizing polyunsaturated fatty acids (10,11).

As regards the lipid solvent soluble fluorescent products, our experiment as well as some other reports indicates
that chloroform-methanol extractable fluorescence in biological tissues also increases as a function of age. A major portion of fluorescent lipid peroxidation products in biological
systems was found in this solubility class and was reported to
be derived from the reaction of malonaldehyde, a peroxidation
product of polyunsaturated lipids, with amino acids. But, as
it was not clear whether TBARS could change and whether similar fluorescent products could be produced in the course of
peroxidation and polymerization of pure unsaturated fatty
acids, the following experiment was carried out.

Pure polyunsaturated fatty acids, including linoleic,
linolenic and arachidonic acids, were allowed to oxidize at
25°C under oxygen and irradiation by ultraviolet light. At
various stages of oxidation, samples of approximately 30 mg
were taken and dissolved in 2 : 1 chloroform-methanol in a
volume-to-weight ratio of 20 : 1. The chloroform-methanol ex-
tracts were processed for estimation of peroxide value, TBARS,
and fluorescence intensity. Peroxide value and TBARS were
measured by modified Dahle's methods (12). Fluorescence measurements were made on a Hitachi 204-S spectrophotofluorometer
at an excitation wavelength of 360 nm and an emission wavelength of 420 nm with quinine sulfate (1 µg/ml 0.1 N H_2SO_4) as
a standard.

TBARS of linolenic and arachidonic acids were found to
vary linearly with peroxide value throughout the course of the
oxidation, while linoleic acid yielded no TBARS. As grade of
unsaturation increased, the molar yield of TBARS increased.
These results were in accord with the observations by Dahle
and co-workers (12).

Fig. 5 indicates the interrelationship between TBARS,
peroxide value and fluorescence intensity during the course of
the oxidation of arachidonic acid. TBARS and peroxide value

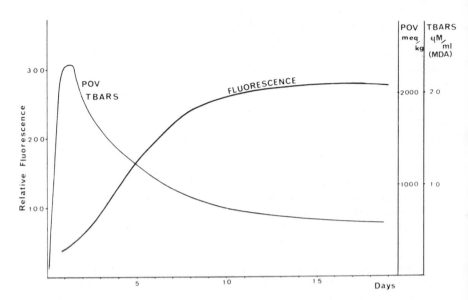

Figure 5.   Relationship between peroxide value (POV), thiobarbituric acid reacting substance (TBARS) and fluorescence intensity in the course of oxidation of arachidonic acid.

increased fairly rapidly in the early stage of oxidation and then decreased.   On the other hand, fluorescence intensity began to increase in the time period that TBARS tended to decrease.   Then fluorescence intensity reached a plateau level and remained more or less constant thereafter.   Oxidized arachidonic acid showed excitation maximum at 360 nm and emission maximum at 420 nm after 24 hours of oxidation, but excitation and emission maxima shifted to longer wavelength in the later stage showing 380 nm and 440 nm, respectively.   Fluorescence excitation and emission spectra as well as fluorescence intensity remained almost constant throughout the remainder of the examination (20 days).

The mechanism and nature of such fluorescent products remain obscure, but it is quite possible that these fluorophores are produced from polymerized lipid peroxides.   It seems unlikely that such products occupy a considerable portion of lipid solvent soluble fluorescent products in biological tissues, as biological systems contain protein in close association with oxidizing lipids.   However, these results may indicate that for measurement of lipid peroxidation damage to biological systems, estimation of not only TBARS, but also fluorescence is necessary.

III. VASCULAR CHANGES INDUCED BY LIPID PEROXIDE

The second part of this paper deals with vascular changes induced by lipid peroxide.

Since the report of Glavind, who demonstrated the accumulation of lipid peroxide in atherosclerotic aortae, the correlation between lipid peroxide and atherosclerosis has been suggested, we have had little information as to the relationship between the two. On the other hand, it has recently been observed that prostacyclin (PGI$_2$) is formed in vascular tissues, especially in the endothelial cells, and that some lipid peroxides inhibit synthesis of this substance.

The purpose of this study was to ascertain whether vascular damage could be induced by 15-hydroperoxy arachidonic acid (15-HPAA), which is one of such lipid peroxides that inhibit synthesis of PGI$_2$.

15-HPAA which we used was prepared and purified by Prof. Nakano and Dr. Sugioka, Gunma University.

First, this substance was examined to see if it could actually inhibit PGI$_2$ synthesis. PGI$_2$-like substance was estimated by its activity to inhibit platelet aggregation (13). Rabbits were anesthetized with intravenous injection of sodium pentobarbital and descending aortae were removed. After rinsing in cold saline, the aorta was cut into approximately 2 mm long segments. These aorta rings were first incubated in solution A (0.25 ml/10 mg wet weight) for 30 minutes. Solution A was a mixture of 15-HPAA (0 - 1000 γ/ml Na$_2$CO$_3$) and pH 9.0 borate-buffered saline (0.1 M borate buffer pH 9.0/0.15 M NaCl, 1 : 9 v/v) in a volume-to-volume ratio of 1 : 9. The aorta rings were then incubated in the same amount of pH 9.0 borate-buffered saline (solution B) for 60 minutes. Each 10 μl of solution A or solution B in which the aorta ring had been incubated was added to 215 μl platelet rich plasma 1 minute before the addition of 25 μl collagen solution (3 μg/ml), and the platelet aggregation was recorded by NKK aggregometer.

As shown in Fig. 6, 15-HPAA strongly inhibited PGI$_2$ synthesis, because solution B contained PGI$_2$ newly formed in the aorta ring during 60 minutes' incubation after addition of 15-HPAA. But this substance showed no effect on the activity of PGI$_2$ which had already been formed before the addition of 15-HPAA.

15-HPAA was then given to rabbits by intravenous injection. The rabbits were killed at various intervals after injection and fixed by in vivo fixation with 1% glutaraldehyde solution in phosphate buffer. The aortae were processed for scanning and transmission electron microscopy. In chronic experiment, each rabbit was given 15-HPAA at each dose level of 1 mg/g body weight intravenously, once every two weeks for

Shunsaku Hirai *et al.*

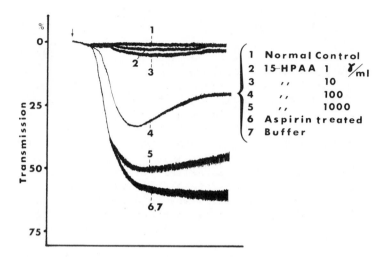

Figure 6.   Effect of incubation fluids of 15–HPAA treated
aortic samples on platelet aggregation.

Figure 7.   Scanning electron micrograph of the inner surface
of the aorta of normal control rabbit showing smooth and regularly waved surface.   Compare with Fig. 8A and B.   Scale indicates 5 μ length.

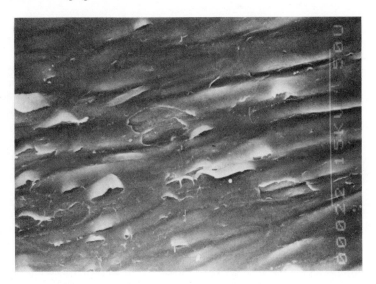

Figure 8A. Scanning electron micrograph of the inner surface of the aorta of the rabbit which received 15-HPAA by intravenous injection showing marked exfoliation of the intima. Scale indicates 50 μ length (acute experiment).

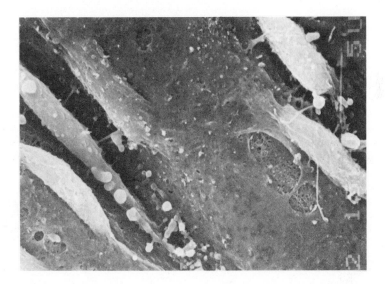

Figure 8B. Higher magnification of the inner surface of the same aorta as shown in Fig. 8A showing damaged areas in detail. Multiple exfoliations and adhesion of platelets to damaged sites can be observed. Scale indicates 5 μ length.

ten weeks and sacrificed one week after the last injection. In acute experiment, each rabbit received 3 mg/g body weight of 15-HPAA by single intravenous injection and killed after one hour of injection.

As demonstrated in Fig. 7, scanning electron micrograph of the inner surface of the aorta of normal control rabbit showed smooth and regularly waved surface. While the inner surface of the aorta of the rabbit which received 15-HPAA showed marked exfoliation of the intima with occasional adhesion of platelets to the damaged areas (Fig. 8A and B).

On the other hand, the activity of the damaged aorta to produce PGI_2, estimated by the method described above, showed only slight decrease even in the acute experiment and almost no decrease in the chronic experiment in spite of such severe morphological changes. It seems unlikely that these damages were secondary to inhibition of PGI_2 synthesis. As it has recently been reported that hydroperoxide of linoleic acid can also induce similar vascular changes, it seems quite probable that some lipid peroxides can produce vascular damages.

At present, it is not clear whether such vascular changes are correlated with atherosclerosis or not, but these results may suggest the importance of lipid peroxide in the pathogenesis of vascular damages including atherosclerosis.

IV. CONCLUDING REMARKS

From the various experiments which have been described, it seems reasonable to conclude that there is a close relationship between lipid peroxide and aging process.

Aging phenomena can be divided into two categories; normal and pathological. Accumulation of lipofuscin age pigments is an example of the former, while atherosclerosis has been considered as a pathological aging phenomenon of the vascular tissue. Therefore, lipid peroxide may probably be correlated with both categories of aging phenomena.

Further studies on this problem are needed because more diverse age-related phenomena might be correlated with lipid peroxide as it is suggested by Harman's free radical theory of aging.

REFERENCES

1. Hirai, S., and Yoshikawa, M. (1966) Jap. J. Geriat. 3, 256.
2. Yoshikawa, M., and Hirai, S. (1967) J. Gerontol. 22, 162.

3. Takeuchi, N., Tanaka, F., Katayama, Y., Matsumiya, K.,
    and Yamamura, Y. (1976) Exp. Gerontol. 11, 179.
4. Osawa, N., and Ibayashi, H. (1969) Clin. Endocrinol. 17,
    42.
5. Uchiyami, M., and Mihara, M. (1978) Analyt. Biochem. 86,
    271.
6. Fletcher, B. L., Dillard, C. J., and Tappel, A. L. (1973)
    Analyt. Biochem. 52, 1.
7. Tappel, A. L., Fletcher, B. L., and Deamer, D. (1973) J.
    Gerontol. 28, 415.
8. Csallny, A. S., and Ayaz, K. L. (1976) Lipids 11, 412.
9. Chio, K. S., and Tappel, A. L. (1969) Biochemistry 8,
    282.
10. Tappel, A. L. (1955) Arch. Biochem. Biophys. 54, 266.
11. Hirai, S., and Yoshikawa, M. (1970) Proc. 4th Internat.
    Symp. on Vit. E, p. 288.
12. Dahle, L. K., Hill, E. G., and Holman, R. T. (1962) Arch.
    Biochem. Biophys. 98, 253.
13. Okuma, M., Yamori, Y., Ohta, K., and Uchino, H. (1979)
    Prostaglandins 17, 1.

PEROXIDE-MEDIATED METABOLIC ACTIVATION OF CARCINOGENS

Peter J. O'Brien

Department of Biochemistry
Memorial University of Newfoundland
St. John's, Newfoundland, Canada

A number of pathological conditions stemming primarily
from exposure to toxic chemicals are believed to be manifested
through a process of lipid peroxidation. Peroxidation of the
lipids of animal tissues in vivo may be induced by carbon
tetrachloride (1) or ethanol (2). Lipid peroxidation in vivo
is increased during lung damage by oxygen toxicity and liver
damage by various drugs, herbicides and pesticides, particularly those which deplete intracellular GSH levels. The metals
iron, copper or lead also enhance lipid peroxidation in vivo
(3). The peroxidation is enhanced by a Vitamin E or a selenium deficient diet (4). Dietary fatty acid hydroperoxides
can be toxic to the gastrointestinal tract and can be carcinogenic (5,6). Antioxidants or selenium markedly protect from
chemical carcinogens including polycyclic aromatic hydrocarbons, nitrosamines, nitroquinoline, urethane, uracil mustard
and arylamines (7). Free radicals are also often associated
with chemical carcinogenesis (8).

During an investigation into the intracellular mechanisms
for lipid peroxide decomposition, I discovered some time ago
that the cytosol formed one product whereas microsomes formed
a wide range of products. The cytosolic enzyme was identified
as glutathione peroxidase, and was found to catalyse the reduction and thus detoxification of the hydroperoxide by GSH
to hydroxy fatty acids (9). The microsomal enzyme was later
identified as cytochrome P450 and involved the peroxide decomposition to harmful lipid free radicals which interacted to
form a complex range of products (10). Cytochrome P420 was six
fold more effective than cytochrome P450. Cytochrome P450 but
not cytochrome P420 is also a particularly powerful peroxidase
with tertiary hydroperoxides (11). The electron donors were
however oxidised to products similar to the mixed function
oxidase activity rather than to the complex range of free

radical mediated products that characterise peroxidase reactions. The significance of this activity was realised when it
was found that tertiary hydroperoxides could substitute for
NADPH, NADPH:cytochrome P450 reductase and O_2 in the mixed
function oxidase reaction [reviewed (12)]. Thus the tertiary
hydroperoxides carried out N-dealkylation, O-dealkylations,
side chain oxidations and aromatic ring hydroxylation (13).
A similar donor specificity with different animal species was
observed. The induction by phenobarbital or 3 methylcholanthrene was accompanied by the same differences in products
formed. Furthermore an epoxide intermediate was demonstrated
during acetanilide and phenanthrene hydroxylation. Clearly
the active hydroxylating species of cytochrome P450 is analogous to a peroxidase compound I which is unusual in being
able to carry out an oxene transfer besides the usual electron transfer (12). The range of products however may differ
between the two systems. Thus whilst cytochrome P450 seems to
carry out principally a 2e oxidation reaction in contrast to
the 1e oxidation reaction of other peroxidases, it is now
realised that cytochrome P450 can also carry out 1e reactions
so that any 1e oxidised product formed would be reduced back
by the NADPH and reductase and result in a poor stoichiometry
of NADPH oxidised to drug hydroxylated. Furthermore the 2e
oxidised product could undergo a further 1e oxidation by
either system.

     In the following the peroxide mediated metabolic activat-
ion of various carcinogens catalysed by cytochrome P450,
methemoglobin, and prostaglandin synthetase will be compared.

I. POLYCYCLIC AROMATIC HYDROCARBON METABOLISM

A. *Cytochrome P450 catalysed*

     As shown in Table I, the rate of total benzopyrene meta-
bolism by cumene hydroperoxide was very similar to that with
the mixed function oxidase using microsomes from normal or
phenobarbital treated rats. The 3OH and 9OH phenols were however markedly decreased at higher hydroperoxide concentrations
due to further oxidation to the quinones. 80% of the products
were quinones with hydroperoxide in contrast to 30% with the
mixed function oxidase. However 3MC microsomes were less
effective with cumene hydroperoxide in catalysing benzopyrene
metabolism and in forming epoxides as shown by decreased 3OH,
9OH and diols. Capdevila et al (14) have recently shown that
cumene hydroperoxide can efficiently catalyse the oxidation
of 3OH to 3,6-quinone by cytochrome P450. Cytochrome P450 also

TABLE I. Benzopyrene metabolism by rat liver microsomes
[phenobarbital (PB) and 3-methylcholanthrene
(3MC) induced]

		NADPH	ROOH
		(n mol/min/mgm)	
Total metabolism	PB	5.2	5.8
	3MC	5.8	2.8
		(% products)	
Quinones	PB	33	80
	3MC	26	67
3OH, 9OH	PB	40	5
	3MC	44	20
Diols	PB	28	15
	3MC	30	13

Incubation mixture contained 20 μM [^{14}C] benzopyrene,
100 μM cumene hydroperoxide or NADPH generating system,
rat liver microsomes 1 mgm/ml, 50 mM Tris-HCl buffer
pH 7.5 at 25°C.

catalysed the oxidation of 6OH and 9OH benzo(a)pyrene to the
quinones. As shown in Figure 1a, both 1e and 2e oxidation mechanisms explain the variety of products formed by the cumene
hydroperoxide/cytochrome P450 catalysed hydroxylation and the
mixed function oxidation of benzopyrene. Cytochrome P450 can
catalyse both a 1e and a 2e oxidation mechanism.

The 6 position of benzo(a)pyrene is chemically more reactive than the other sites on this hydrocarbon. As shown in Figure 1, the 1e oxidation reaction forms a mixture of equal
amounts of 6,12-, 1,6- and 3,6-benzo(a)pyrene diones [15]. An
initial mild oxidation forms a radical cation[16] which reacts
with H_2O to form 6-hydroxy benzo(a)pyrene. The latter readily
undergoes autoxidation in aqueous buffer to produce first the
6-oxo radical, and then diones and superoxy radicals. A mechanism has been proposed in which oxygen couples directly with
the 6-oxo radical producing ketoperoxyl radicals. These radicals form diones via hydroperoxides or diperoxides [17]. The
rapid oxidation of 6OH benzo(a)pyrene in liver homogenate is
partly nonenzymic (see Figure 1b) [18] and P450 catalysed.

In contrast, the cytochrome P450 catalysed oxidation of
benzo(a)pyrene to 3OH-benzo(a)pyrene is probably mediated by a
2,3 epoxide. However the 3OH-benzopyrene is stable [17] and

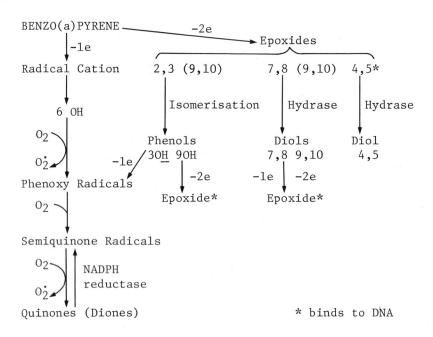

Figure 1a. *Mixed function oxidation of benzo(a)pyrene*

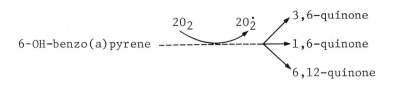

Figure 1b. *Autoxidation of 6-hydroxybenzopyrene*

requires cytochrome P450 to catalyse its 1e oxidation to 3,6-benzo(a)pyrene dione. Enhanced 1,6-dione could also be explained by a similar process.

The lower quinones and higher epoxides and diols in the mixed function oxidase system reflects the reduction by the NADPH/cytochrome P450 reductase system of the radical cation making more benzopyrene available for epoxide formation. Reduction of the 3-and 9-phenoxy radicals would also increase phenol levels. The formation of the highly mutagenic epoxide and epoxide diol which many investigators believe could explain benzopyrene carcinogenicity would be markedly decreased in the hydroperoxide catalysed reaction.

TABLE II. The oxidation of benzopyrene to quinones

	Fluorescence	TLC products p mol/5 min		Quinone Ratio (from HPLC)		
	3OH BP	Polar	Quinones	1,6	3,6	6,12
Microsomes −NADPH	+	980	750	1.65	2.2	1.0
Microsomes −CHP[a]	+	670	1220	2.1	2.35	1.0
Microsomes −LEHPO[b]	−	600	670	1.2	1.0	1.0
met Hb−LEHPO	−	2480	4270	1.1	1.0	1.0
Platelets −arachidonate	−	605	700	1.1	1.0	1.0

[a]CHP − cumene hydroperoxide [b]LEHPO − linolenate hydroperoxide
Incubation mixture (1 ml) contained 20 µM [^{14}C] benzo(a)-
pyrene, 5 µM heme or rat liver microsomes with 1 µM P450, 250
µM CHP or LEHPO or 1 mM NADPH. Arachidonate (1 mM) was added
to platelet microsomes (1 mgm/ml).

B. *Lipid peroxide catalysed*

Various polycyclic aromatic hydrocarbon carcinogens including benzopyrene, 20-methylcholanthrene (19), 3-methylchol-
anthrene (20) and 9,10-dimethyl-1,2-benzanthracene are very
effective in inhibiting the autoxidation of unsaturated lipids.
The benzopyrene was converted to the products identified as
3,6-quinone and 1,6-quinone (19,21).
As shown in Table II, linoleic acid hydroperoxide in the
presence of a heme catalyst results in a very rapid oxidation
of benzo(a)pyrene to quinones (22). The assay for aryl hydrocarbon hydroxylase activity was however negative (22) indicating no 3OH−benzo(a)pyrene was formed. The latter is believed
to be formed via the 2,3-epoxide. Presumably only cytochrome
P450 can carry out the 2e oxidation required for epoxide formation. The three quinones were formed in equal amounts indicating no 3OH or 9OH formation and thus no epoxide formation.
It is generally thought that benzo(a)pyrene 7,8-dihydro-
diol-9,10-epoxide is the most active metabolite in

mutagenicity and in binding to nucleic acids. Furthermore different P450 species seem to differ in their specificity eg. LM2 in rabbits was effective in 3OH formation whereas LM4 was effective in diol epoxide formation (23). Butylated hydroxyanisole, a phenolic antioxidant that inhibits the neoplastic effects of benzopyrene seems to act on benzopyrene metabolism by increasing 3OH and decreasing 4,5-epoxide and diol epoxide formation (24) within four hours after administration to mice. Presumably the antioxidant alters microsomal metabolism by diminishing activation reactions leading to the formation of ultimate carcinogenic metabolites. In vitro the antioxidant inhibited the overall metabolism of benzopyrene without changing the metabolite pattern (24). Other investigators have reported a marked increase in the activities of the detoxifying enzymes glutathione S-transferase, UDP-glucuronyltransferase and epoxide hydrase (25). Induction of "LM2" or selective inhibition of "LM4" by antioxidant metabolites could explain the protective effect.

Quinones seem to be rapidly metabolised in vivo. Thus whilst microsomes form large amounts of quinones, the trachea formed only a small amount of quinones (26). Lind et al (27) have demonstrated that BP quinones after reduction with DT diaphorase are conjugated with UDP-glucuronic acid in the presence of microsomal UDP-glucuronyltransferase. However most of the metabolites of BP quinones are not sensitive to hydrolysis by glucuronidase and it is possible that BP quinones are also polyhydroxylated via "recycling" to water soluble derivatives. Liver cells incubated with benzo(a)-pyrene resulted in a release of 40% of quinones, 23% of 3OH and 56% of diols (28).

The question arises as to whether the quinone pathway results in detoxification. The quinones formed (particularly the 6,12-quinone) however also markedly decrease the covalent binding of 7,8-diol 9,10-epoxide to DNA (29). The addition of UDPGA decreases the quinones by trapping the phenols as glucuronides or by conjugating the reduced quinones. This results in enhanced DNA modification! The quinones presumably inhibit the mixed function oxidase by diverting the electrons from the reductase to O_2 and away from cytochrome P450 (29). Ts'o et al (30) however have shown that 6OH benzo(a)pyrene spontaneously reacts with nucleic acid forming covalent bonds and causing strand breakage. Lorentzen and Ts'o (31) have also found similar effects with the quinones following reduction by microsomes and NADPH. In this case hydroxyl radicals seem to be responsible and we have confirmed that hydroxyl radicals were formed (32). The covalent binding of benzo(a)pyrene semiquinone radicals to DNA has been reported in vitro by Kodama et al (33). However the quinones are not mutagenic in the Ames test presumably because the bacteria can readily detoxify the activated oxygen species. Recently Lorentzen et al (34) showed that the quinones are highly cytotoxic and inhibitory to cellular DNA and RNA synthesis. This was prevented by the removal of oxygen. They suggest that the quinones owe their activity to oxidation-reduction cycles involving semiquinones, hydroquinones and molecular oxygen. The enzymes responsible for the reduction of benzopyrene quinones have been shown to be liver microsomal NADPH:cytochrome P450 reductase (35) and cytosolic DTdiaphorase (27). The 6,12-quinone is less soluble than the other quinones. Anaerobic reduction of all 3-quinones by microsomes/NADPH followed first order kinetics with rate constants for the respective BP quinones in the order 3,6>1,6> 6,12 (29). The Km for 3,6-quinone is 1.3 μM. The 6,12-quinone however has a lower Km and the semiquinones are more autoxid-isable. We have shown that the DT diaphorase is much less act-ive than the P450 reductase in producing hydroxyl radicals presumably because hydroquinones rather than the semiquinones are formed. The induction of the diaphorase by 3-methylchol-anthrene is clearly a detoxification mechanism and necessary for glucuronide formation (27). We have also shown that xanth-ine oxidase can reduce benzopyrene quinones resulting in hydroxyl radical formation (32). The oxidase could be import-ant in the intestinal mucosa. The rapid conversion of benzo-pyrene to quinones by lipid peroxides/hematin or arachidonate/ prostaglandin synthetase therefore whilst not producing muta-genic epoxides could result in the generation of cytotoxic and mutagenic hydroxyl radicals. Hydroxyl radicals were also form-ed prior to quinone formation by the oxidation of 3OH benzo-pyrene indicating semiquinone formation precedes quinone formation (Figure 1a).

C. *Prostaglandin synthetase catalysed*

Marnett et al (36) have shown that sheep vesicular gland prostaglandin synthetase-arachidonate readily oxidises benzo-(a)pyrene to the quinones. No diols were found.As shown in Table II, arachidonate and platelets were also found to carry out this oxidation. Sivarajah et al (87) have also confirmed that human platelets will catalyse the binding of benzo(a)-pyrene to exogenous DNA. If the somatic mutation theory for the initiation of the onset of atherosclerosis is correct, then the enhanced atherosclerosis risk with cigarette smoking could be explained by platelet catalysed benzo(a)pyrene oxida-tion. Marnett et al (37) have also shown that prostaglandin synthetase can also activate benzopyrene 7,8-diol to mutagens.

II. N-DEALKYLATION OF AZO DYES

   Carcinogenic azo dyes such as p-dimethylaminoazobenzene
(DAB) and monomethylazobenzene (MAB) were very effective in
inhibiting the autoxidation of unsaturated lipids (20). The
products were identified as methylaminoazobenzene and carbon
dioxide (38). Such a conversion also occurs in diets of dye
dissolved in cottonseed oil and mixed with ground brown rice
(39).
   As shown in Table III, the oxidative N-demethylation of
DAB, MAB and tertiary amines also occurs with linoleate hydro-
peroxide/heme. Kadlubar et al (40) found N-demethylation of
dimethylaniline and aminopyrine with cumene hydroperoxide/
catalase. We have shown that a blue violet violene radical
cation (570 nM) is formed during the reaction (41). It is
readily reduced by NADPH/microsomes so that the 1e oxidation
pathway is not observed with the mixed function oxidase. The
aminopyrine free radical has also been detected with electron
spin resonance (42). Presumably the aminopyrine is oxidised in

TABLE III. Peroxide-mediated metabolism of aromatic amines

Tertiary or secondary amines	N-Dealkylation $\longrightarrow$ HCHO[c]		
	Microsomes NADPH	P450 ROOH[a]	Heme ROOH[b]
N,N-Dimethylaminoazobenzene	13	60	45
Methylaminoazobenzene	6	36	32
N,N-Dimethylaniline	3	35	121
Aminopyrine	3	6	82
Benzphetamine	3	4	3
	N-Dealkylation $\longrightarrow$ Acrolein		
Cyclophosphamide	+	++	++

[a]Cytochrome P450/cumene hydroperoxide [b]Methemoglobin/lino-
leate hydroperoxide [c]Expressed as n mol HCHO formed/min/mgm
protein.
Incubation mixture contained 1 mM amine, 0.5 mM cumene hydroperoxide or linoleate hydroperoxide or an NADPH generating
system, 1 mgm microsomes/ml or 1 µM methemoglobin or 0.2 µM
cytochrome P450 in 0.1 M phosphate buffer at pH 7.5 at 37°C.

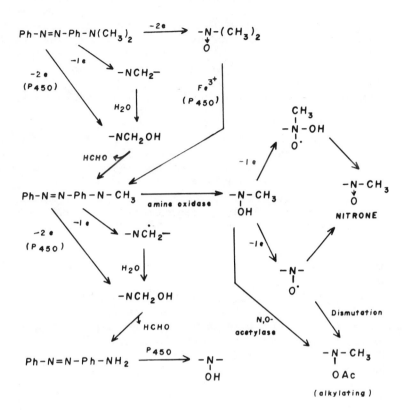

Figure 2. *Metabolic activation of p-dimethylaminoazobenzene*

a one electron transfer step because aminopyrine is a peroxidase donor. This 1e mechanism is shown in Figure 2 for the N-demethylation of DAB and MAB. Recently we have shown that the MAB radical reacts with DNA and binds to protein.

Liver microsomes are believed to carry out the demethylation, 4'-hydroxylation and azo reductive cleavage of DAB in vivo (43). However the enzymology involved has not been fully investigated. The mechanism for the N-demethylation probably involves a C-hydroxylation via a 2e oxidation by cytochrome P450 to a hydroxymethyl intermediate as proposed for various secondary and tertiary amines. Evidence that cytochrome P450 is involved rather than amine oxidase is the remarkable effectiveness of cumene hydroperoxide in catalysing the N-demethylation of DAB (see Table III). Kadlubar et al (44) have also shown that cumene hydroperoxide/microsomes catalyse the N-demethylation of MAB. Amine oxidase, unlike the cytochrome P450 containing mixed function oxidase cannot utilise cumene hydro-

peroxide instead of oxygen. Evidence that P448 was the species
of P450 involved was our finding that 3MC (4 mgm/100g) injected intraperitoneally induces the N-demethylase about threefold
A similar increase in ring hydroxylation of aminoazobenzene
also occurred. The protective effect of 3MC on DAB induced
carcinogenesis could be best explained by the increase in
N-demethylase and ring hydroxylation.

   As shown in Figure 2, an alternative mechanism for the N-
demethylation of tertiary amines, eg. dimethylaniline has been
suggested to involve an N-oxide intermediate. Recently a reconstituted cytochrome P448 system has been shown to mediate
N-oxidation of N,N-dimethylaniline but in the microsomes,
amine oxidase is also responsible for 56% of the N-oxidation
(45). DAB N-oxide which can be readily formed by DAB with
peroxyacids is easily decomposed in the presence of heme compounds to MAB with some DAB and 3OH-MAB also formed (46). We
have shown that microsomes can also catalyse this reaction
suggesting that oxidised P450 can dealkylate the N-oxide.
These products may indicate that the N-oxide can raise the
heme to a higher oxidation state and then act as its own
hydrogen donor. Pyridine N-oxide has been shown to catalyse
such a reaction in the same way as hydroperoxides (47).
However with microsomes and NADPH, 4OH-DAB was formed instead
of 3OH-MAB so that an N-oxide intermediate is unlikely.

   The in vivo binding of an MAB metabolite to proteins on
alkaline treatment releases 3-methylmercapto-4-MAB similar to
that obtained by reacting N-benzoyloxy MAB with methionine or
protein. This suggests that N-hydroxylation of MAB occurs and
the NOH MAB could be the active carcinogen. Microsomal amine
oxidase is believed to catalyse the N-hydroxylation of MAB and
cytochrome P450 is believed to catalyse the N-hydroxylation of
aminoazobenzene (44). Thus lipid peroxidation or the hydroperoxide-cytochrome P450 system is not likely to activate MAB
to NOH MAB.

   Another N-dealkylation example is cyclophosphamide, a
widely used anti cancer drug. It is believed that N-dealkylation results in the release of acrolein which causes the
toxic side effects of sterility, infection, alopecia and
hemorrhagic cystitis (48). The alkylating properties of the
nitrogen mustard moiety of cyclophosphamide are believed to be
responsible for the antitumor properties. The N-dealkylation
involves an initial 4-hydroxylation probably mediated by cytochrome P450. As Fe^{2+}/H_2O_2 could also carry out the reaction
(49), it is likely that all the above 1e peroxide oxidising
systems can activate cyclophosphamide.

III. AROMATIC ARYLAMINE HYDROXYLATION

A. *N-hydroxylation of secondary amides*

It is now established that the carcinogen activation reaction undergone by aromatic amines is hydroxylation on the
nitrogen to form the corresponding hydroxylamino derivatives.
Gorrod (50) has proposed that N-hydroxylation of secondary
amides occurs by a cytochrome P450 pathway whereas free amines
would be oxidised by an amine oxidase. The first step in the
metabolic activation of acetylaminofluorene to a carcinogen
involves N-hydroxylation via a cytochrome P448 dependent monooxygenase (51). As shown in Table IV, cumene hydroperoxide is
very poor in catalysing the microsomal catalysed N-hydroxylation or ring hydroxylation of acetylaminofluorene presumably
because the species of P448 involved is a poor peroxidase with
cumene hydroperoxide. NOH AAF does not react with proteins or
nucleic acids but requires activation. Floyd et al (52) have
shown that lipid hydroperoxides at very low concentrations, in
the presence of heme or cytochrome P420,oxidise the NOH AAF by
a 1e process to nitroxy free radicals. The latter dismutate to

TABLE IV. Peroxide mediated metabolism of aromatic amines

Amines	N-Hydroxyln (-2e)		Ring hydroxyln^c (-2e)		Oxidn (-1e)
	Micros. NADPH	P450 ROOH^a	Micros. NADPH	P450 ROOH^a	Heme ROOH^b
a) Secondary amides					
Acetylaminofluorene 3MC	++*		1.2	0.1	
Acetanilide PB			0.5	0.16	
3MC			3.6	0.17	
b) Secondary amines					
Methylaminoazobenzene	+*	−			+
β-Naphthylamine	+*	−			+
c) Primary amines					
Aminoazobenzene	+	+	2.5	30.1	+
Aniline			0.5	50.1	+
Benzidine	+*				+
o-Aminoazotoluene	+*				+

^{a}Cytochrome P450/cumene hydroperoxide ^{b}Cytochrome P420 or
methemoglobin/linoleate hydroperoxide ^{c}n mol/min/mgm protein
Uninduced or 3MC (3 methylcholanthrene) or PB (phenobarbital)
induced liver microsomes were used.
*Possible carcinogenic pathway

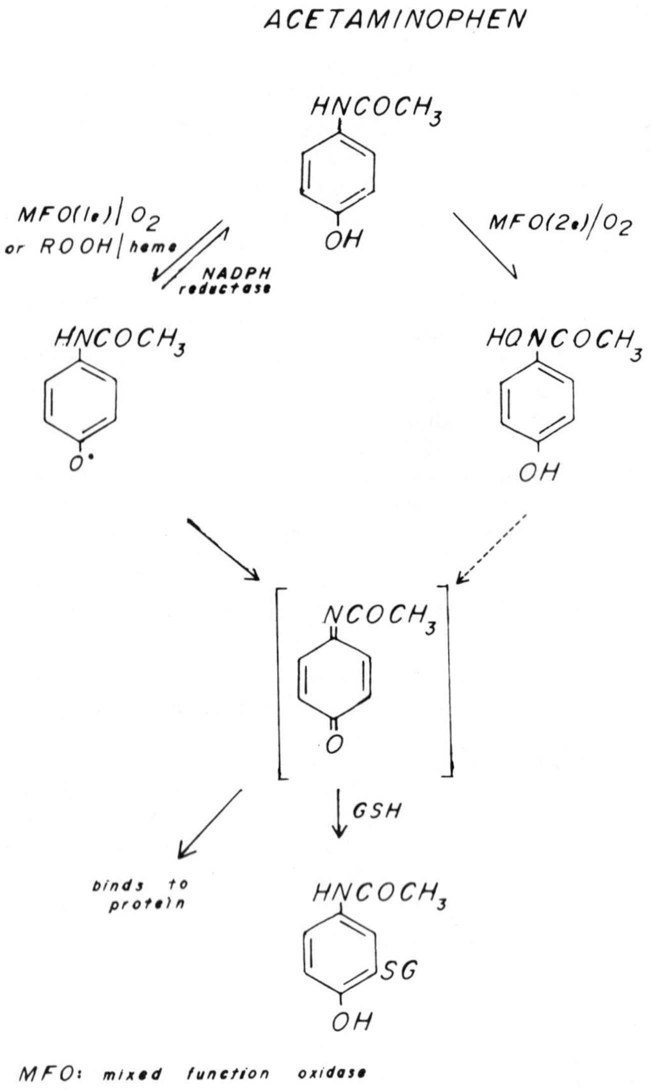

ACETAMINOPHEN

Figure 3. *Metabolic activation of acetaminophen*

the carcinogens nitrosofluorene and N-acetoxyacetyl amino-
fluorene. However the nitroxy free radical was reduced by ascorbate suggesting a protective effect for ascorbate against
amine carcinogenesis. In the salmonella mutagenesis system
however ascorbic acid did not affect NOH AAF mutagenicity (53).
The authors suggested a microsomal N,O-acyltransferase cataly-

ses the formation of the electrophilic arylnitrenium ion
species. The ultimate hepatic carcinogen in the rat appears to
be the sulfate ester (54). However the sulfotransferase is not
found in several target tissues such as mammary gland and
Zymbals gland (55) so that a nitroxy free radical is a possibility in these tissues.

Although the mixed function oxidase does not activate the
NOH AAF, cumene hydroperoxide and microsomes readily form the
nitroxy radical (56). Clearly the nitroxy radical is reduced
back to the NOH AAF by the NADPH/reductase system so that the
le oxidation pathway is not apparent.

B. *le oxidation of secondary amides*

The 3MC induced cytochrome P448 seems to be involved in
the aryl hydroxylation of acetanilide to acetaminophen. Acetaminophen then undergoes further oxidation to a reactive intermediate which covalently binds to tissue macromolecules.
Significant tissue necrosis and covalent binding occur only
after liver glutathione has been depleted to approximately
30% of the control level. The electrophilic intermediate can
be trapped as a thiol adduct.

As shown in Figure 3, the reactive intermediate has been
suggested to be formed by N-hydroxylation (57). However the
NOH acetaminophen was synthesised and found not to react with
glutathione to form the glutathione adduct illustrated.
Furthermore the rate of reaction with cumene hydroperoxide-
microsomes was several orders of magnitude faster than NADPH-
microsomes in the formation of the adduct (58). As the other
le oxidising systems such as lipid peroxides/heme or prostaglandin synthetase/arachidonate were very effective at carry-
ing out the reaction, it is suggested that a phenoxy radical
was formed which reacted with glutathione to form the adduct.
This indicates that the mixed function oxidase carries out a
le oxidation of acetaminophen, but that the NADPH/reductase
reduces the phenoxy radical in an attempt to detoxify the
acetaminophen. The peroxide catalysed metabolic activation of
acetaminophen is therefore analogous to that with benzopyrene
phenols. Ring hydroxylation is normally thought to be a carcinogen detoxifying mechanism but it would clearly be interesting to check for glutathione adducts on oxidation of 4OH-
aminoazobenzene and 7OH-acetylaminofluorene. Recently the
peroxidase mediated oxidation of the carcinogen diethylstil-
bestrol resulted in the formation of p-semiquinone and quinones with substantial binding to DNA (59).

BENZIDINE

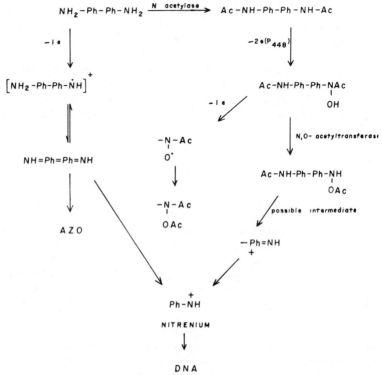

Figure 4. *Mechanisms for benzidine metabolic activation*

C. *The oxidation of primary and secondary arylamines*

As shown in Table IV, the microsomal catalysed N-hydroxylation of the carcinogenic secondary amines methylaminoazobenzene and β-naphthylamine is catalysed by microsomal amine
oxidase and is not catalysed by microsomes-cumene hydroperoxide .

Microsomes-cumene hydroperoxide catalysed the N-hydroxylation of the primary amine aminoazobenzene (44) but poorly
catalysed the N-hydroxylation of acetylaminofluorene. The
ring hydroxylation of the primary amines in contrast was much
more effective than the mixed function oxidase. The 1e oxidising systems linoleate hydroperoxide-cytochrome P420 or
methemoglobin did not carry out the N-hydroxylation or ring
hydroxylation but did carry out a 1e catalysed oxidation to a
free radical. The free radical formed by the 1e oxidation of
benzidine using prostaglandin synthetase/arachidonate readily

binds to DNA and tRNA (60). A mechanism for the reactive
intermediates involved are shown in Figure 4. It is suggested
that the 1e pathway results in the formation of a semiquinoid
blue product which is then oxidised to an orange diphenoquin-
onediimine (61). The latter could then give rise to the same
alkylating nitrenium species believed to be the ultimate
carcinogen following the N-hydroxylation of benzidine. The
primary amine o-aminoazotoluene is also probably oxidised by
similar pathways. The latter compound is also oxidised by
lipid peroxidation as it prevents lipid peroxidation (20).

IV. PEROXIDE-CATALYSED CARCINOGEN ACTIVATION AND BLADDER CANCER

A good case can be made that prostaglandin synthetase activity of the kidney is responsible for bladder carcinogenesis.
Although the metabolic activation of several arylamines has
been shown to involve N-hydroxylation, the mixed function oxidase activity seems to be located in the kidney cortex (62) and
dose not in any case carry out N-hydroxylation of amines (63).
In the case of β-naphthylamine it is believed the N-hydroxylation occurs in the liver where it is released as the N-glucuro-
nide. The urine acidity could cause hydrolysis to the NOH arylamines which then form the highly reactive electrophilic arylnitrenium ions (64). Prostaglandin synthetase is however located in the bladder and in the last cells of the kidney medulla
in contact with urine prior to its entry into the urinary space
(62). Furthermore prostaglandin synthetase is responsible for
the activity of kidney medullary microsomes and arachidonate
in oxidising benzidine to products which bind to DNA and tRNA
(60). Nitrofuran N4(5-nitro-2-furyl)-2-thiazolyl is also oxidised (65). Benzopyrene is oxidised to products that bind tRNA
(66) and the 7,8-diol to products which bind to DNA (37). We
have also found that NOH N-acetyl-2-aminofluorene is activated
to nitrosofluorene and dimethylaminoazobenzene is activated to
methylaminoazobenzene. Acetaminophen is also activàted to
reactive intermediates that conjugate GSH (58). Except for the
last compound all of the above compounds induce bladder cancer.
Furthermore recent epidemiologic studies show elevated bladder
cancer among hairdressers and beauticians (67,68). Ames et al
(69) have shown that three primary amine hair dye components
(2,5-diaminotoluene, p-phenylenediamine, 2,5-diaminoanisole)
become strongly mutagenic after oxidation by hydrogen peroxide.
The skin dyes are absorbed by the skin and appear in the urine.
We have also shown that these dyes are excellent peroxidase
donors for prostaglandin synthetase (70). Azo dyes are believed to be detoxified by reductive cleavage by the liver microsomes and the products excreted in the urine. We have also

preliminary evidence that prostaglandin synthetase can reverse
this. Thus dimethyl p-phenylenediamine is oxidised to a yellow
azo dye and aniline is oxidised to azobenzene.

There has been found a strong association between human
cancer and increased prostaglandin formation (71). A number of
malignant tumors produce high levels of prostaglandins (72).
Furthermore tumor promoters stimulate cultured kidney cells to
produce prostaglandins (73). Antioxidants are known to be effective in vivo in protecting against carcinogenesis (74) and
in inhibiting prostaglandin synthetase. However anti-inflammatory agents that markedly inhibit prostaglandin synthetase
should now be tested.

V. ACTIVATION OF HYDRAZINES

Metabolic activation of hydrazine moieties of the anti-
tuberculosis drug, isoniazid may be responsible for the
hepatitis side effects (75). Liver injury also results from
the metabolic activation of hydrazine moiety of the anti-
depressant iproniazid (76). 1,2-Dimethylhydrazine, 1,1-di-
methylhydrazine, methylhydrazine and the anticancer drug procarbazine are also carcinogenic following metabolism (77).
Isoniazid is carcinogenic and mutagenic in animal tests.
Recently, cytochrome P450 has been implicated in the activation of these hydrazines by an initial N-hydroxylation step
(80). Thus propane formation from iproniazid was induced by
phenobarbital in vivo, inhibited in vitro by CO and SKF 525A
and decreased in vivo by $CoCl_2$ (80). Similar conclusions
were reached for the N-oxidation of procarbazine to azoprocar-
bazine (81). Furthermore cytochrome P450 also seems to
catalyse the oxidation of azo to azoxyprocarbazine (81) and the
N-dealkylation of azoxy to the aldehyde p-formyl-N-isopropyl-
benzamide (82). The other product in the last reaction is
believed to be the methyl diazonium ion suggested as the
carcinogen. Our research indicates that lipid peroxides or
cumene hydroperoxide can oxidise procarbazine to azo procarbazine and azoxyprocarbazine by a 2e oxidation in the absence
of microsomes. Microsomes or heme enhanced the peroxide effect. Cu^{2+} or Mn^{2+} carries out a metal catalysed autoxidation
to the azoxyprocarbazine by a 1e oxidation (83), which involves
considerable O_2 uptake with H_2O_2 formation and DNA degradation
(84). Isoniazid and iproniazid during autoxidation catalysed
by transition metals induce unscheduled DNA synthesis in
cultured human fibroblasts (78) probably as a result of H_2O_2
formation (79). The N-oxidation of procarbazine to azo and
azoxyprocarbazine can therefore be catalysed by both a 1e and
2e pathway. The 1e pathway can be distinguished from the 2e

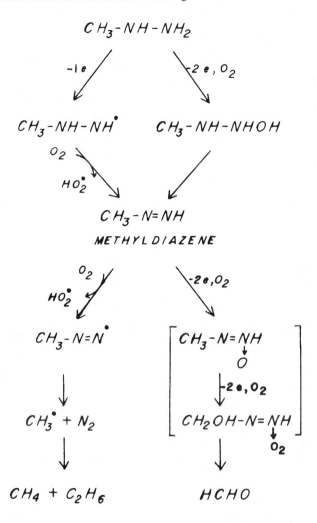

Figure 5. *Metabolic activation of methylhydrazine*

pathway by the uptake of O_2, and formation of H_2O_2 and hydroxyl radicals associated with this pathway. The conversion of
azoxy to the aldehyde involves a N-dealkylation reaction via
a 1e or a 2e pathway. It is suggested that this last reaction
results in the formation of the ultimate carcinogen the methyl
diazonium ion.

The oxidation of methylhydrazine to methane was also
greatly catalysed by H_2O_2 or cumene hydroperoxide or linolenic
acid hydroperoxide/hematin. Microsomes with NADPH or cumene
hydroperoxide were poor at catalysing methane formation but

formed HCHO instead.   In Fig. 5 it is suggested that methyldiazene is formed by either the 1e or 2e oxidation pathway.
Autooxidation or 1e oxidation of the methyldiazene could form
methane whereas N-oxidation and N-dealkylation by amine oxidase or mixed function oxidase could form HCHO.   75% of the
N-methylhydrazine in vivo is respired as methane and 25% as
CO_2 (85).   Phenobarbital increases CO_2 formation in vivo (86).

ACKNOWLEDGMENTS

This research was supported by the National Cancer
Institute of Canada.

REFERENCES

1. Riley, C.A., Cohen, G. and Lieberman, M. Science 183, 208
    (1974).
2. Koster, U., Albrecht, D. and Kappus, H. Toxicol. Appl.
    Pharmacol. 41, 639 (1977).
3. Ramstoeck, E.R., Hoekstra, W.G. and Ganther, H.E. Toxicol.
    Appl. Pharmacol. 54, 251 (1980).
4. Hafeman, I.G. and Hoekstra, W.G. J. Nutr. 107, 666 (1977).
5. Cutler, M.G. and Hayward, M.A. Nutr. Metab. 16, 87 (1974).
6. Cutler, M.G. and Schneider, R. Food Cosmet. Toxicol. 11,
    443 (1973).
7. Wattenburg, L.W. Adv. in Cancer Res. 28, 197 (1979).
8. Ts'o, P.O.P., Caspary, W.J. and Lorentzen, R.J. In "Free
    Radicals in Biology" Vol. III, (W. Pryor, ed), p. 251.
    Academic Press, New York (1980).
9. Little, C. and O'Brien, P.J. Biochem. Biophys. Res. Com-
    mun. 31, 145 (1968).
10. Little, C. and O'Brien, P.J. Can. J. Biochem. 47, 493
    (1969).
11. Hrycay, E.G. and O'Brien, P.J. Arch. Biochem. Biophys.
    147, 14 (1971).
12. O'Brien, P.J. In "Microsomes, Drug Oxidations and Chemi-
    cal Carcinogenesis" (A.H. Conney, M.J. Coon, R.W.
    Estabrook, J.R. Gillette, H.V. Gelboin and P.J. O'Brien,
    eds), p. 263. Academic Press, New York (1980).
13. O'Brien, P.J. Pharmacol. Therap. 2A, 517 (1978).
14. Capdevila, J., Estabrook, R.W. and Prough, R.A. Arch.
    Biochem. Biophys. 200, 186 (1980).
15. Nagata, C., Tagashira, Y. and Kodama, H. In "Chemical
    Carcinogenesis" Part A, (P.O.P. Ts'o and J. DiPaolo, eds),
    p. 87. Marcel Dekker, New York (1974).

16. Girke, S. and Wilk, M. In "Chemical Carcinogenesis" Part A, (P.O.P. Ts'o and J. DiPaolo, eds), p. 183. Marcel Dekker, New York (1974).

17. Lorentzen, R.J., Caspary, W.J., Lesko, S.A. and Ts'o, P.O. P. Biochemistry 14, 3970 (1975).

18. Lesko, S., Caspary, W., Lorentzen, R. and Ts'o, P.O.P. Biochemistry 14, 3978 (1975).

19. Mueller, G.C., Miller, J.A. and Rusch, H.P. Cancer Res. 5, 401 (1945).

20. Rusch, H.P. and Kline, B.E. Cancer Res. 1, 465 (1941)

21. Berenblum, I., Crowfoot, D., Holiday, E.R. and Schoental, R. Cancer Res. 3, 151 (1943).

22. Rahimtula, A. submitted for publication.

23. Deutsch, J., Leutz, J.C., Yang, S.K., Gelboin, H.V., Chiang, Y.L., Vatsis, K.P. and Coon, M.J. Proc. Natl. Acad. Sci. 75, 3123 (1978).

24. Lam, L.K.T., Fladmoe, A.V., Hochalter, J.B. and Wattenberg, L.W. Cancer Res. 40, 2824 (1980).

25. Cha, Y.N. and Bueding, E. Biochem. Pharmacol. 28, 1917 (1979).

26. Mass, M.J. and Kaufman, D.G. Biochem. Biophys. Res. Commun. 89, 885 (1979).

27. Lind, C., Vadi, H. and Ernster, L. Arch. Biochem. Biophys. 190, 97 (1978).

28. Burke, M.D., Vadi, H., Jernstrom, B. and Orrenius, S. J. Biol. Chem. 252, 6424 (1977).

29. Shen, A.L., Fahl, W.E., Wrighton, S.A., Fefeonte, C.R. Cancer Res. 39, 4123 (1979).

30. Ts'o, P.O.P., Caspary, W., Cohen, B., Leavitt, J., Lesko, S., Lorentzen, R. and Schenchtman, L. In "Chemical Carcinogenesis" Part A, (P.O.P. Ts'o and J. DiPaolo, eds), p. 113. Marcel Dekker, New York (1974).

31. Lorentzen, R. and Ts'o, P.O.P. Biochemistry 16, 1467 (1977).

32. Hawco, F., Hulett, L. and O'Brien, P.J. In "Microsomes, Drug Oxidations and Chemical Carcinogenesis" (A.H. Conney, M.J. Coon, R.W. Estabrook, J.R. Gillette, H.V. Gelboin and P.J. O'Brien, eds), p. 263. Academic Press (1980).

33. Kodama, M., Ioki, Y. and Nagata, C. Gann 68, 253 (1977).

34. Lorentzen, R.J., Lesko, S.A., McDonald, K. and Ts'o, P.O.P. Cancer Res. 54, 3194 (1979).

35. Capdevila, J., Estabrook, R.W. and Prough, R.A. Biochem. Biophys. Res. Commun. 83, 1291 (1978).

36. Marnett, L.J., Reed, G.A. and Johnson, J.T. Biochem. Biophys. Res. Commun. 79, 569 (1977).

37. Marnett, L.J., Reed, G.A. and Dennison, D.J. Biochem. Biophys. Res. Commun. 82, 210 (1978).

38. Rusch, H.P. and Miller, J.A. Proc. Soc. Exp. Biol. Med. 68, 140 (1945).

39. Kensler, C.J., Magill, J.W. and Suguira, K. Cancer Res. 7, 95 (1947).
40. Kadlubar, E.F., Morton, K.C. and Ziegler, D.M. Biochem. Biophys. Res. Commun. 54, 1255 (1973).
41. Rahimtula, A.D., Hawco, F. and O'Brien, P.J. In "Microsomes, Drug Oxidations and Chemical Carcinogenesis" (A.H. Conney, M.J. Coon, R.W. Estabrook, J.R. Gillette, H.V. Gelboin and P.J. O'Brien, eds), p. 419. Academic Press, New York (1980).
42. Sayo, H. and Hosokawa, M. Chem. Pharmacol. Bull. 2351 (1980).
43. Mueller, G.C. and Miller, J.A. J. Biol. Chem. 179, 535 (1948).
44. Kadlubar, F.F., Miller, J.A. and Miller, E.C. Cancer Res. 36, 1196 (1976).
45. Hlavica, P. and Hulsmann, S. In "Microsomes, Drug Oxidations and Chemical Carcinogenesis" (A.H. Conney, M.J. Coon, R.W. Estabrook, J.R. Gillette, H.V. Gelboin and P.J. O'Brien, eds), p. 217. Academic Press, New York (1980).
46. Terayama, H. Gann 55, 195 (1963).
47. Stohrer, G. and Brown, G.B. J. Biol. Chem. 244, 2495 (1969).
48. Levy, L. and Harris, R. Biochem. Pharmacol. 26, 1015 (1977).
49. Alarcon, R.A. and Merenhofer, J. Nature: New Biol. 233, 250 (1971).
50. Gorrod, I.W. Chem. Biol. Interact. 7, 289 (1973).
51. Thorgeirsson, S.S., Jallow, D.J., Sasame, H.A., Green, I. and Mitchell, J.R. Mol. Pharmacol. 9, 398 (1973).
52. Floyd, R.A., Soong, L.M., Walker, R.N. and Stuart, M. Cancer Res. 36, 2761 (1976).
53. Schut, H.A.J., Wirth, P.J. and Thorgeirsson, S.S. Mol. Pharmacol. 14, 682 (1978).
54. DeBairn, J.R., Rowley, J.Y., Miller, E.C. and Miller, J.A. Cancer Res. 37, 1461 (1977).
55. Irving, C.C., Janss, D. and Russell, L.T. Cancer Res. 31, 387 (1971).
56. Floyd, R.A. In "Free Radicals in Biology" Vol. IV, (W. Pryor, ed), p. 187. Academic Press, New York (1977).
57. Hinson, J.A., Nelson, S.D. and Mitchell, J.R. Mol. Pharmacol. 13, 625 (1977).
58. Moldeus, P. and Rahimtula, A.D. Biochem. Biophys. Res. Commun. 96, 469 (1980).
59. Metzler, M. and McLachlan, J.A. Biochem. Biophys. Res. Commun 85, 874 (1978).
60. Zenser, T.V., Mattammal, M.B. and Davis, B.B. Cancer Res. 40, 114 (1980).

61. Kroupa, J. and Matrka, M. Coll. Czech. Chem. Commun. 36, 2729 (1971).
62. Zenser, T.V., Mattammal, M.B. and Davis, B.B. J. Pharmacol. Exp. Therap. 207, 719 (1978).
63. Seal, V.S. and Guttmann, H.R. J. Biol. Chem. 234, 648 (1959).
64. Kadlubar, F.F., Miller, J.A. and Miller, E.C. Cancer Res. 37, 805 (1977).
65. Zenser, T.V., Mattammal, M.B. and Davis, B.B. J. Pharmacol. Exp. Therap. 214, 312 (1980).
66. Marnett, L.J. and Reed, G.A. Biochemistry 18, 2923 (1979).
67. Wynder, E.L., Orderdonk, J. and Mantel. N. Cancer 16, 1388 (1963).
68. Dunham, L.J., Rabson, A.S., Stewart, H.L., Frank, A.S. and Young, J.L. J. Natl. Cancer Inst. 41, 683 (1968).
69. Ames, B.N., Kammen, H.O. and Yamasaki, E. Proc. Natl. Acad. Sci. 72, 2423 (1975).
70. O'Brien, P.J. and Rahimtula, A.D. Biochem. Biophys. Res. Commun. 70, 832 (1976).
71. Papanicolaon, N., Mourtokalakis, T., Tcherdakoff, P., Bariety, J. and Milliery, P. Prostaglandins 10, 405 (1975).
72. Humes, J.L. and Strausse, H.R. Prostaglandins 5, 183 (1974).
73. Levine, L. and Hassid, A. Biochem. Biophys. Res. Commun. 79, 477 (1977).
74. Wattenberg, L. Adv. in Cancer Res. 28, 197 (1978).
75. Mitchell, J.R., Thorgeirsson, U.P., Black, M. and Timbrell, J.A. Clin. Pharmacol. Therap. 18, 70 (1975).
76. Snodgrass, W.P., Potter, W.Z., Timbrell, J.A. and Mitchell, J.R. Clin. Res. 22, 323A (1974).
77. Fishbein, L. In "Studies in Environmental Science 4 Potential Industrial Carcinogens and Mutagens" (L. Fishbein ed), p. 311. Elsevier Publ. Amsterdam (1979).
78. Whiting, R.F., Wei, L. and Stich, H.F. Biochem. Pharmacol. 29, 842 (1980).
79. Whiting, R.F., Wei, L., and Stich, H.F. Mutation Res. 62, 505 (1979).
80. Nelson, S.D., Mitchell, J.R., Timbrell, J.H., Snodgrass, W.R. and Corcoran, G.B. Science 193, 901 (1976).
81. Dunn, D.L., Lubet, R.A. and Prough, R.A. Cancer Res. 39, 4555 (1979).
82. Wiebkin, P. and Prough, R. A. Cancer Res. 40, 3524 (1980).
83. Weinkam, R.J. and Shiba, D.A. Life Sciences 22, 937 (1978).
84. Berneis, K., Bollag, W., Kofler, M. and Luthy, H. Experientia 21, 318 (1965).
85. Dost. F.N., Reed, D.J. and Wang, C.H. Biochem. Pharmacol. 15, 1325 (1966).

86. Dewald, B., Baggiolini, M. and Aebi, H. Biochem. Pharmacol. 18, 2179 (1969).
87. Sivarajah, K., Anderson, M. W. and Eling, T. E. Life Sciences 23, 2571 (1978).

LIPID PEROXIDATION AND OXIDATIVE DAMAGE TO DNA[1]

Bruce N. Ames
Monica C. Hollstein[2]
Richard Cathcart[3]

Department of Biochemistry
University of California
Berkeley, California

Our enthusiasm for the field of lipid peroxidation has evolved in the last few years from at least five different interests in the area of mutagenesis and carcinogenesis: a) polyhalogenated and other carcinogens that are not mutagens in our Salmonella mutagenicity test, b) promoters in carcinogenesis, c) the role of antioxidants such as Vitamin E and ascorbate, and of selenium, in preventing cancer induced by various carcinogens, d) the effect of carotene and Vitamin A as anti-carcinogens, e) the connection between ageing, spontaneous cancer, and oxygen metabolism. We are becoming convinced that an understanding of lipid peroxidation, and of oxidative damage to DNA, and the organism's defenses against them, will lead to practical ways for both minimizing cancer incidence and increasing life span.

I. HALOGENATED CARCINOGENS AND LIPID PEROXIDATION

An extremely high percentage of chlorinated and brominated chemicals are carcinogens in animal cancer tests and thus represent an extremely suspect class of chemicals.

[1]This work was supported by DOE contract DE-AT03-80EV70156 to B.N.A. and by NIEHS Center Grant ES01896.
[2]M.C.H. is supported by NIEHS Training Grant ES07075.
[3]R.C. -is supported by NIH Postdoctoral Fellowship ES05191.

In the U.S. about 10 x 10^9 kilos (20 billion lbs.) of Cl_2 is made each year. About 4 billion lbs go into bleaching paper pulp and this results in a wide variety of chlorinated chemicals in rivers and lakes. About 1 billion lbs is used for treating water which results in chloroform and other halomethanes in municipal drinking supplies. The more organic material in the water that is chlorinated the greater amount and variety of chlorinated organics in the water. For example, New Brunswick, N.J. drinking water contains about 50 μg $CHCl_3$ per liter and also about 40 μg $CHCl_2Br$ per liter (1). Whenever there is bromide ion present during water chlorination (as in sea water), brominated organics, which are in general considerably more carcinogenic than their chlorinated analogs, are formed as well. About 12 billion lbs of Cl_2 go into chlorinated hydrocarbons of various types for industrial use. We have previously pointed out (2) the variety of chlorinated pesticides and PCBs which have been found to accumulate in human body fat and human mother's milk, particularly DDE (a DDT derivative) and PCB's which strongly bio-concentrate because of their high lipid solubility. This is an enormous biological experiment, as organic chemicals containing chlorine and bromine are not used in natural mammalian biochemical processes and, with one exception, do not appear to have been normally present in the human diet until the onset of the modern chemical age. The one exception is the ingestion of brominated and chlorinated organics in the diet of Japanese and other people from eating seaweed. We suspect that these halogenated organics in seaweed may contribute to cancer.

Halogenated carcinogens such as PCB's, dieldrin, DDE, and DDT, CCl_4, $CHCl_3$ are not mutagens in our Salmonella mutagenicity test or in other short-term tests for mutagenicity (3,4,5,6). Though some chlorinated carcinogens are mutagens in our test, and in the other short-term tests, these are usually either a) direct alkylating agents, such as methyl bromide and methyl chloride, or b) carcinogens with double bonds activated through epoxides, such as vinyl chloride, or c) bifunctional halogens such as ethylene dibromide, ethylene dichloride, or dibromochloropropane, which appear to form active sulfur-half-mustards with glutathione.

Our mutagenicity test makes use of a rat liver homogenate as a first approximation of mammalian metabolism and a set of histidine-requiring mutants of Salmonella bacteria as an extremely sensitive indicator of DNA damage (2,7). The test detects about 85% ± 5% of carcinogens as mutagens with a large number and a wide variety of carcinogens of

different structures having been tested (3). Thus the question of the lack of mutagenicity of the chlorinated carcinogens such as CCl_4, $CHCl_3$, TCDD (dioxin), DDT, dieldrin, and
PCB's in the Salmonella test, coupled with the question of
why these chlorinated chemicals are carcinogens, has been
puzzling us for many years and has led us to the field of
lipid peroxidation.

Carbon tetrachloride was one of the first of the chlorinated organics shown to be effective in causing lipid peroxidation in rodents. This appears to be due to a homolytic
cleavage by cytochrome P450 type enzymes with the formation
of a $CCl_3 \cdot$ radical which sets off lipid peroxidation, primarily in the liver (8). Many other chlorine or bromine
containing organics have been shown to cause lipid peroxidation, and the liver, testes and brain appear to be the main
tissues affected. TCDD has been shown to be particularly
effective in this regard (9). It appears highly likely that
the carcinogenicity of these halogen containing chemicals is
due to their ability to form radicals (10) which cause lipid
peroxidation. Lipid peroxidation, being a chain reaction,
causes the production of a considerable amount of active
forms of oxygen such as $HO \cdot$ and ROOH which can damage DNA
(see below). In addition, other carcinogens not detected as
mutagens in our test such as lead and cadmium and substituted hydrazines may be active by causing lipid peroxidation.

II. WORK IN PROGRESS ON IMPROVING THE TEST SYSTEM FOR OXI-
DANTS OF DNA

If, in fact, lipid peroxidation is at the root of the
carcinogenicity of the halogenated compounds discussed, two
possible reasons suggest themselves for why the chlorinated
carcinogens are not mutagenic in the Salmonella test. One
explanation is that a free radical is actually formed from
CCl_4, or other compounds, in the in vitro system using the
rat liver homogenate, which does seem to be the case (11).
The extremely short-lived free radicals, however, would not
make it from the microsomal membrane, with its cytochrome
P450, to the bacterial membrane with the bacterial DNA.
Thus we have initiated some experiments on membrane fusion
to try to fuse the liver microsomes to the surface of the
bacterial membrane so as to generate lipid peroxidation in
the vicinity of the bacterial DNA. This work (with Lynne
Haroun) appeared promising, but we have temporarily halted
it to concentrate instead on developing better tester

strains for oxidative damage to DNA as we believe this is also a problem.

To explore the possibility that we could develop better base-pair-substitution tester strains for agents causing oxidative damage to DNA such as HO·, $^1O_2$, and ROOH) we screened about a hundred histidine-requiring mutants (containing the pKM101 plasmid) for reversion by a variety of potential oxidative type mutagens which would not require liver activation. We have found that streptonigrin, which is believed to generate superoxide (12) and hydrogen peroxide are both considerably more mutagenic on new strains which we have developed than on the standard tester strain TA100 or hisG46 pKM101 (TA92), a derivative containing the same histidine mutation that is in TA100. It is already clear that two different new tester strains are needed for these mutagens as the histidine mutation we found to be optimum for streptonigrin, hisG428, is not reverted very well by hydrogen peroxide, while the strain selected for maximum response to reversion by hydrogen peroxide, hisC3108, is not the most effective strain for detection of streptonigrin mutagenicity. The hisC3108 mutation is at a UGA (umber) chain – terminating triplet while the hisG428 mutation is at a UAA (amber) chain – terminating triplet. Thus it seems likely that both of these agents may be hitting A:T base pairs rather than G:C base pairs (Figures 1 and 2), though the context also appears to be important. The hydroperoxides tested, cumene hydroperoxide and t-butyl hydroperoxide, are quite mutagenic on hisG428 pKM101 while they are not appreciably mutagenic on the standard tester strains. Methyl ethyl ketone peroxide is also somewhat mutagenic on both new strains.

The histidine mutation hisG46 used in our TA100 tester strain has been sequenced by Wayne Barnes in St. Louis (personal communication) and been shown to contain −CCC− (pro) instead of the −CTC− (leu) in the wild type. Thus hisG46 may primarily detect mutagens damaging a G:C base pair.

Singlet oxygen possibly mutates tester strains at a G:C base pair. Mutagenesis has been reported by both methylene blue plus light, which generates $^1O_2$, and by light alone which could be active through interaction with endogenous flavins (13,14). We are now comparing a set of tester strains to determine the base pair specificity. Oxygen itself is mutagenic in a series of oxygen-sensitive mutants of TA100 (15).

Various hydroperoxide derivatives of thymine and thymine glycols can be formed from lipid peroxidation and photoreactions. Some of these have been shown to be highly mutagenic on the TA100 tester strain by Wang et al (16).

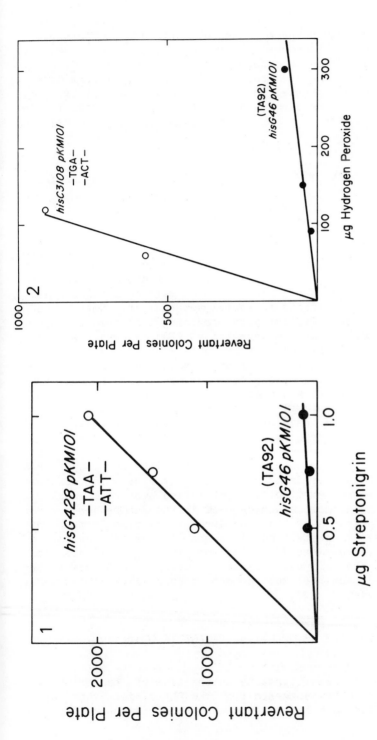

FIGS. 1 and 2. Comparison of various histidine auxotrophs to reversion by streptonigrin or hydrogen peroxide. Reversion was determined on petri plates incorporating mutagen and bacteria directly into the top agar (without S-9 mix), as previously described (7). Spontaneous revertants have been subtracted.

With the sequencing of a good part of the histidine
operon by Barnes (personal communication) it is now feasible
to determine any induced base pair change in the hisG, hisD,
and hisC genes quite rapidly. We hope to determine the
actual base pair changes induced by the various active forms
of oxygen. We are examining the chemistry in parallel (see
below) as a minor chemical adduct may be the major mutagenic
adduct.

III. THE MUTAGENICITY AND CARCINOGENICITY OF MALONALDEHYDE

Malonaldehyde, one of the most commonly assayed products
of lipid peroxidation, was shown to react with DNA a number
of years ago (17). It was also shown to cause tumors in
mice by Shamberger et al. (18) in a skin painting study
using croton oil as a promoter. In addition, it has been
reported to be mutagenic in our Salmonella test (19).
Nevertheless, questions as to its mutagenicity have arisen
because of the difficulty in preparing pure malonaldehyde.
A later report on its mutagenicity (20) showed that the
mutagenic activity previously observed was due to the
mutagenic activity of partial hydrolysis products, such as
methoxy acrolein. Malonaldehyde itself was shown to have
only a trace of mutagenic activity and even this may have
been due to residual impurities. We (M. Fiss and B. N.
Ames, unpublished) have also examined the mutagenicity of
pure malonaldehyde. We prepared it by the enzymatic procedure of Summerfield and Tappel (21) and found the pure
material to be without mutagenic activity on hisD3052 and
the standard set of Salmonella tester strains, in contrast
to the impure chemically synthesized material which was
mutagenic. The malonaldehyde used in the Shamberger cancer
study was prepared by the procedure giving rise to mutagenic
impurities. Thus it is possible that the carcinogenicity of
malonaldehyde was caused by its impurities. It is also possible that the Salmonella tester strains did not detect the
particular type of DNA damage (cross links?) caused by
malonaldehyde.

IV. THE CHEMISTRY OF OXIDATIVE DAMAGE TO DNA

A considerable literature in the area of radiation and
its effects on DNA suggests that the DNA damage by radiation
is mostly due to effects of $HO\cdot$ and other oxidative species

(22,23). Lipid peroxidation caused by radiation, or by
other radical forming agents, would also generate considerable hydroxyl radicals. The chemistry of DNA and deoxynu-
cleoside damage by HO· has been investigated fairly
thoroughly by a number of laboratories. Much recent work
has been done by Teoule and Cadet and their colleagues (24).
Isomeric thymine glycols are among the products produced,
and a DNA repair system for thymine glycol has been
described (25,26,27). We have also been investigating the
oxidation products of the various deoxy- and ribonucleosides
incubated with hydroxyl radicals and the kinetics of this
process by using HPLC (Figure 3). We find that all of the
deoxynucleosides are sensitive to destruction by HO· and
that there is considerable depurination and depyrimidina-
tion.
     Singlet oxygen is an extremely reactive substance that
can damage the deoxynucleosides. It is produced by various
photo reactions when dyes, light and O_2 are present, as a
toxic byproduct of photosynthesis, and it may be produced by
other mechanisms, possibly including radiation (28). Some
work in the area of DNA damage has been done previously by
the group of Teoule and Cadet and others (29). Simon and
Van Vunakus (30) did some interesting early work in this
field. We have also done a study of the kinetics and destruction of the various nucleoside bases by singlet oxygen
using HPLC (Figure 4). Deoxyguanosine is the most sensitive
base to destruction by the $^1O_2$ (Figure 5). Unlike HO·,
singlet oxygen does not appear to depurinate deoxyguanosine.
     Our interest in the chemistry of oxidative damage to DNA
is to complement our studies on oxidative mutagenesis. We
also wish to have available the various oxidation products
for studies on DNA repair and for a project on determining
oxidative DNA damage in humans.

V. THE ORGANISM'S DEFENSES AGAINST OXIDATIVE DAMAGE TO DNA

     Numerous papers at this meeting have discussed the
variety of metabolic defenses against superoxide (31),
hydroperoxides (32), singlet oxygen, and the variety of radicals and peroxides generated from lipid peroxidation and
oxidative metabolism. It seems likely that the appropriate
level of these defense systems will be of paramount importance in minimizing spontaneous and induced cancer due to
oxidative damage (33). That this is a major factor in
cancer is supported by various lines of evidence (34,35,36)
including the ability of a variety of substances which are

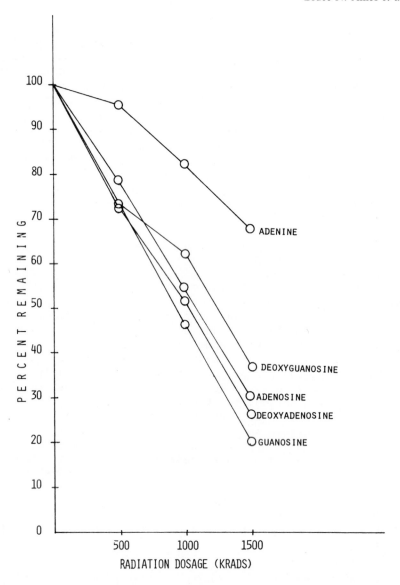

FIG. 3.   3 mM solutions of adenine, adenosine, deoxy-
adenosine, guanosine, and deoxyguanosine in water were irradiated with increasing dosages of gamma radiation from a
cobalt-60 source.   The percent of these compounds remaining
after irradiation was measured using High Pressure Liquid
Chromatography equipped with a Waters Radial Pak-A Cartridge
and a Spectra Physics automated peak integrator.   The buffers
used for the HPLC were 0.25 M Ammonium Acetate, pH 5.0 and
60/40 acetonitrile.

5

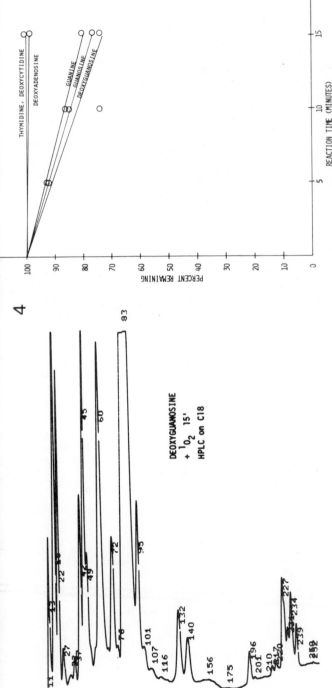

FIGS. 4 and 5. Destruction of deoxyguanosine by singlet oxygen. Singlet oxyten was generated using Sensitox II, 10 mg/l in a deoxyguanosine solution, 0.5 mg/l with 50 mM tris buffer, pH 8.0. Oxygen was bubbled into the mixture continuously during illumination by a tungsten-halogen lamp with a UV-cutoff filter. After the appropriate time, the Sensitox was removed by passage of the solution through a millipore filter. An aliquot was chromatographed using a Waters Radial Pak-A cartridge and Waters HPLC system. Details of the separation methods will be described in future publications. The dye sensitized reaction has been shown dependent on visible light, oxygen and Sensitox II.

part of these defense systems to act as anti-carcinogens in
animal cancer tests with carcinogens such as hydrazines,
chlorinated compounds, quinones, and polycyclic hydrocarbons, which give rise to radicals (10,37) and lipid peroxidation. These anti-carcinogens include ascorbate (38,39),
an anti-oxidant; vitamin E (40), an anti-oxidant and free
radical trap; β-carotene and Vitamin A derivatives (41),
radical and O_2 traps (42,53); polyamines (43,44), antioxidants; and selenium (45), a component of glutathione
peroxidase. Some evidence from human epidemiology also suggests that dietary modifying factors, presumably including
the above substances, may strongly modify the frequency of
lung cancer in cigarette smokers (46). Many of the mutagens
in cigarette smoke, such as the nitrogen oxides, would cause
lipid peroxidation and oxidative damage (47).

   In addition to lipid peroxidation caused damage to DNA
we suspect the promotion stage in carcinogenesis may involve
damage to the cell membrane by lipid peroxidation (34) and
be counteracted by the above defense systems. Phorbol
esters and a variety of other promoters such as TCDD,
phenols, halogens of various sorts, wounding of tissues, and
asbestos are known to cause lipid peroxidation or produce
oxygen radicals.

   Other degenerative diseases such as atherosclerosis and
diabetes and loss of mental activity may well have a muta-
tional component (48) and be related to oxidative DNA damage
as well.

   If ageing, as well as spontaneous cancer, is primarily
due to accumulated damage to DNA, mainly from oxidative damage (35,36,49,50,51,52), then the optimization of the
defense mechanisms for oxidative damage may prolong life
span. The evidence on the effect of antioxidants in pro-
longing life span has been reviewed (49,52).

ACKNOWLEDGMENTS

   We are indebted to E. Schwiers for her experiments on
developing new tester strains and to M. Fiss and T. Philbert
for experiments on the chemistry of oxidative damage to
nucleosides.

REFERENCES

1. Youssefi, M., Faust, S. D., and Zenchelsky, S. T. Clinical Chemistry 24, 1109 (1978).
2. Ames, B. N. Science 204, 587 (1979).
3. Ames, B. N., and McCann, J. Cancer Research 41, 4192 (1981).
4. Hollstein, M., McCann, J., Angelosanto, F. A., and Nichols, W. W. Mutation Research 65, 289 (1979).
5. McCann, J., Choi, E., Yamasaki, E., and Ames, B. N. Proc. Natl. Acad. Sci. USA 72, 5135 (1975).
6. McCann, J., and Ames, B. N. Proc. Natl. Acad. Sci. USA 73, 950 (1976).
7. Ames, B. N., McCann, J., and Yamasaki, E. Mutation Research 31, 347 (1975).
8. Kornbrust, D. J., and Mavis, R. D. Molec. Pharm. 17, 408 (1980).
9. Albro, P. W., Corbett, J. T., Harris, M., and Lawson, L. D. Chem.-Biol. Interactions 23, 315 (1978).
10. Mason, R. P. in "Reviews in Biochem. Toxicol." (E. Hodgson, J. R. Bend, and R. M. Philpot, eds.), p. 151. Elsevier, North Holland (1979).
11. Poyer, J. L., McCay, P. B., Lai, E. K., Janzen, E. G., and Davis. E. R. Biochem. Biophys. Res. Comm. 94, 1154 (1980).
12. White, J. R., Vaughan, T. O., and Yeh, W. S. Fed. Proc. 30, 114 (1971).
13. Gutter, B., Speck, W. T., and Rosenkranz, H. S. Mutation Research 44, 177 (1977).
14. Jose, J. G. Proc. Natl. Acad. Sci. USA 76, 469 (1979).
15. Bruyninckx, W. J., Mason, H. S., and Morse, S. A. Nature 274, 606 (1978).
16. Wang, S. Y., Hahn, B. S., Batzinger, R. P., and Beuding, E E. Biochem. Biophys. Res. Comm. 89, 259 (1979).
17. Brooks, B. R., and Klamerth, O. L. Europ. J. Biochem. 5, 178 (1968).
18. Shamberger, R. J., Andreone, T. L., and Willis, C. E. J. Nat. Canc. Inst. 53, 1771 (1974).
19. Mukai, F. H., and Goldstein, B. D. Science 191, 868 (1976).
20. Marnett, L. J., and Tuttle, M. A. Canc. Res. 40, 276 (1980).
21. Summerfield, F. W., and Tappel, A. L. Biochem. Biophys. Res. Comm. 82, 547 (1978).
22. Held, K. D., and Powers, E. L. Int. J. Rad. Biol. 36, 665 (1979).
23. Myers, L. S. in "Free Radicals in Biology" (W. A. Pryor, ed.), Vol. 4, p. 95. Academic Press, New York (1980).

24. Bonicel, A., Mariaggi, N., Hughes, E., and Teoule, R. Rad. Res. 83, 19 (1980).
25. Cerutti, P. A. in "DNA Repair Mechanisms" (P. C. Hanawalt, E. D. Friedberg, and C. F. Fox, eds.), p. 1. Academic Press, New York (1978).
26. Targovnik, H. S., and Hariharan, H. S. Rad. Res. 83, 360 (1980).
27. Demple, B., and Linn, S. Nature 287, 203 (1980).
28. Clough, R. L. J. Am. Chem. Soc. 102, 5242 (1980).
29. Cadet, J., and Teoule, R. Photochem. Photobiol. 28, 661 (1978).
30. Simon, M. I., and Van Vunakis, H. Arch. Biochem. Biophys. 105, 197 (1964).
31. Fridovich, I. Science 201, 875 (1978).
32. Chance, B., Sies, H., and Boveris, A. Physiol. Rev. 59, 527 (1979).
33. Tappel, A. L. in "Free Radicals in Biology" (W. A. Pryor, ed.), Vol. 4, p. 1. Academic Press, New York (1980).
34. Demopoulos, H. B., Pietronigro, D. D., Flamm, E. S., and Seligman, M. L. J. Env. Path. Toxicol. 3, 273 (1980).
35. Oberley, L. W., Oberley, T. D., and Buettner, G. R. Med. Hypotheses 6, 249 (1980).
36. Totter, J. R. Proc. Natl. Acad. Sci. USA 77, 1763 (1980).
37. Ts'o, P. O., Caspary, W. J., and Lorentzen, R. J. in "Free Radicals in Biology" (W. A. Pryor, ed.), Vol. 3, p. 251. Academic Press, New York (1977).
38. Cameron, E., Pauling, L., and Leibowitz, B. Cancer Research 39, 663 (1979).
39. Kallistratos, G., and Fasske, E. J. Canc. Res. Clin. Oncol. 97, 91 (1980).
40. Cook, M. G., and McNamara, P. Cancer Research 40, 1329 (1980).
41. Peto, R., Doll, R., Buckley, J. D., and Sporn, M. B. Nature 290, 201 (1981).
42. Foote, C. in "Free Radicals in Biology" (W. A. Pryor, ed.), Vol. 2, p. 85. Academic Press, New York (1976).
43. Vanella, A., Pinturo, R., Rapisarda, A., Di Silvestro, I., and Rizza, V. Ital. J. Biochem. 29, 133 (1980).
44. Kallistratos, G., and Fasske, E. Folia Biochim. Biol. Graeca 17, 1 (1980).
45. Daoud, A. H., and Griffin, A. C. Canc. Lett. 9, 299 (1980).
46. Hirayama, T. Nutrition Cancer 1, 67 (1979).
47. Tappel, A. L. Executive Health 16, 1 (1980).
48. Trosko, J. E., and Chang, C-c. Med. Hypotheses 6, 455 (1980).
49. Harman, D. Age 1, 145 (1978).
50. Leibovitz, B. E., and Siegel, B. V. J. Geront. 35, 45 (1980).

51. Tolmasoff, J. M., Ono, T., and Cutler, R. G. Proc. Natl.
     Acad. Sci. USA 77, 2777 (1980).
52. Harman, D., and Eddy, D. E. Age 2, 109 (1979).
53. Packer, J. E., Mahood, J. S., Mora-Arellano, V. O.,
     Slater, T. F., Willson, R. L., and Wolfenden, B. S.
     Biochem. Biophys. Res. Comm. 98, 901 (1981).

Note added in proof: A more recent discussion of natural
anti-oxidants, and the description of a new anti-oxidant,
uric acid, which has appeared during primate evolution, is
discussed in Ames, B. N., Cathcart, R., Schwiers, E.,
Hochstein, P. Proc. Natl. Acad. Sci. USA 78, 6858 (1981).

Index